Sustainable Nanotechnology and the Environment: Advances and Achievements

ACS SYMPOSIUM SERIES **1124**

Sustainable Nanotechnology and the Environment: Advances and Achievements

Najm Shamim, Editor
U.S. Environmental Protection Agency
Washington, DC

Virender K. Sharma, Editor
Florida Institute of Technology
Melbourne, Florida

Sponsored by the
ACS Division of Industrial and Engineering Chemistry

American Chemical Society, Washington, DC

Distributed in print by Oxford University Press

Library of Congress Cataloging-in-Publication Data

Sustainable nanotechnology and the environment : advances and achievements / Najm Shamim, editor, U.S. Environmental Protection Agency, Washington, DC, Virender K. Sharma, editor, Florida Institute of Technology, Melbourne, Florida ; sponsored by the ACS Division of Industrial and Engineering Chemistry.
pages cm. -- (ACS symposium series ; 1124)
Includes bibliographical references and index.
ISBN 978-0-8412-2784-2 (alk. paper)
1. Nanotechnology--Environmental aspects. 2. Nanostructured materials--Environmental aspects. 3. Sustainable engineering. 4. Sanitary engineering--Technological innovations. I. Shamim, Najm, editor of compilation. II. Sharma, Virender K., editor of compilation. III. American Chemical Society. Division of Industrial and Engineering Chemistry.
TD196.N36S87 2013
338.9'27--dc23

2012049658

The paper used in this publication meets the minimum requirements of American National Standard for Information Sciences—Permanence of Paper for Printed Library Materials, ANSI Z39.48n1984.

Distributed in print by Oxford University Press

PRINTED IN THE UNITED STATES OF AMERICA

Foreword

The ACS Symposium Series was first published in 1974 to provide a mechanism for publishing symposia quickly in book form. The purpose of the series is to publish timely, comprehensive books developed from the ACS sponsored symposia based on current scientific research. Occasionally, books are developed from symposia sponsored by other organizations when the topic is of keen interest to the chemistry audience.

Before agreeing to publish a book, the proposed table of contents is reviewed for appropriate and comprehensive coverage and for interest to the audience. Some papers may be excluded to better focus the book; others may be added to provide comprehensiveness. When appropriate, overview or introductory chapters are added. Drafts of chapters are peer-reviewed prior to final acceptance or rejection, and manuscripts are prepared in camera-ready format.

As a rule, only original research papers and original review papers are included in the volumes. Verbatim reproductions of previous published papers are not accepted.

ACS Books Department

Contents

Indexes

Preface

It has been ten years since Barbara Karn initially organized the symposium on Green Nanotechnology at the ACS Meeting under the Industrial & Engineering Chemistry Division. At the time, Barbara along with Tina Masciangioli, Wie-xian Zhang, Vicki Colvin, and Paul Alivisatos successfully put together an initial pioneering symposium consisting of relevant presentations and the fruits of this endeavor were published in the ACS Symposium book, entitled "Nanotechnology and the Environment: Applications and Implications". Later, Dr. Barbara Karn along with Dr. Stanislaus S. Wong continued the process of arranging these symposia till 2011. It was our great pleasure to organize the 10th symposium in 2012, particularly honoring Drs. Barbara Karn and Stanislaus Wong for their excellent contributions to the field of green nanotechnology.

Ten years has seen a tremendous advancement in the nanotechnology field, more so on the applications side than in nanoscience itself, although scientific progress in both has not been lagging behind.

We thought it fit to maintain the tradition and put together the contributions and papers of this latest symposium as chapters and to publish these as the following Symposium Book. We would be deluding ourselves if we thought that this symposium entirely captures the tremendous the length and breadth of scientific advancements in all disciplines related to nanotechnology. Neither one book nor many books can achieve this goal. But we are certain about one thing: This book makes a serious effort at bringing forth and synergistically combining the concepts of green chemistry, sustainability and nanotechnology and to some degree, should motivate scientists at all levels to think clearly and seriously about creating and optimizing novel and sustainable green approaches to nanotechnology.

The chapters in this book can be divided into three broad categories: 1) Advancement in research on pollution control through the green chemistry principles of nanotechnology; 2) Emergence of nanomaterials in widespread applications in various scientific fields, including but not limited to sensors and catalysts; 3) Extension of research into nanotechnology and green nanotechnology at a rapid pace. Review articles on the individual aspects of these diverse and complementary topics have become important resources for researchers, industry leaders, and regulators, both nationally and internationally. This book contains a few chapters associated with these particular themes, and provides glimpses of the many difficulties and challenges faced by those who seek to not only understand but also regulate the new nanomaterials.

Nanotechnology represents a unique field of science, and necessitates new and novel sustainable approaches to create usable end products for the market place with the primary goal of yielding less adverse effects upon both human health and the environment.

Acknowledgments

We would like to thank all the authors who graciously agreed at the ACS Conference in San Diego (March 2012) to convert their presentations to full chapters for this book. More so, they did not mind our repeated and sometimes nagging emails about completing the work in time. Without their research and efforts this book would be naught.

The ACS Editorial Staff: What an outstanding group of folks. Mr. Hauserman, Tim Marney, Mrs. Arlene Furman, and Mrs. Kathleen Squibb. The cooperation and outstanding professionalism kept Virender and I in line. The ACS even accepted our little excuses with a smile (We think).

We are indeed very thankful to the reviewers for this book who kindly donated their time and efforts to help us in the review process.

Virender thanks the support of Center of Ferrate Excellence at Florida Institute of Technology.

On a personal note, V. K. Sharma is happy that he met Dr. Barbara Karn in the University of Miami in 1987 and cherishes her friendship for the last 25 years. Both of us are very happy that they met and wish her the very best for enjoying the lovely moments of her life.

Chapter 1

Ten Years of Green Nanotechnology

Barbara Karn*,[1] and Stanislaus S. Wong*,[2]

[1]Program Director for Nano EHS, National Science Foundation, 4201 Wilson Blvd., Arlington, VA, 22230
[2]Professor of Chemistry, Department of Chemistry, State University of New York at Stony Brook, Stony Brook, New York, 11794; Joint appointment with the Condensed Matter Physics and Materials Sciences Department, Brookhaven National Laboratory, Upton, New York, 11973
***E-mail: bkarn@nsf.gov (B.K.); stanislaus.wong@stonybrook.edu (S.S.W.)**

This Chapter examines a brief history of Green Nanotechnology during the last ten years. The field is both multi-disciplinary and inter-disciplinary, and, while it has made great strides for the past ten years, it is still evolving. One very important and positive aspect of Nanotechnology with respect to its latent impact upon research and applications is its deepening connection to both Green Chemistry and Sustainability. Although much work has gone into developing and strengthening this fruitful relationship, it has not caught up with other technical advances in Nanotechnology. Nevertheless, the convergence of three important fields of science, namely Nanotechnology, Green Chemistry, and Sustainability, could result in a novel and vital holistic area of study, namely Green Nanotechnology, with the potential for fewer societal issues, fewer hazardous reagents to deal with, and the generation of very efficient products associated with a reduced environmental footprint.

Introduction

"*D'où venons nous? Que sommes nous? Où allons nous*?" reads the inscription on Gaugin's 1897 painting. The quote is an appropriate framework for an introduction to Green Nanotechnology, examining its past, present and future.

Where Do We Come From? (*D'où venons nous*?)

Green Nanotechnology is a subset of nanotechnology itself. It couples the rigor of a legitimate scientific discipline with the fervent excitement and awesome responsibility for developing new opportunities for enhancing sustainability.

History

Nanotechnology is a relatively new discipline. Studies at the nanoscale were enabled by the practical development of instruments in the 1980s by talented visionaries including but not limited to Binnig, Roher, Gerber, and Quake working at IBM and beyond (*1*, *2*). These included the scanning tunneling microscope and the atomic force microscope, which allowed scientists, for the first time, to "see" atoms, and furthermore, permitted them to controllably manipulate individual atoms and molecules. In doing so, this new and unprecedented capability enabled researchers to actually spatially move these chemical entities around on a substrate in a well-defined manner with the potential for not only probing novel properties but also conducting localized reactions.

Research at the nanoscale in the 1990s was supported by both government and industry. Research program directors at several government agencies noted that special phenomena were occurring at a size range, roughly between one and 100 nm. Realizing the game changing nature of research into matter at this length scale, these program directors convinced the science managers under the Clinton administration to form a National Nanotechnology Initiative (NNI) so as to coordinate funding among federal US agencies.

The NNI defines nanotechnology as the understanding and control of matter at the nanoscale, at dimensions between approximately 1 and 100 nanometers, where unique phenomena enable novel applications. Encompassing nanoscale science, engineering, and technology, nanotechnology involves imaging, measuring, modeling, and manipulating matter at this length scale. From the beginning, nanotechnology was different. In 1992, Eric Drexler stated: "Molecular nanotechnology promises a fundamental revolution in the way we make things. By bringing precise control to the molecular level…it can serve as a basis for (making) manufacturing processes cleaner, more productive, and more efficient than those known today." Nanotechnology also represented a paradigm shift in recognizing that special and often unexpected properties occur at the nanoscale due to the ultrasmall size of the material.

The NNI concept was formalized through the Nanotechnology Research and Development Act of 2003. The NNI vision denoted a future in which the ability to understand and control matter at the nanoscale can potentially lead to a revolution in understanding the underlying technology and industry that benefit society. In fact, the development of nanotechnology in its broadest and most altruistic sense seeks not only to improve the comprehension of nature through an increase in fundamental knowledge but also to provide yet another tool in tackling pressing basic global concerns of immense moral impact. As examples, the latter include providing for a clean water supply; ensuring food security and decent human shelter; improving transportation, power, and energy systems; and reducing the weight of pollution and of greenhouse gases in industrial manufacturing processes in order to preserve a clean environment (*3*). To this end, not surprisingly, a lot of effort has been expended towards developing nanomaterials for the destruction of environmental pollutants as well as for remediation purposes. Other groups have sought to understand the role of the structure, shape, bandgap, size, morphology, and surface chemistry of the nanoparticles themselves in determining their formation and reactivity in a range of aquatic, air, and soil environments (*4*).

The four goals of the NNI include:

(1) Advancing world-class nanotechnology research and development;
(2) Fostering the transfer of new technologies into products for commercial and public benefit;
(3) Developing and sustaining educational resources, a skilled workforce, and the supporting infrastructure and tools to advance nanotechnology;
(4) Supporting the responsible development of nanotechnology. (*5*)

Green Nanotechnology 'sustains' the fourth goal. By either eliminating or minimizing harmful polluting substances in the synthesis of nanomaterials or using the products of nanotechnology to eliminate or minimize these pollutants in current chemical processes, Green Nanotechnology enables nanotechnology itself to develop in a responsible and more sustainable manner.

The efforts of the NNI federal expenditures work in concert with similar commercial outlays. As a result, many applications of nanotechnology are currently the focal point of extensive commercialization efforts with the potential for developing future markets that might be worth trillions of dollars (*2*). Sectors incorporating nanotechnology include but are not limited to food/agriculture, metallurgical additives and alloys, semiconductors/electronics/ computers, personal care products, energy/batteries, textiles, military applications, sporting goods, construction, medicine, as well as the automotive and chemical industries.

Commercial revenue impacted by nanotechnology spans three levels of the value chain – nanomaterials, nanointermediates, and nano-enabled products. Nanomaterials are those purposefully engineered materials that fit the NNI definition having dimensions between 1 and 100 nanometers and exhibiting unique properties, different from and potentially better than what is characteristic of the bulk. Carbon nanotubes, metal and metal oxide nanoparticles, quantum dots, graphene, dendrimers, and ceramic nanomaterials represent typical examples.

Nanointermediates incorporate nanomaterials which possess nanoscale features, but they are not the final products. Examples include solar cells, drug delivery systems, agricultural chemicals, sensors, lubricants, and catalysts. These nano-enabled products signify the end of the value chain, incorporating designed materials and intermediates that will be used as key components and constituents of aircraft, pharmaceuticals, phosphors, computer chips, sporting goods, automobiles, and so forth.

Each year, the NNI publishes a budget indicating expenditures of each participating agency in what are called Program Component Areas (PCA) (Table 1) (*5*). The 7th PCA deals with the environmental health and safety of nanotechnology. That is, the focus is on discovering and addressing potential negative implications of nanotechnology before they can obstruct its sustainable development and progress. In this light, the NNI has held several workshops that have dealt with nanotechnology as well as its combined environmental and societal impact, both positive and negative. Early on, international meetings dealing with the responsible development of nanotechnology initiated productive conversations and set the stage for more detailed dialogues concerning both the potential applications and probable implications of nanotechnology with respect to the environment.

Table 1. NNI Program Component Areas (PCAs)

No.	*Title*	*Description*
1	*Fundamental Nanoscale Phenomena and Processes*	Discovery and development of fundamental knowledge pertaining to new phenomena in the physical, biological, and engineering sciences that occur at the nanoscale. Elucidation of scientific and engineering principles related to nanoscale structures, processes, and mechanisms.
2	*Nanomaterials*	Research aimed at the discovery of novel nanoscale and nanostructured materials and at a comprehensive understanding of the properties of nanomaterials (ranging across length scales, and including interface interactions). R&D leading to the ability to design and synthesize, in a controlled manner, nanostructured materials with targeted properties.
3	*Nanoscale Devices and Systems*	R&D that applies the principles of nanoscale science and engineering to create novel, or to improve existing, devices and systems. Includes the incorporation of nanoscale or nanostructured materials to achieve improved performance or new functionality. To meet this definition, the enabling science and technology must be at the nanoscale, but the systems and devices themselves are not restricted to that size.
4	*Instrumentation Research, Metrology, and Standards for Nanotechnology*	R&D pertaining to the tools needed to advance nanotechnology research and commercialization, including next-generation instrumentation for characterization, measurement, synthesis, and design of materials, structures, devices, and systems. Also includes R&D and other activities related to development of standards, including standards for nomenclature, materials, characterization and testing, and manufacture.

Continued on next page.

Table 1. (Continued). NNI Program Component Areas (PCAs)

No.	*Title*	*Description*
5	*Nanomanufacturing*	R&D aimed at enabling scaled-up, reliable, and cost-effective manufacturing of nanoscale materials, structures, devices, and systems. Includes R&D and integration of ultra-miniaturized top-down processes and increasingly complex bottom-up or self-assembly processes.
6	*Major Research Facilities and Instrumentation Acquisition*	Establishment of user facilities, acquisition of major instrumentation, and other activities that develop, support, or enhance the Nation's scientific infrastructure for the conduct of nanoscale science, engineering, and technology R&D. Includes ongoing operation of user facilities and networks.
7	*Environment, Health, and Safety*	Research primarily directed at understanding the environmental, health, and safety impacts of nanotechnology development and corresponding risk assessment, risk management, and methods for risk mitigation.
8	*Education and Societal Dimensions*	Education-related activities such as development of materials for schools, undergraduate programs, technical training, and public communication, including outreach and engagement. Research directed at identifying and quantifying the broad implications of nanotechnology for society, including social, economic, workforce, educational, ethical, and legal implications.

Beginning in 2003, the symposium at the American Chemical Society (ACS) meeting on nanotechnology and the environment has been held annually. From 2006 onwards, it was primarily dedicated towards Green Nanotechnology. Green Nanotechnology can be seen as an offshoot of Green Chemistry and Green Engineering, and has evolved with the contributions of a number of groups both nationally as well as worldwide.

What Are We? (*Que sommes nous*?)

There are two goals to Green Nanotechnology with the ultimate objective of ensuring the responsible development of nanotechnology. This community is multidisciplinary, multi-organizational, and multinational. Specifically, Green Nanotechnology encourages collaborations and enables a vision that uses scientific research to fundamentally move toward sustainability. In terms of rendering nanomanufacturing processes less energy and resource intensive, the principles of Green Chemistry applied to nanotechnology involve, if at all possible, (a) the use of cost-effective, nontoxic precursors; (b) minimization of carcinogenic reagents and solvents (if possible, through utilization of aqueous solvents in order to bypass potentially toxic, acidic, or basic analogues); (c) reduction of experiments carried out with either pyrophoric compounds or unstable precursors to avoid risk; (d) use of relatively few numbers of reagents, i.e. atom economy, coupled with a conscious effort to circumvent the generation of greenhouse gases; (e) minimization of reaction steps leading to a reduction in waste, reagent use, and power consumption; (f) development of reactions to generate high-purity materials with little if any byproducts through high-yield processes; (g) ambient temperature and pressure synthesis, if at all possible, so as to preclude the need for either vacuum or high temperature processes; and (h) efficiency of scale-up (*2*). Parallel efforts in developing adequate characterization facilities as well as in promoting computer-aided nanomaterials modeling and process design tools have been of utmost importance in furthering all of these 'green' objectives with the aim of creating environmentally benign building blocks and reducing the environmental footprint of manufacturing protocols (*3*). Optimally, green nanomanufacturing should use less material, less water, and less energy, and produce less waste as compared with conventional processes.

Green Nanotechnology is an application that pays attention to implications. It is based on lowering the risk of using nanomaterials, limiting the risk of producing nanomaterials, and using nanomaterials to lower the risk of producing unwanted chemical intermediates and end-products. Green Nanotechnology maintains a commitment to rigor with applications related to sustainability as well as awareness about and prevention of possible harmful consequences from its successful implementation. Moreover, while not a focal point of the current Chapter, responsible development of Green Nanotechnology necessarily involves balanced and measured considerations of toxicity and cost-benefit risk assessments. Indeed, the explicit goal of all of these analyses is to ensure a smart process design of materials that will eliminate hazards throughout the material's life cycle and potentially deleterious impacts on the environment (*2*).

In this Chapter, we are more interested in actual novel synthesis processes for developing nanoscale materials relevant for light harvesting, photo-oxidation, solar photovoltaics, solar water splitting, fuel cells, catalysis, and energy storage (*6*, *7*). Incidentally, as a tangible example of the emphasis in the community on finding worthwhile applications for these types of nanostructures, Fe_3O_4 magnetic nanoparticles, which can be recovered by magnetic separation and subsequently recycled multiple times, have been used to catalyze the coupling of aldehydes, alkynes, and amines to bioactive intermediates such as propargylamines at high yield (*8*). More in line with our spotlight on novel 'green' synthesis techniques, characteristic examples include recent successes in the development of template-assisted methods as well as the demonstration of proven protocols such as ultrasound and microwave irradiation specifically tailored for the production of novel, high-quality, multifunctional materials. In addition to technique development, there has been a growing emphasis on creating novel approaches towards greener synthesis to either (i) enhance the environmental friendliness of "green synthetic techniques," ideally or (ii) serve as an alternative, when "green synthetic techniques" have not been practical. In this light, a number of "fundamentally green approaches" have included strategies such as substituting common reagents and solvents with more benign replacements such as biological reagents, ionic liquids, supercritical fluids, and even solvent-free methods.

We have recently and extensively reviewed all of these issues (*9*) and simply highlight a set of different but typical examples herein, illustrating the diversity and ingenuity of many of these approaches. In terms of novel method development, an industrially relevant low cost environmentally benign CO_2-based hydrometallurgical process has been developed to generate general classes of CO_2-philic compounds useful for applications in catalysis, utilizing CO_2-soluble organometallic complexes (including zinc derivatives) that readily dissolve in carbon dioxide at a low pressure and temperature of 860 psi and 25°C, respectively (*10*). In addition, the rapid universal synthesis of nanocrystals of a microporous metal-organic framework (MOF) has been published using an ultrasonic technique; these MOFs have been used for the size-selective sensing of organoamines through the measurement of their emission spectra (*11*). Because of the ubiquity of carbon nanotubes and the popularity of their assembly using chemical vapor deposition techniques, it is noteworthy that carbon nanofibers at purity levels of 98.5% can now be produced in bulk with as much as a 8200 weight percent yield on water-soluble, nontoxic sodium chloride supports (*12*).

For oxides in particular, monodisperse Cu_2O nanocubes with an average edge length of 500 nm have been produced by reducing freshly prepared $Cu(OH)_2$ with $N_2H_4 \cdot H_2O$ in water at room temperature with the analogous nanoboxes synthesized by etching these cubes with acetic acid (*13*). Cu_2O nanocubes ranging in size from 25 to 200 nm have been synthesized by reducing copper (II) salt in water with ascorbic acid in air in the presence of polyethylene glycol (PEG) and sodium hydroxide, with PEG concentration playing a major role in determining the size variation (*14*). Rare earth hydroxide nanorods measuring 10-150 nm in diameter and 30-600 nm in length have been generated without either templates or surfactants by a simple precipitation method (e.g. through reaction of a metal nitrate with ammonia solution under vigorous stirring) under ambient conditions with dimensions controlled by adjusting the aging time (e.g. 0 to 96 h) and the precipitation agents; the corresponding oxide rods were obtained by calcination (*15*). By analogy with a method of creating doped titania nanoparticles (~ 5 nm) by hydrolyzing sol gel precursors under 348 K for 24 h at ambient pressure (*16*), a vapor-diffusion sol-gel technique, wherein water is slowly diffused into a mixture of bimetallic alkoxides yielded solid-solution sub-15 nm nanocrystals of $Ba_xSr_{1-x}TiO_3$ in quantitative yield on the gram scale, was developed to create nanoparticles at low temperatures, ambient pressure, and neutral pH (*17*). The reaction of germinate ions with $NaBH_4$ in an aqueous medium at room temperature for 6 to 48 hours resulted in highly hydrogenated Ge that could be transformed into crystallized Ge after an air-drying treatment at 20-250°C for 4 hours. Interestingly, a reaction time of 12 h yielded ultra-small particles measuring 5 to 20 nm, whereas reaction times of 24 h favored the formation of worm-like structures with diameters of 10 to 80 nm and lengths of up to 1000 nm (*18*).

Regarding other semiconducting species, complementary to oxides, ~10 nm capped CdS nanoparticles have been successfully manufactured by stirring a solution of cadmium nitrate with an aqueous solution of soluble starch at room temperature, adjusting pH conditions, and ultimately adding in an aqueous solution of sodium sulfide. After stirring for 5 h and aging for an additional 12 h, a yellow precipitate could be isolated after acetone washing and low-temperature drying steps (*19*). High-quality CdSe nanoparticles, capped with 3-mercaptopropionic acid and ranging in size from 1.58 to 3.42 nm, have been ambiently synthesized through the mediation of a hydrazine hydrate-Se complex intermediate (*20*). CdS quantum dots ranging in size from 1 to 3 nm have also been fabricated from the ambient reaction of H_2S with R-2-thiotetrahydrothiazole-4-carboxylic acid, itself *in situ* produced from the reaction of *L*-cysteine with carbon disulfide in water (*21*). Interestingly, these CdS quantum dots exhibited high stability even under weakly acidic conditions at pH 6.5. Incidentally, the spontaneous formation of rose-like Se crystal (with petals possessing typical thicknesses of 400 nm) in aqueous solution at room temperature has been reported, based on the oxidation of Na_2Se in the presence of solutions of thioglyceraol derivatives at pH 11 in a dark ambient atmosphere (*22*).

For metals, sodium borohydride and silver nitrate were mixed together to produce yellow colloidal Ag nanoparticles, measuring 12 ± 2 nm in diameter (*23*). A facile method was recently developed for preparing uniform Ag nanoparticles with sizes of less than 10 nm, in which aniline was used to assist in the reduction of silver nitrate to Ag nanoparticles in the presence of dodecylbenzenesulfonic acid as a stabilizer; upon addition of excess NaOH, nanoparticles measuring 8.9 ± 1.1 nm were formed in ~ 2 min at 90°C in 94% yield and these remained stable for more than one year at ambient temperatures (*24*). Interestingly, citrate ions have been used as reducing agents, stabilizers, and complexing agents in the presence of surfactants to form Ag nanoplates, measuring from 100 to 350 nm under ambient conditions (*25*). Highly stable, starch-stabilized aqueous dispersions of 1-8 nm nanoparticles of Ag, Au, and Au-Ag alloys of varying compositions have also been synthesized in water using glucose as the reducing agent and starch as the protecting agent by microwave heating precursors for as little as 60 s in some cases (*26*). Macroscale quantities of bimetallic Au-Ag alloy nanoclusters have been successfully achieved through simple mortar grinding of metal precursors in the presence of chitosan biopolymer, which was used as both a stabilizing and reducing agent (*27*). Finally, bio-assisted syntheses of Ti nanospheres (40-60 nm) (*28*), anisotropic Au nanotriangles (25 to 35 nm) (*29*), and Pt nanocrystals (including 4 nm polyhedra and 7-8 nm cubes) (*30*) have been initiated using *Lactobacillus*, *Aspergillus clavatus*, and single peptide sequences, respectively.

Where Are We Going? (*Où allons nous*?)

The discussion above has outlined some history and provided a flavor for the current state of science for Green Nanotechnology. It is necessary to continue to expand the knowledge base and to transfer this knowledge into practical commercial use. Below we suggest some possible pathways forward.

(1) Integrate Green Nanotechnology procedures, concepts, and philosphies into the current curriculum for students in chemistry, physics, materials science, and engineering. Embracing the broad principles and techniques of Green Chemistry and Green Engineering, these ideas and

methodologies can form a robust platform and a powerful pedagogical foundation from which to expand and include nanomaterials design, synthesis, and ultimately, processing. In general, these paradigms should become part of and guide the conventional wisdom in the field. In particular, crystallographic and surface characteristics of nanomaterials need to be carefully considered, as these affect the nature of the synthesis and increase the number and variety of techniques available for sustainable synthesis.

(2) Continue to maintain the "weapon of openness." Green Nanotechnology does not exist in a vacuum. Exchange of information can be expanded through dedicated journals, special symposia, and focused sessions in a variety of professional societies. Industries that manufacture nanomaterials and nanomaterial-containing products must be kept informed of the most effective practical green processes as they bring new products and processes on line. A robust system of information exchange needs to expand, develop, and fundamentally thrive.

(3) Incorporate life cycle thinking into nanomaterial processes and products. The impacts of resource extraction, energy use, as well as the generation of pollutants from cradle to grave, i.e. from factory to consumer to ultimate disposal, require thoughtful consideration, when new nanomaterials are designed and used.

(4) Lastly, the ultimate goal of Green Nanotechnology is to aid sustainability. While no single technology alone can ensure sustainability, newer technologies, especially nanotechnology, can proactively design out pollution, thereby "doing it right in the first place," while producing useful, usable, and viable products that promote sustainable development and provide the mainstay for the future global economy. There is an open window of opportunity right now, and we need to take advantage of it through thoughtful and productive use of Green Nanotechnology.

References

1. Rugar, D.; Hansma, P. *Phys. Today* **1990** (October), 23–30.
2. Binnig, G.; Rohrer, H.; Gerber, C.; Weibel, E. *Phys. Rev. Lett.* **1983**, *50*, 120–123.
3. Diallo, M.; Brinker, C. J.Nanotechnology for Sustainability: Environment, Water, Food, And Climate. In *Nanotechnology Research Directions for Societal Needs in 2020: Retrospective and Outlook*; Roco, M., Mirkin, C. A., Hersam, M. C., Eds.; Springer Publishing: New York, 2011; Chapter 5, pp 157−187.
4. Roco, M. *J. Nanopart. Res.* **2003**, *5*, 181−189.
5. *National Nanotechnology Initiative Strategic Plan*, National Science and Technology Council Committee on Technology, Subcommittee on Nanoscale Science, Engineering, and Technology: Washington, DC, February 2011; pp 29−32.

6. Zach, M.; Hagglund, C.; Chakarov, D.; Kasemo, B. *Curr. Opin. Solid State Mater. Sci.* **2006**, *10*, 132–143.
7. Abrams, B. L.; Wilcoxon, J. P. *Crit. Rev. Solid State Mater. Sci.* **2005**, *30*, 151–182.
8. Zeng, T.; Chen, W.-W.; Cirtiu, A.; Moores, A.; Song, G.; Li, C.-J. *Green Chem.* **2010**, *12*, 570–573.
9. Patete, J. M.; Peng, X.; Koenigsmann, C.; Xu, Y.; Karn, B.; Wong, S. S. *Green Chem.* **2011**, *13* (3), 482–519.
10. Stallone, K. B.; Bonner, F. J. *Green Chem.* **2004**, *6*, 267–270.
11. Qiu, L.-G.; Li, Z.-Q.; Wu, Y.; Wang, W. T.; Xu, T.; Jiang, X. *Chem. Commun.* **2008**, 3642–3644.
12. Ravindra, R.; Bhat, B. R. *J. Nanoeng. Nanomanuf.* **2012**, *2*, 31–35.
13. Wang, Z.; Chen, X.; Liu, J.; Mo, M.; Yang, L.; Qian, Y. *Solid State Commun.* **2004**, *130*, 585–589.
14. Gou, L.; Murphy, C. J. *J. Mater. Chem.* **2004**, *14*, 735–738.
15. Xin, Y.; Wang, Z.; Qi, Y. X.; Zhang, Z.; Zhang, S. *J. Alloys Compd.* **2010**, *507*, 105–111.
16. Xu, J.; Ao, Y.; Fu, D.; Yuan, C. *J. Colloid Interface Sci.* **2008**, *328*, 447–451.
17. Bier, C. W.; Cuevas, M. A.; Brutchey, R. L. *J. Mater. Chem.* **2010**, *20*, 5074–5079.
18. Jing, C.; Zang, X.; Bai, W.; Chu, J.; Liu, A. *Nanotechnology* **2009**, *20*, 505607/1–8.
19. Wei, Q.; Kang, S.-Z.; Mu, J. *Colloids Surf., A* **2004**, *247*, 125–127.
20. Kalasad, M. N.; Rabinal, M. K.; Mulimani, B. G. *Langmuir* **2009**, *25* (21), 12729–12735.
21. Mo, Y.; Tang, Y.; Gao, F.; Yang, J. X.; Zhang, Y. *Ind. Eng. Chem. Res.* **2012**, *51*, 5995–6000.
22. Deng, D.-W.; Yu, J.-S.; Pan, Y. *Eur. J. Inorg. Chem.* **2008**, 1129–1134.
23. Solomon, S. D.; Bahadory, M.; Jeyarajasingam, A. V.; Rutkowsky, S. A.; Boritz, C.; Mulfinger, L. *J. Chem. Educ.* **2007**, *84* (2), 322–325.
24. Yang, J.; Yin, H.; Jia, J.; Wei, Y. *Langmuir* **2011**, *27*, 5047–5053.
25. Jiang, X. C.; Chen, C. Y.; Chen, W. M.; Yu, A. B. *Langmuir* **2010**, *26* (6), 4400–4408.
26. Raveendran, P.; Fu, J.; Wallen, S. L. *Green Chem.* **2006**, *8* (1), 34–38.
27. Murugadoss, A.; Kai, N.; Sakurai, H. *Nanoscale* **2012**, *4*, 1280–1282.
28. Prasad, K.; Jha, A. K.; Kulkarni, A. *Nanoscale Res. Lett.* **2007**, *2*, 248–250.
29. Verma, V. C.; Singh, S. K.; Solanki, R.; Prakash, S. *Nanoscale Res. Lett.* **2011**, *6*, 16/1–7.
30. Forbes, L. M.; Goodwin, A. P.; Cha, J. N. *Chem. Mater.* **2010**, *22*, 6524–6528.

Chapter 2

Green Synthesis of Nanomaterials: Environmental Aspects

J. Virkutyte*[,1] and R. S. Varma*[,2]

[1]Pegasus Technical Services, Inc., E. Hollister Street, Cincinnati, Ohio 45219, U.S.A.
[2]Sustainable Technology Division, National Risk Management Research Laboratory, U.S. Environmental Protection Agency, MS 443, 26 West M.L.K. Drive, Cincinnati Ohio 45268, U.S.A.
***E-mail: Virkutyte.Jurate@epa.gov (J.V.); Varma.Rajender@epa.gov (R.S.V.)**

There is always a search for reliable and environmentally friendly processes to manufacture metal and metal oxide nanoparticles minimizing or even eliminating the use of hazardous chemicals. The only way to develop these "green" processes is to adapt benign synthesis approaches that use mild reaction conditions and non-toxic reaction precursors. This chapter will summarize the work of our research group on the "green" synthesis and functionalization of nanoparticles used for the environmental remediation as well as present the summary of the "state of the art" in the exploitation of various environmentally-friendly synthesis approaches and the use of benign reaction precursors and conditions to synthesize metal and metal oxide nanoparticles.

Introduction

Recently, there was a significant increase in commercial and research applications of nanotechnology, translating into more than $9 billion in investments over the past several years (*1*).

Nanotechnology can be defined as 'engineering at a small scale', and is generally applied to a great number of research and development areas ranging from medicine to manufacturing and even to textiles and/or consumer products

(*2*). A nanometer is one billionth of a meter (10^{-9} m), which is roughly ten times the size of an individual atom. Usually nanotechnology includes the use of functional structures, designed from atomic or molecular scope with at least one characteristic dimension measured in nanometers. This tiny size is responsible for the unique physical, chemical, and biological properties as well as various phenomena. Unfortunately, phenomena and properties may not be easily predicted in comparison to the ones that observed at larger scales. These new properties are not only caused by the small size but by the new phenomena, such as quantum confinement. A typical example of this is the color of light emitting from semiconductor nanoparticles depending on their sizes.

Nanotechnology can provide us with knowledge about materials and devices that will and already have impacted many fields. By adopting nanoscale properties as a tunable physical variable, it is possible to greatly increase the range of performance of existing chemicals and materials. Furthermore, alignment of linear molecules in an ordered array on a substrate surface (self-assembled monolayers) can function as a new generation of chemical and biological sensors (*3*). In addition, switching devices and functional units at the nanoscale can improve computer storage and operation capacity by a factor of a million (*4*). Nanostructured ceramics and metals can greatly improve mechanical properties, in both ductility and strength.

For all these applications, it is vitally important to control matter at atomic or molecular levels, which means tailoring the fundamental properties, phenomena, and processes exactly at the scale where the basic properties are initiated. Therefore, to achieve these goals and realize all these effective applications, it is necessary to acquire new knowledge, create and apply new tools, and try new approaches.

The current research mainly focuses on nanoscience that attempts to discover new materials and novel phenomena, develop new characterization tools, and fabricate nanodevices with a great array of properties and applications. Therefore, synthesis of nanomaterials is one of the most active areas in nanotechnology. At present, there are numerous methods and approaches to synthesize nanomaterials exhibiting required characteristics. One of the most difficult challenges in the synthesis of nanomaterials is to manufacture the structures with desired and controlled properties in large amounts for various industrial applications (*3*). Therefore, techniques are highly warranted for atomic and molecular control of material building blocks, which can be assembled, used, and tailored for manufacturing nanodevices/nanomaterials with multi-functionality for a number of applications.

In order to successfully synthesize nanodevices and nanomaterials, there is a need to establish nano-manufacturing standards to achieve the effective and validated product and process performance. These challenge must follow three directions (*5*):

1. Atomic-scale synthesis: develop and assemble the technologies that require standards which are atomically precise. These include the work directed at solving artifact integrity, precision placement, dimensional metrology, and manufacturing issues.

2. Molecular-scale manipulation and assembly: identify and address the fundamental measurements, control and standardization issues related to manipulation, and assembly of nanodevices or nanomaterials using optical, physical, or chemical methodologies.
3. Micro-to-millimeter-scale synthesis technologies: develop the technologies that require positioning, manipulating, assembling, and manufacturing across nanometer-to-millimeter length scales.

Furthermore, to successfully manufacture nanomaterials and nanodevices one should not only solve the process standardization issues but address various research challenges that are associated with the surface phenomena. For example,outer layer atoms in the particle may have a different composition and, therefore, chemistry from the rest of the particle. Furthermore,that nanoparticle surface may be prone to environmental changes such as redox conditions, pH, ionic strength, and the presence of microorganisms and natural organic matter. In addition, the small size and large surface to volume ratios may lead to chemical as well as physical differences in nanoparticle properties, including mechanical, biological and sterical catalytic activity, thermal and electrical conductivity, optical absorption, and melting point (*6*).

It is generally accepted that the intended nanodevice/nanoparticle application defines its composition, e.g. if a nanoparticle is going to be used in the presence of biological systems, such as drug carrier in the human body, functional groups will be attached to its surface to prevent from aggregation or agglomeration (*2*). Also, various coatings and other surface active materials should be used that form transient van der Waals interactions with the surface of the target nanoparticle and exist in equilibrium with the free surfactant molecule. On the contrary, if the intended use is in electronics industry, nanoparticles should be synthesized to enhance the strength and hardness of materials and to exhibit enhanced electrical properties by controlling the arrangements within the nanoclusters. If the nanoparticles are to be used for environmental remediation and catalysis, attaching numerous functional groups, doped with ions and anions as well as varied size and structure, could increase their activity. Finally, if the nanoparticles are used in applications that require increased superconductivity and magnetism, they can be manufactured utilizing relevant mechanisms and surface functionalization approaches (*7*).

Nanoparticles consists of three layers (Figure 1): 1) a surface that can easily be functionalized, 2) a shell that may be added according to the desired properties of the final products, and 3) the core that can be synthesized utilizing various methods, reaction conditions, and precursors (*2*). The nanoparticle surface can be functionalized with various polymers, metals, metal oxides, small molecules, surfactants, and other compounds that are necessary to achieve the required composition and be used for the target application. The nanoparticles can be charged ($SiO^{-}Me^{+}$) or uncharged (citrate, polyethylene glycol (PEG), which depends on the use of nanomaterials. Moreover, the shell is generally made of inorganic material that has a different structure than the core, e.g. iron oxide on iron nanoparticle, titanium dioxide on titania, quantum dots, and polystyrene-polyaniline nanoparticles (*2*). It is important to indicate that the core is usually referred as the nanoparticle itself where its physicochemical properties are nearly always controlled by the properties of the core. However, in the case of fate and transport of nanoparticles in environment, both, the core and shell properties rather than the core alone will likely dominate these. On the other hand, the risks associated with the occurrence of nanoparticles in the environment are always related to the surface, core, and shell without emphasizing any of the components of the nanoparticle.

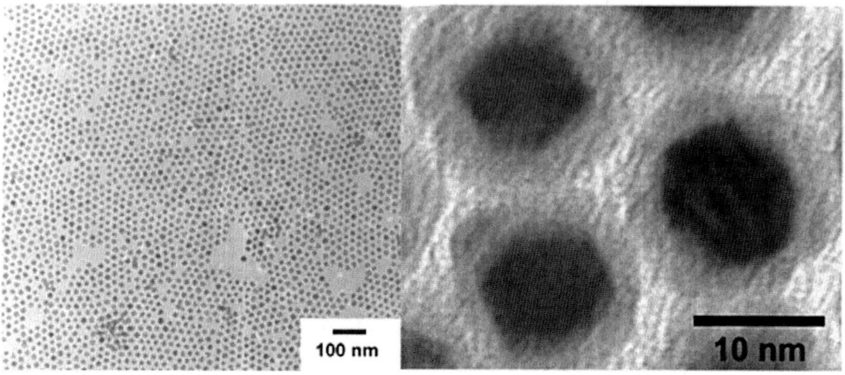

Figure 1. Low and high resolution images of Pt@Fe_2O_3 core−shell nanoparticles. Reprinted with permission from (8). Copyright (2003) American Chemical Society.

There are two main methods to synthesize nanomaterials: 1) top-down and 2) bottom-up approaches. Top-down approaches include lithographic techniques, etching, sputtering, and grinding in a ball mill among a few. Bottom-up approaches include "growing" nanoparticles from simple molecules - reaction precursors. It is possible to control the size and the shape of the nanoparticles using target precursors' concentrations, reaction conditions (temperature, pH), and functionalization of the surfaceas well as templates utilizing a bottom-up approach.

The functionalization of nanoparticle surface is one of the most important aspects of the synthesis, especially when the nanoparticles or nanomaterials are synthesized for a certain application. For instance, the high chemical activity with a large surface is usually the main reason for aggregation, which is an irreversible and highly undesirable process, because aggregation significantly diminishes reactivity through the reduced specific surface area and the interfacial free energy (*9*). Therefore, to avoid aggregation, nanoparticles have to be functionalized with certain chemicals or minerals to increase their stability during storage, transportation, and application.

Currently, conventional functionalization methods include the addition of dispersant molecules such as surfactants and/or polyelectrolytes to the nanoparticle surface (*10*). Unfortunately, these dispersant molecules not only alter the chemistry of the surface, but also produce large amounts of wastes (*1*). Therefore, to avoid producing wastes and subsequently contaminating the environment, it is necessary to search and develop environmentally benign synthesis approaches and functionalization pathways to create biocompatible stabilizing agents.

There is numerous research evidence available on the "greener" synthesis and functionalization of the nanomaterials using polyphenols, citric acid, vitamins (B, C, D, K), biodegradable polymers, and silica.

This chapter will summarize the work of our research group on the "green" synthesis and functionalization of nanoparticles used for the environmental remediation. It alsopresents the summary of the "state of the art" in the exploitation of various environmentally-friendly synthesis approaches, the use of benign reaction precursors, and conditions to synthesize metal and metal oxide nanoparticles.

Bio-Inspired Nanoparticles

To succeed in the nanotechnology field, one needs to develop reliable synthesis protocols that deal with different chemical compositions, shapes, and sizes and controlled monodispersity of metal and metal oxide nanoparticles. Conventionally, metal and metal oxide nanoparticles are produced using widely available physical and chemical methods. However, the use of generally accepted but highly reactive reducing agents such as hydrazine hydrate and sodium borohydride may lead to undesired toxicity issues and may not be environmentally friendly. Therefore, it is important to develop an effective and trustworthy "green" chemistry process for the synthesis of bio-nanomaterials. The use of bio-world for the synthesis of nanoparticles is not extraordinary, because many of organisms, both unicellular and multicellular are known to produce inorganic nanomaterials either intra-or extra cellularly (*11*). Some of the examples of bio-organisms synthesizing inorganic materials include magnetotactic bacteria (which synthesize magnetite nanoparticles), diatoms (which synthesize siliceous materials), and S-layer bacteria (which produce gypsum and calcium carbonate layers). Therefore, these examples from nature have led to the development of biomimetic approaches for the synthesis of various nanomaterials for a great array of purposes.

Figure 2 shows an example of bio-inspired synthesis at mild conditions and temperatures. In summary: (i) various organisms can be used as eco-friendly and sustainable precursors to manufacture stable and efficient nanoparticles while reducing waste products that leads to atomically precise molecular manufacturing, ii) synthesis of nanoparticles with tailored functionalities and the subsequent use of these nanoparticles as catalysts for greater efficiencies of various processes by minimizing or even eliminating the use of toxic materials, iii) "green" synthesized nanoparticles and nanomaterials can be effectively utilized to reduce or even eliminate environmental pollution, and iv) nanomaterials can be utilized for efficient alternative energy production, e.g. solar and fuel cells (*12*).

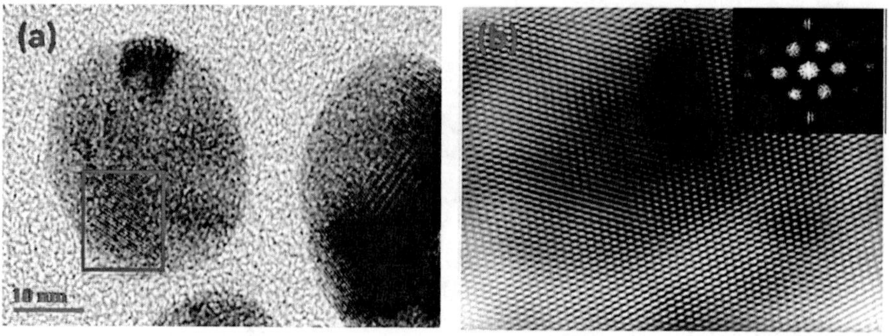

Figure 2. a) TEM images of Azadirachta indica extract mediated silver nanoparticles prepared at room temperature and corresponding (b) FFT (inset of b) typical SAED patterns. Reprinted with permission from (13). Copyright (2012) American Chemical Society.

However, if organisms are used to produce nanoparticles, several aspects must be taken into account (*6*):

1. The most effective organism: it is important to know and effectively use the important intrinsic properties of target organisms, e.g. enzymatic activity and biochemical pathways, to synthesize metal and metal oxide nanoparticles. For instance, there are plants that can accumulate and detoxify metals and, therefore, are considered the best candidates to act as the reducing and capping agents for nanoparticle synthesis.
2. Optimal conditions for cell growth and enzymatic activity: it is important to optimize the conditions, such as the amount of light, nutrients, inoculum size, pH, agitation speed, buffer strength, and temperature for the most effective cell growth.
3. Optimal reaction conditions: for the larger production of nanoparticles, the reaction yield and the production rate are the most important parameters that should be taken into account and optimized by manipulating substrate and biocatalyst concentrations, electron donor type and amount, pH, exposure time, temperature, buffer strength, mixing speed, the amount of light, etc. Other secondary factors should also be taken into consideration, such as the energy expenditure from microwave, ultrasound, and visible light irradiation.

Bio-Inspired Synthesis of Nanoparticles

The potential of organisms to be used for nanoparticle production ranges from simple prokaryotic cells to eukaryotic fungi or even plants. Importantly, the nanoparticle synthesis using microbial organisms is a "green" chemistry approach that connects nanotechnology and microbial biotechnology, therefore giving birth to a new area of science,bionanotechnology (*14*). Unfortunately, most of the nanoparticles synthesized using simple bio-pathways are not monodispersed and result in a slow rate of the synthesis. Therefore, to overcome these limitations, numerous factors such as microbial cultivation approaches and the extraction techniques should be taken into account; methods and/or approaches should be developed and optimized as well as new methods, such as photo-biological, must be used. More details can be found in the recent review article by (*15*).

Fungi and Bacteria

Fungi have enzymes, proteins, and reducing components on its cell surface. Therefore, fungi have numerous advantages over other organisms in the synthesis of metal and metal oxide nanoparticles (*15*). Intracellular biosynthesis includes the formation of nanoparticles through the reduction by enzymes (reductase) found in the cell wall, in the cytoplasmic membrane, and on the inner surface of the fungal cell.

Various metals, such as silver and gold, can be obtained using mild reaction conditions in the presence of fungi as the reducing and stabilizing agent (Figure 3).

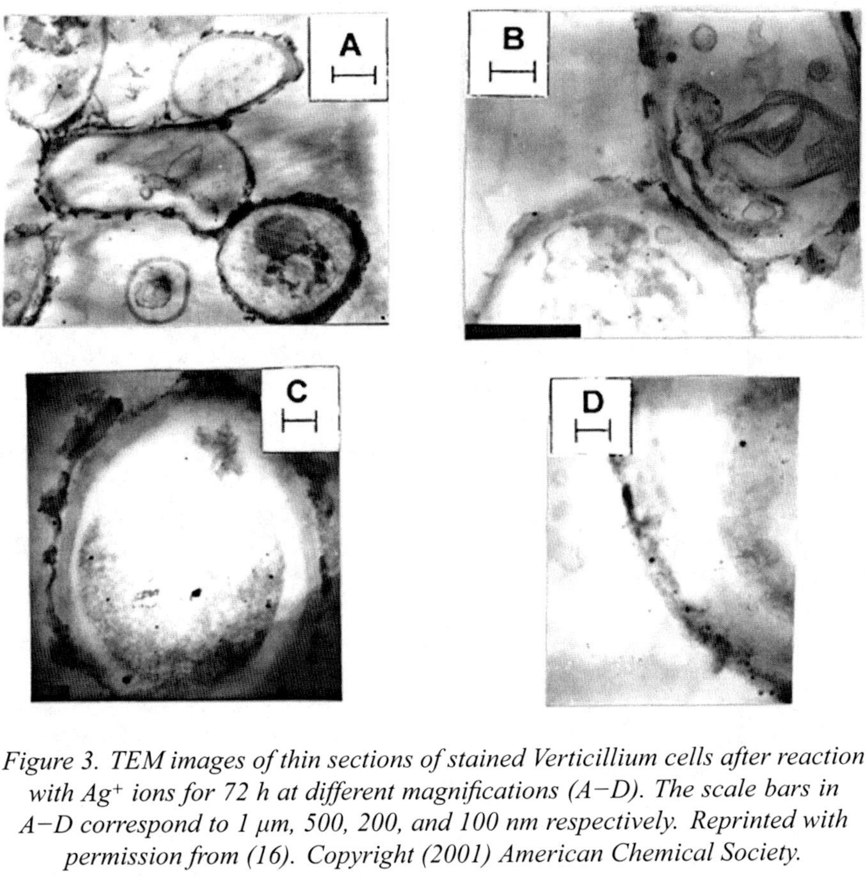

Figure 3. TEM images of thin sections of stained Verticillium cells after reaction with Ag^+ ions for 72 h at different magnifications (A−D). The scale bars in A−D correspond to 1 μm, 500, 200, and 100 nm respectively. Reprinted with permission from (16). Copyright (2001) American Chemical Society.

For instance, Au nanoparticles were synthesized by the fungus *Cylindrocladium floridanum*. It was found that in 7 days, the fungus accumulated fcc (111)-oriented crystalline gold nanoparticles (SPR band of UV vis spectrum at 540 nm) on the surface of the mycelia (*17*). The activity of gold nanoparticles was tested against 4-nitrophenol. The process followed a pseudo-first order kinetic model with a reaction rate constant of 2.67×10^{-2} min^{-1} with 5.07×10^{-6} mol/dm^3 of gold nanoparticles with a size of ca. 25 nm. The reaction rates significantly increased when the gold concentration increased from 2.54×10^{-6} to 12.67×10^{-6} mol/dm^3 (ca. 25 nm) with reduced Au nanoparticle size from 53.2 to 18.9 nm, respectively.

In addition, polydispersed Ag nanoparticles (5 to 40 nm in size) were synthesized using the fungus *Trichoderma viride* at room temperature with a maximum absorbance at 420 nm on ultraviolet-visible spectra (*18*). Antimicrobial activity was tested against four bacterial strains—*Salmonella typhi* (gram-negative rods), *Escherichia coli* (gram-negative rods), *Staphylococcus aureus* (gram-positive cocci), and *Micrococcus luteus* (gram-positive cocci). Interestingly, there was a synergistic effect observed in antimicrobial activity of ampicillin, kanamycin, erythromycin, and chloramphenicol in the presence of silver nanoparticles. Ag nanoparticles (30-50 nm in size) were also synthesized in the presence of *Geotricum sp* (*12*). As can be seen in the FTIR spectra, the bands of amide (I) and (II) in the protein act as the stabilizers on the nanoparticle surface. *Fungus Verticillium* (from Taxus plant) was used to manufacture silver nanoparticles (25 ± 12 nm in size) at room temperature (*16*). An important finding of the study was that the silver ions used as the precursors of the synthesis did not exhibit toxicity to fungal cells; therefore, cells continued to multiply even after the biosynthesis of the silver nanoparticles.

Prokaryotic bacteria and actinomycetes have been most extensively researched for the synthesis of metallic nanoparticles. The ability of bacteria to survive and grow in stressful situations (e.g. when heavy metals are present at high concentrations) may be attributed to the specific mechanisms of resistance, which include inactivation and complexation of metals, impermeability to metals, and the lack of specific metal transport systems (*19*).

One of the reasons to use these unicellular microorganisms for nanoparticle synthesis is their relative ease of manipulation (*20*) Thus, the synthesis of well-dispersed Au nanoparticles (8 nm in size) was performed in the presence of the prokaryotic microorganism *Thermomonospora sp.* (*21*). *Brevibacterium casei* was used to manufacture Ag and Au (10 to 50 nm in size) nanoparticles (*22*). Yet again, FTIR data confirmed that the presence of proteins was responsible for the reduction and stabilization of the nanoparticles.

Pd-ferrimagnetic nanoparticles exhibit properties that could be used for a wide range of applications. Unfortunately, conventional synthesis methods generally result in high environmental and economic costs. However, the use of Fe (III)-reducing bacterium *Geobacter sulfurreducens* was found to be effective in manufacturing stable and non agglomerated Pd nanoparticles on biomagnetite (20 to 30 nm in size, 10 mol % of Pd) with reduced synthesis costs and easy recovery of the catalyst (*23*). These nanoparticles were effective in the Heck coupling reaction of iodobenzene to ethyl acrylate or styrene with a complete conversion to ethyl cinnamate or stilbene within 90 and 180 min, respectively. *Lactobacillus sp.* and *Sachharomyces cerevisae* can be used to synthesize TiO_2 nanoparticles (8 to 35 nm in size) at room temperature (*24*). Such synthesis was possible due to the pH-sensitive membrane bound oxido-reductases and to the carbon source dependent fH2 (formin homology proteins) in the culture solution.

De Corte and colleagues used *Shewanella oneidensis* in the presence of H_2 to manufacture bio-Pd/Au nanoparticles (Figure 4) that were effective in the dehalogenation (78% in 24 h) of environmental contaminants, such as diclofenac and trichlorethylene (*25*). Unfortunately, these nanoparticles failed to efficiently catalyze the degradation of various recalcitrant halogenated compounds.

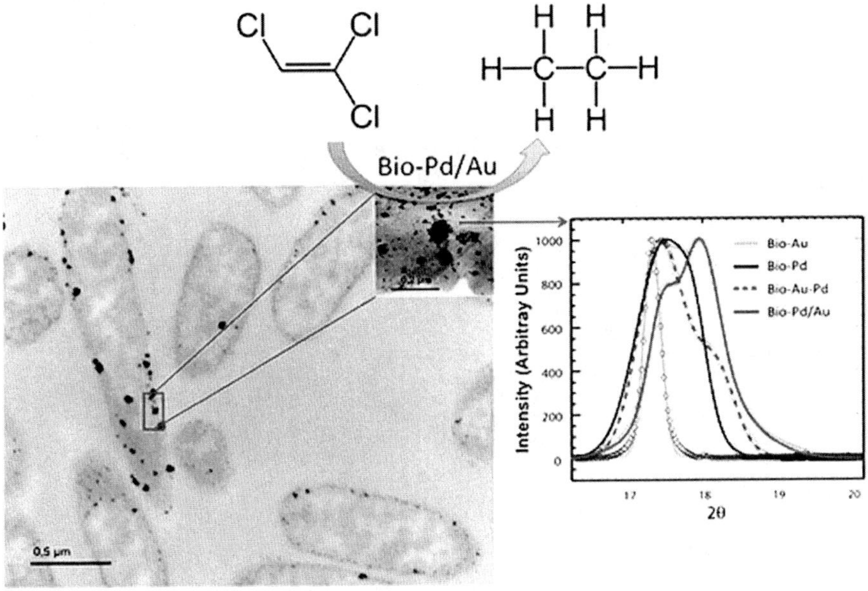

Figure 4. Biologically produced bio-Pd/Au nanoparticles by Shewanella oneidensis in the presence of H_2. Reprinted with permission from (25). Copyright (2011) American Chemical Society.

Platinum nanoparticles were synthesized using metal-reducing bacterium, *Shewanella algae*, at room temperature within 60 min at pH 7 from $PtCl_6^{2-}$ ions as the reaction precursor (*26*). Pt nanoparticles of about 5 nm were located in the periplasmic space, which is a preferable cell surface location for easy recovery of biogenic nanoparticles. Spherical gold nanoparticles and nanoplates were also synthesized using *Shewanella algae* as the reducing and stabilizing agent in the presence of H_2 as the electron donor and a pH of 2.8 at room temperature (*27*). The formation of well-dispersed nanoparticles was observed within 1 hour with a mean size of 9.6 nm. On the contrary, gold nanoplates with an edge length of 100 nm started to form within 6 h and more than 60% of the total nanoparticle population was nanoplates with an edge length of 100 to 200 nm within 24 h. The yield of gold nanoplates manufactured with *S. algae* extract was four times higher than that prepared with resting cells of *S. algae*.

Yeasts and Algae

Of all the eukaryotes, yeasts are probably the most studied and applied in bioprocesses. Yeasts are eukaryotic microorganisms in the kingdom of fungi, with more than one thousand species. Unfortunately, there is not sufficient data to discuss in detail the efficiency and the probable mechanisms of yeasts to manufacture metal and metal oxides. However, several studies indicate that yeasts extracts could be used to manufacture nanoparticles. For instance, a silver-tolerant yeast strain MKY3 was used to extracellularly synthesize silver nanoparticles with sizes ranging from 2 to 5 nm (*28*). Gold and silver nanoparticles with sizes from 20 to 100 nm and 10 to 20 nm, respectively at pHs from 4 to 10 were synthesized in the presence of *Saccharomyces cerevisae* broth in 24 and 48 hours, respectively, at room temperature (*29*). Distinct surface plasmon peaks were observed at 540 nm and 415 nm for gold and silver nanoparticles respectively.

Another yeast species, *Hanensula anomala* (Figure 5) was used to reduce the gold salt in the presence of amine-terminated polyamidoamine dendrimer and cysteine as the stabilizers (*30*).

Figure 5. Gold nanoparticles preparation using yeast species Hansenula anomala. Reprinted with permission from (30). Copyright (2011) American Chemical Society.

Candida guilliermondii is another suitable candidate to manufacture silver and gold nanoparticles that exhibit distinct surface plasmon peaks at 530 and 425 nm, respectively, with fcc structures as well as 50-70 nm and 10-20 nm in sizes (*31*). Silver and gold nanoparticles were tested against five pathogenic bacterial strains. The highest efficiency for both gold and silver nanoparticles was observed against *Staphylococcus aureus*, indicating suitability of these nanoparticles in an environmental remediation field. Furthermore, marine yeasts (*Pichia capsulata)* derived from the mangrove sediments were utilized for the synthesis of stable silver nanoparticles (1.5 mM $AgNO_3$, 0.3% NaCl, pH of 6.0 and incubated at 5 °C for 24 h, UVvis absorption peak at 430 nm) (*32*). *Yarrowia lipolytica* acted as a reducing and stabilizing agent in the synthesis of gold nanoparticles and nanoplates by varying concentrations of chloroauric acid at pH 4.5 (*33*). Herein authors reported that in the presence of 109 cells ml^{-1} and 0.5 or 1.0 mM of the gold salt, the mixture turned purple or golden yellow, respectively, indicating the formation of Au nanopartcles. Nanoparticles with various sizes were synthesized when 1010 cells ml^{-1} were incubated with 0.5, 1.0, or 2.0 mM chloroauric acid. On the contrary, nanoplates were synthesized in the presence of 3.0, 4.0, or 5.0

mM $HAuCl_4$ and 1011 cells ml^{-1}. Obviously, an increase in salt concentration and a fixed number of cells resulted in the increase in the nanoparticle size. However, an increase in cell numbers while using a constant gold salt concentration resulted in the significant decrease in nanoparticle size. As confirmed by FTIR data, carboxyl, hydroxyl, and amide groups on the surface of the cell were responsible for the synthesis.

There are only a limited number of reports available on the use of algae in the bio-synthesis of metal and metal oxide nanoparticles. Algae are a diverse group of photosynthetic simple, typically autotrophic organisms, ranging from unicellular to multicellular forms. For instance, *Spirulina platensis* has been used to synthesize silver and gold nanoparticles with sizes ranging from 7 to 16 nm and 6 to 10 nm, respectively and surface plasmon absorbance peaks at 424 nm and 530 nm in 120 h at 37 °C (*34*). When both silver and gold salts were mixed together during the synthesis protocol, core (Au)-shell (Ag) nanostructures (17 to 25 nm, bimetallic 50:50 ratio) were observed. For the core-shell nanoparticles, absorption peaks were observed at 509, 486, and 464 nm at 75:25, 50:50, and 25:75 (Au:Ag) mol concentrations, respectively. According to FTIR data, proteins were responsible for the reduction and stabilization of the metal and core-shell nanoparticles. Furthermore, marine microalgae *Tetraselmis suecica* was used to synthesize gold nanoparticles (79 nm in size) from chloroauric acid solution with a surface plasmon absorbance peak at 530 nm (*35*).

The formation of single-crystalline triangular and hexagonal gold nanoplates was observed when an aqueous solution of chloroauric acid was treated with the extract of the unicellular green algae *Chlorella vulgaris* at room temperature (*36*). A sustained synthesis of colloidal gold (9 nm in size, ca. 16 wt % of cellular gold and ca. 80 % metallic gold) in the presence of *Klebsormidium flaccidum* green algae was demonstrated proving the intracellular reduction process (*37*). Figure 6 shows the use of algae to produce nanocrystalts with various shapes.

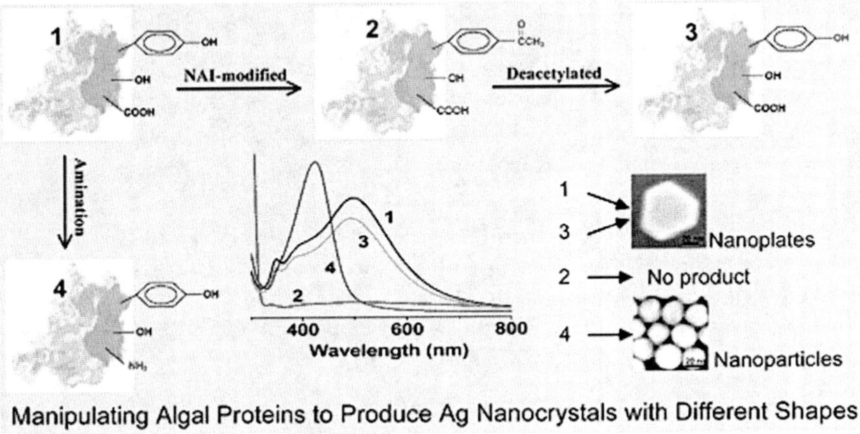

Figure 6. Schematics of the use of algae to produce silver nanoparticles. Reprinted with permission from (38). Copyright (2011) American Chemical Society.

Plants

Various plants can be used to accumulate and detoxify heavy metals (*6*), and, therefore, they might be tested as potential candidates to synthesize metal nanoparticles. Recently, the use of plant extracts and living plants has received an increased attention as a simple, viable, and effective approach to manufacturing nanoparticles as well as a good alternative to conventional chemical and physical synthesis methods. Currently, there are numerous plants and fruits that can be utilized for the reduction and stabilization of mono, bi, and tri-metallic nanoparticles in "one-pot" synthesis.

For instance, $AgNO_3$ and garlic (*Allium sativum*), as the Ag^+ and sulfur sources, respectively, were used to manufacture a visible light activated Ag- TiO_2 nanocatalyst, as can be seen in Figure 7 (*39*).

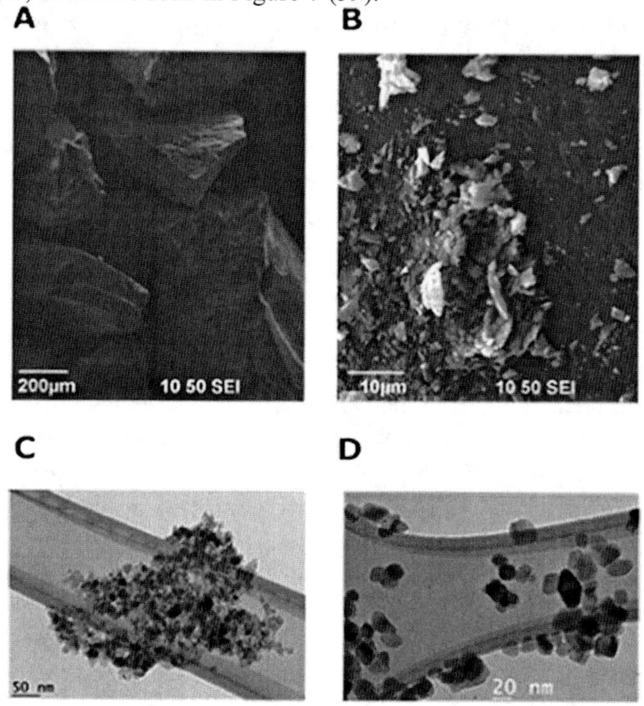

Figure 7. Selected SEM (a) Ag, garlic loaded-TiO2 calcined at 450 °C and (b) Ag, garlic loaded-TiO2 calcined at 700 °C and TEM (c) Ag, garlic loaded-TiO2 calcined at 450 °C, (d) Ag, garlic loaded-TiO2 calcined at 700 °C images of as-prepared catalysts (39). Reproduced by permission of The Royal Society of Chemistry (2012).

Such a catalyst does not change from anatase to rutile phase even at high (700°C) calcination temperatures, which is an important characteristic of an effective catalyst to be used in the environmental field. The catalyst was tested against the degradation of methyl orange and dichlorophenol under the visible light irradiation. It was found that the process followed first order reaction kinetics with reaction rates ranging from 9.5×10^{-4} to 19.4×10^{-3} min^{-1}. The dye decomposed in 3 h and more than 70% dichlorophenol degraded in 4 h. Demineralization of the solution was in the range of 40 to 90%. It is important to point out that a complete elimination of *Vibrio fischeri* was accomplished with 1 mg L^{-1} of nanocatalyst regardless of the calcination temperature. Ag-and garlic loaded TiO_2 was used for up to 5 consecutive runs without the significant loss in the photocatalytic activity and could be regenerated by adopting a simple procedure of washing with water and acetone as well as drying at 200 °C for 2 h.

It was reported that silver and gold nanoparticles can be manufactured inside living plants - alfalfa (*Medicago sativa*) (*40–44*). During the studies, alfalfa plants were grown in a silver and gold salt-rich environment; subsequently, nanoparticle synthesis was achieved within plants by ion uptake from solid media. For instance, crystalline icosahedral (ca. 4 nm) and fcc twinned (6-10 nm) gold nanoparticles were formed when $AuCl_4^-$ was treated with alfalfa plants (*43*, *44*). By varying the pH, crystalline gold nanoparticles with cluster radius of 6.2 Å at pH 5 and 9 Å at pH 2 were formed, suggesting that another layer of gold may have been deposited onto the nanoparticle surface at the lower pH (*42*). In addition, anisotropic polyhedra and nanoplates (30 to 60 nm in size and morphologies including 30 nm decahedra and 15 nm icosahedra nanoparticles in isopropanol extracts) were formed when 3 mM of $AuCl_4^-$ solution was reduced with *Medicago sativa* at room temperature and pH 3.5 and 3 (*45*). Interesting, when water was used as an extracting agent instead of isopropanol, triangular gold nanoplates were formed with 500 nm to 4 um in size.

In addition, *Citrus sinensis* peel was used to prepare silver nanoparticles at room and 60 °C temperatures with sizes ranging from 35 nm at 25 °C to ca. 10 nm for the synthesis at 60 °C (*46*). These silver nanoparticles were effectively used to destroy bacterial contamination, such as *E. coli*, *Ps. aeruginosa* (Gram-negative), and *S. aureus* (Gram-positive). Dried mahogany leaves (*Swietenia mahogany JACQ.)* that are rich in polyhydroxy limonoids were utilized to rapidly prepare silver, gold, and Ag/Au alloys (*47*). When the target metal salts were treated with leaf extracts, the competitive reduction of Au (II) and Ag (I) occurred resulting in the formation ofbimetallic Au/Ag alloy nanoparticles. *Hibiscus rosa sinensis* was used to produce stable silver and gold nanoparticles at room temperature (*48*).

In addition to plant extracts, *Indian propolis,* which is full of hydroxyflavonoids, and its chemical constituents pinocembrin and galangin are used to produce stable silver and gold nanoparticles with varied morphologies at pH 10.6 (*49*). *Anacardium occidentale* leaf extract, rich in polyols, can be used to synthesize silver, gold, Au/Ag alloy, and Au core-Ag shell nanoparticles at room temperature without any other reducing and stabilizing agents or any harsh reaction conditions (*50*). As suggested by XRD data, these nanoparticles were highly crystalline and predominantly in the cubic phase. *Camellia sinensis* extract, which contained polyphenols, was used to prepare silver and gold colloidal nanoparticles with average sizes of 40 nm that exhibited highly efficient single photon-induced luminescence that could be manipulated by changing concentrations of metal ions and the quantity of the reducing agent (*51*).

Ocimum sanctum (Tulsi) leaf extract demonstrated its applicability to reduce and stabilize silver nanoparticles with sizes ranging from 4 to 30 nm in 8 minutes (*52*). The resulting silver nanoparticles exhibited significant antimicrobial activity, which could be utilized in medical or environmental applications. The stem latex of a medicinal *Euphorbia nivulia* containing peptides and terpenoids can be effectively exploited to manufacture silver and copper nanoparticles with sizes ranging from 5 to 10 nm at room temperature in the presence of microwave irradiation for 30 sec (*53*). These nanoparticles were effectively used to diminish gram-positive and gram-negative bacterial contamination. Furthermore, gum kondagogu (*Cochlospermum gossypium*) was used to synthesize Au, Ag, and Pt nanoparticles at 45 to 75 °C in water (*54*). It was found that the -OH groups present in the gum matrix were responsible for the reduction of the target metal salt. Thus, highly crystalline with face centered cubic geometry Au, Ag, and Pt colloidal nanoparticles with average sizes of 7.8 ± 2.3 nm, 5.5 ± 2.5 nm, and 2.4 ± 0.7 nm, respectively, were formed and were extremely stable without undergoing any subsequent oxidation.

Over the last few years, sunlight-induced nanoparticle synthesis methods were found to be effective in producing stable metal nanoparticles. Ethanol extract of *Andrachnea chordifolia* was used to produce Ag nanoparticles (*55*). Interestingly, the smallest (3.4 nm) silver nanoparticles were formed when silver salt solution was irradiated for 5 min by direct sunlight radiation, whereas larger nanoparticles (3 to 30 nm in size) were formed when the solution was left in the dark for about 48 h. *Azadirachta indica* leaves extract that contained a significant amount of terpenoids was tested to produce silver nanoparticles (*56*). After 2 hours, stable spherical nanoparticles with an average diameter of 20 nm and crystallite size of 36.6 nm were formed. Unfortunately, after 4 hours of the reaction, agglomeration was observed indicating that synthesis approach may have to be optimized or improved.

Coleus amboinicus Lour (Indian Borage) is a medicinal plant that contains phytochemicals such as carvacrol (monoterpenoid), caryophyllene (bicyclic sesquiterpene), patchoulene, and flavanoids (quercetin, apigenin, luteolin, salvigenin, and genkwanin) and therefore, could be used to manufacture silver and gold nanoparticles (*17*, *57*, *58*). There was a significant shift in the shape of nanoparticles from anisotrophic nanostructures like triangle, decahedral, and hexagonal to isotrophic spherical nanoparticles when the leaf concentration was increased (*57*). Furthermore, these nanoparticles ranged from 2.6 to 79.8 nm; this difference in sizes highly depended on the concentration of the leaf extract. The crystalline nature of the nanoparticles was confirmed by XRD with peaks corresponding to (1 1 1), (2 0 0), (2 2 0), and (3 1 1) planes and bright circular spots in the selected-area electron diffraction (SAED). However, during gold nanoparticle synthesis, mainly spherical, truncated triangle, hexagonal, and decahedral nanoparticles were formed with sizes ranging from 4.6 to 55.1 nm (*57*). FTIR data proved that aromatic amines, amide (II) groups, and secondary alcohols were responsible for the reduction and stabilization of silver and gold nanoparticles.

Roots can be used to synthesize metal nanoparticles. For example, Korean red ginseng root was used to manufacture stable gold nanoparticles in cysteine, histidine, saline, and sodium chloride at various pHs with sizes from 2 to 40 nm without the addition of other chemicals or stabilizing agents (*59*). Such a wide distribution in sizes could be attributed to the difference in kinetics of nanoparticle formation if compared to a more conventional $NaBH_4$-induced method. It is generally accepted that $NaBH_4$ is a strong reducing agent that can rapidly reduce metal ions and nucleate a number of particles by depleting the gold salt before large particles can be formed. On the contrary, ginseng roots-facilitated reduction process is slower in comparison to that of $NaBH_4$, which may result in the formation of various-sized particles. Importantly, gold nanoparticles synthesized with ginseng roots did not aggregate in the solution, which strongly suggested that phytochemicals (ginsenosides polysaccharides, flavones, etc.) acted as coatings providing robust shielding from aggregation. In addition, these coatings proved to be non-toxic as demonstrated through detailed cytotoxicity assays using WST-8 counting kit, performed on normal cervical cells lines.

In addition, rubber latex extracted from *Hevea brasiliensis* provided a suitable alternative to synthesizing spherical colloidal silver nanoparticles with sizes ranging from 2 to 100 nm with a fcc crystalline structure and a characteristic surface plasmonic absorption band at around 435 nm (*60*). It should be noted that lower silver nitrate salt concentrations led to smaller nanoparticle sizes. Yet again, FTIR data confirmed that the reduction of silver ions was predominantly due to the interaction between silver ions and amine groups from ammonia, whereas the stability of the particles resulted from cis-isoprene binding onto the surface of nanoparticles.

Macrotyloma uniflorum can be used to prepare noble metal nanoparticles (*61*, *62*). Thus, well dispersed silver nanoparticles with sizes ranging from 12 to 17 nm and fcc crystalline structure were formed. Moreover, phenolic acids like *p*-hydroxy benzoic acid, 3,4-dihydroxy benzoic acid, *p*-coumaric acid, caffeic acid, ferulic acid, vanilic acid, syringic acid, and sinapic acid were responsible for the reduction and stabilization of silver nanoparticles.

Recently, the use of *Mangifera indica* leaf extract was proved to be effective in the manufacturing well-dispersed silver nanoparticles with surface plasmon resonance at 439 nm (*63*). At pH 8, triangular, hexagonal, and nearly spherical nanoparticles with average sizes of 20 nm were formed. Data suggested that water soluble organics present in the leaf were responsible for the reduction and stabilization of silver ions and silver nanoparticles, respectively.

The ethanolic flower extract of *Nyctanthes arbortristis* (80 °C for 30 min) and water extract of *Terminalia chebula* were used to prepare gold nanoparticles with average sizes of 19.8 nm and overall sizes ranging from 6 to 60 nm (*64*, *65*). It is important to point out that low reaction temperature (ca. 25 °C) favored anisotropy (*64*). FTIR analysis confirmed that the hydrolysable tannins present in the plant extracts were responsible for the reduction and stabilization of gold nanoparticles (*65*). Moreover, aqueous hybrid phenolic-rich sorghum (*Sorghum spp*) was tested in the synthesis of stable Fe and Ag nanocolloids at room temperature (*66*). Hybrid sorghums contain plentiful phenolic compounds and are water soluble, non-toxic, and biodegradable compounds that could be effectively utilized in the "green" synthesis of metal nanoparticles, where they act as both, the reducing and stabilizing agents. Most sorghum species are drought and heat tolerant, making it economical to produce. Ag and Fe nanoparticles synthesized this way were largely uniform with a narrow size distribution and average sizes of 10 and 50 nm, respectively.

Carbohydrates

Carbohydrates are the most abundant class of organic compounds found in living organisms that contains simple sugars, starches, cellulose, etc. Owing to their chemical structure and constituents, these compounds can be used to synthesize nanomaterials where they act as both the reducing and stabilizing agents. Hydroxypropyl starch (Figure 8) was used to synthesize stable Ag nanoparticles with fcc structures without aggregation for more than 6 months and 6 to 8 nm in size (*67*, *68*). It was found that glucose, one of the constituents of the starch, was responsible for the reduction of silver ions. As suggested by FTIR data, silver particles were covered by a starch layer and formed spherical core-shell Ag/starch nanoparticles.

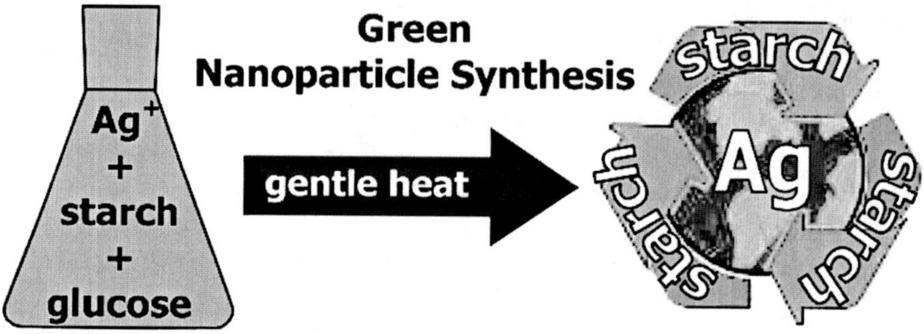

Figure 8. Schematic presentation of Ag nanoparticle preparation in starch and in the presence of glucose solution. Reprinted with permission from (69). Copyright (2003) American Chemical Society.

Hydroxypropyl cellulose can be used to synthesize silver nanoparticles and simultaneously be used as a dopant for visible light active TiO_2 (*70*, *71*). The optimum conditions to synthesize stable silver nanoparticles were pH 12.5, 0.3% hydroxypropyl cellulose solution having a molar substitution of 0.42, and carrying out the reaction at 90 °C for 90 min (*70*). When hydroxypropyl cellulose was used as a dopant, stable nanoparticles were prepared using a "one-pot" synthesis at mild reaction conditions and room temperature. These nanoparticles had an average size of 17 nm and a BET surface area of 133.8 $m^2\ g^{-1}$ that were efficient in degrading organic contaminants in the visible light range (*71*).

Baruwati and Varma (*72*) demonstrated the synthesis of highly dispersed Au, Ag, Pd, and Pt nanoparticles with a narrow size distribution (average size of 5 to 10 nm) in aqueous medium using grape pomace. The grape pomace contained polyphenols that acted as both reducing and capping agents. Authors found that nanoparticles were formed within a few seconds of the reaction when exposed to microwave irradiation at a power level as low as 50 W.

In addition, stable Ag nanoparticles with a relatively narrow size distribution and an average particle size of 25 nm were prepared using monosaccharides (glucose and galactose) and disaccharides (maltose and lactose) by varying the ammonia as the reaction precursor concentration (0.2 to 0.005 mol l^{-1})

(*73*). These as-prepared silver nanoparticles with concentration as low as 1.69 μg ml^{-1} of Ag, showed high antimicrobial and bactericidal activity against Gram-positive and Gram-negative bacteria, including multi-resistant strains such as methicillin-resistant *Staphylococcus aureus*. On the other hand, honey can also be used to manufacture metal nanoparticles. Simple pH adjustment produced highly crystalline nanoparticles of various sizes at room temperature (*74*). For instance, at pH 8.5, predominantly monodispersed and nearly spherical colloidal Ag nanoparticles with an average size of 4 nm were produced. Honey can be used to produce Pt nanoparticles and nanowires with average sizes of 2.2 nm at 100 °C (*75*). However, when the reaction solution was left for extended period of time, crystalline fcc structured Pt nanowires with 5 to 15 nm in lengths were formed. According to FTIR data, nanoparticles were bound to protein through the carboxylate ion group.

Furthermore, polyphenols in tea and coffee extracts as well as epicatechins can be successfully utilized for the production of metal nanoparticles (*76–78*). This way synthesized silver nanoparticles (4 to 100 nm in size in the presence of tea and 15 to 26 nm in the presence of epicatechins), demonstrated benign nature when in contact with human keratinocytes (HaCaTs) used as an *in vitro* model of exposure (*79*). Particle morphology can be seen in Figure 9. In addition, Ag and Pd nanoparticles were manufactured utilizing polyphenols from tea and coffee. The nanoparticles were in the size range of 20–60 nm, depending on the ratios of the reaction precursors and crystallized in face centered cubic symmetry.

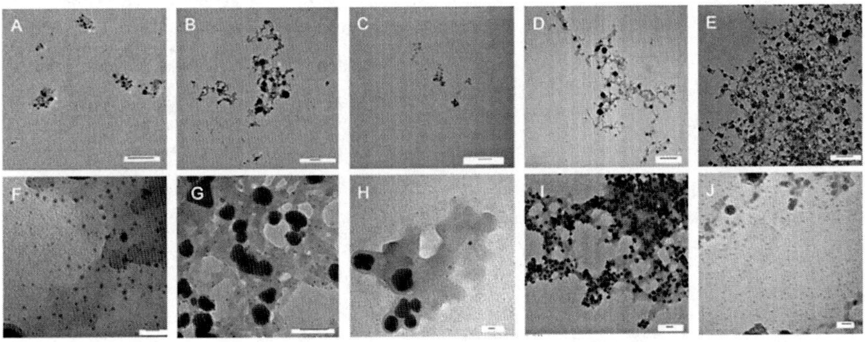

Figure 9. Morphology of the nanoparticles . A) 1 : 1 ratio of water to epicatechin (scale bar: 100nm), B) 2 : 1 ratio of water to epicatechin (scale bar: 100nm), C) 10 : 3 ratio of water to epicatechin (scale bar: 100nm), D) 10 : 1 ratio of water to epicatechin (scale bar: 100nm), E) 20 : 1 ratio of water to epicatechin (scale bar: 100nm), F) 1 : 1 ratio of water to tea extract (scale bar: 20nm), G) 2 : 1 ratio of water to tea extract (scale bar: 100nm), H) 10 : 3 ratio of water to tea extract (scale bar: 20nm), I) 10 : 1 ratio of water to tea extract (scale bar: 100nm), J) 20 : 1 ratio of water to tea extract (scale bar: 20nm) (79). Reproduced by permission of The Royal Society of Chemistry (2010).

Vitamins

Vitamins (B, C, D,and E) can be used to manufacture stable metal nanoparticles applicable for numerous industries and environmental remediation including nano-catalysis. For instance, silver nanoparticles were fabricated from silver salt with vitamin C derivative 6-palmitoyl ascorbic acid-2-glucoside (PAsAG) that worked as both reducing and capping agents in the presence of ultrasound (*80*). It was found that Ag-PAsAG nanocomplex with the average size of 5 nm was subsequently formed and was utilized for the protection of DNA from γ-radiation–induced damages. Nanocomplex inhibited the disappearance of covalently closed circular (ccc) form of plasmid pBR322 with a dose modifying factor of 1.78. Vitamin C was also used to manufacture Au nanodogbones utilizing simple seeded mediated growth method (*81*). The addition of vitamin from 10 to 40 μL significantly influenced the shapes and morphology of nanoparticles. Therefore, when the aspect ratios of nanodogbones were in the range of 2.34 to 1.46, the UV–Vis absorption measurement showed a significant blueshift on the longitudinal surface plasmon resonance band from 713 to 676 nm. Vitamin E was used to synthesize monodispersed silver nanoparticles by simple "one step" procedure in alkaline subphase beneath Vitamin E Langmuir monolayers (*82*). According to the findings, phenolic groups in Vitamin E molecules were converted to a quinone structure; the silver ions were mainly reduced to ellipsoidal and spherical nanoparticles with face centered cubic crystallinity.

Nadagouda and Varma (*83*, *84*) used water soluble, biodegradable, and non-toxic Vitamin B_2 to manufacture Ag and Pd nanoparticles. They discovered that the addition of solvents alters the sizes of the nanoparticles. For instance, the average particle sizes of Ag and Pd were 6.1 ± 0.1nm and 4.1 ± 0.1 nm in ethylene glycol as well as 5.9 ± 0.1 nm and 6.1 ± 0.1 in acetic acid and N-methylpyrrolidinone (NMP), respectively. In the presence of water, Ag and Pd nanoparticles were rod-shaped, whereas in iso-propanol they were wire-shaped with thickness in the range of 10 to 20 nm and several hundred microns in length. On the contrary, in the presence of acetonitrile, nanorods were formed with thicknesses ranging from 100 to 200 nm and lengths of a few microns. Morphologies of Pd nanoparticles can be seen in Figure 10.

Magnetic Nanoparticles

It is vital to manufacture inexpensive and recoverable nanoparticles. Therefore, nanoparticles on a passive or active magnetic core would offer additional advantages in comparison to the un-supported ones. Over the years, magnetic nanoparticles have been utilized to remove heavy metals and destroy dyes, for drug delivery, protein and cell separation, enzyme immobilization, catalysis, biotechnology, magnetic resonance imaging, and environmental remediation because magnetic separation offers unique advantages of high efficiency, cost effectiveness, and rapidity in comparison to other nanoparticles (*86*). Nowadays, magnetic nanoparticles are predominantly used as heterogeneous supports in various catalytic reactions providing easier recoverability with a simple magnet while eliminating the need to employ the solvent before or catalyst filtration after the target process (*87–89*). However, one of the biggest problems that remains unsolved is the intrinsic instability over long periods of time. Usually, magnetic nanoparticles are less than 20 nm in size; therefore they are susceptible to aggregation and agglomeration, which reduces the energy associated with the high surface area to volume ratio. Another drawback is the oxidation of the surface in air, which may result in a complete loss of magnetism and dispersability. Therefore, to avoid all the drawbacks, magnetic nanoparticles should be capped and stabilized using non toxic and environmentally friendly species.

Figure 10. Palladium (Pd) nanobelts, nanoplates and nanotrees using vitamin B1 without any special capping agents, synthesized at room temperature (85). Reproduced by permission of The Royal Society of Chemistry (2009).

To manufacture monodisperse magnetic nanoparticles, anionic surfactants, dispersing agents, proteins, starches, and polyelectrolytes can be added to improve the desired properties for target applications. Recently, bifunctional precious metal-magnetic nanoparticles gained an increased attention due to their potential applications in novel electrical, optical, magnetic, catalytic, and sensing technologies. For example, Fe_2O_3/Ag core–shell magnetic nanocomposite (sizes ranging from 30 to 60 nm) was synthesized using a simple method at room temperature in the presence of sodium hydroxide (NaOH), TSA-doped polyaniline, dimethylsulfoxide (DMSO), and dimethylformamide (DMF) without calcination (*90*).

Over the past years, spinel ferrites (*91–94*) - magnetic $CoFe_2O_4$ nanoparticles in particular, became important constituents in synthetic and biological chemistry as well as magnetic labels of biological systems due to their unique properties, such as strain enhanced coercivity and photo induced magnetic effects (*95*). According to the scientific evidence, several energy intensive techniques could be combined to produce stable spinel ferrites. For instance, a sonochemical and co-precipitation approach in aqueous medium without any surfactant or organic capping agent at 60

°C can be used to manufacture stable spinel cobalt ferrite magnetic nanoparticles with average sizes of 40 to 50 nm (*95*).

"Green" synthesized nano-ferrites can be used for the hydrogen production via water splitting. Thus, nano-ferrites, ethanol as the sacrificial donor, and Pt as co-catalyst were employed to produce hydrogen with the rate of 8.275 μmol h^{-1} and a hydrogen yield of 8275 μmol h^{-1} g^{-1} under visible light compared to 0.0046 μmol h^{-1} using commercial iron oxide that was tested under similar experimental conditions (*96*). It was found that the rate of hydrogen evolution was significantly influenced by the catalyst dose, sacrificial donor, temperature, and the presence of the co-catalyst. Figure 11 shows the schematics of nanoparticle anchoring onto nanoferrite for various applications.

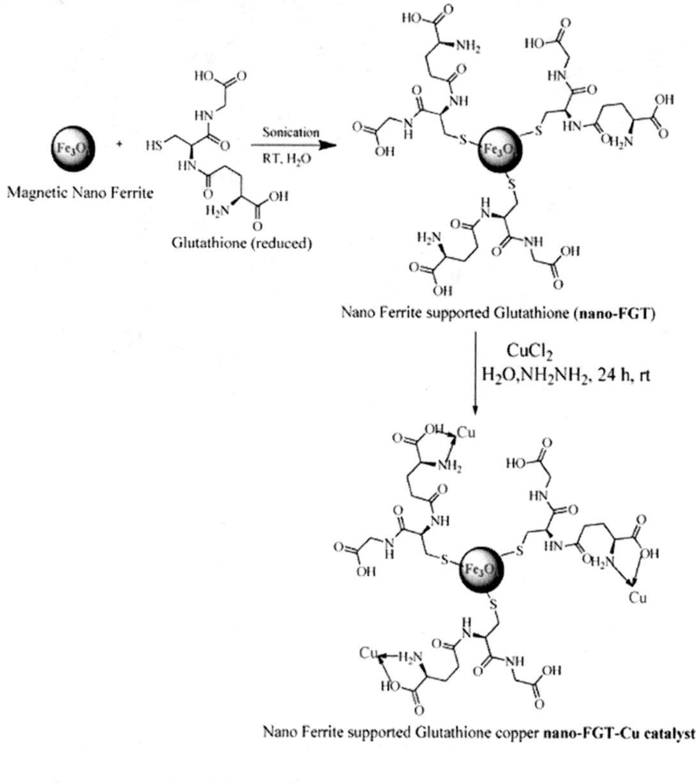

Figure 11. Nano ferrite supported glutathione-copper (nano-FGT-Cu) catalyst (97). Reproduced by permission of The Royal Society of Chemistry (2012).

Hydroxyapatites, which are the main constituent of bones and teeth and are widely used in bone repairs, bone implants, bioactive materials, purification, and separation of biological molecules, can also be used to manufacture magnetic nanocomposites for various industrial applications, including remediation of soils, water, and wastewater (*98*). To further increase their effectivity, functionalization with β-cyclodextrin and pluronic polymer (F127) at room temperature under a nitrogen atmosphere could be accomplished; the final product could be used for drug delivery applications (*99*).

To avoid agglomeration, various porous solids such as clays and zeolites that offer high surface areas and have easily accessible channels for molecules, can be used to disperse nanoparticles (*100*, *101*). As determined by Li et al (*101*), clay supported nanoparticles can be effectively manufactured by exchanging the interlamellar alkali cations with metallic polyoxocations in aqueous solution. Calcination would convert polyoxocations to nanoparticles of corresponding metals anchored on the silicate platelets. The clay layers would partition metal nanoparticles preventing them from sintering, whereas a wide interlayer gallery might prevent pores from being blocked by the metal oxide particles. These clay supported nanoparticles can be used for a great array of applications, especially those that require large surface areas, good thermal stability, and surface acidity. For instance, iron, chromium, cobalt, manganese and cerium pillared clay (laponite) with the surface areas of 510 to 640 $m^2 g^{-1}$ and total pore volume of 0.43 to 0.51 $cm^3 g^{-1}$ was synthesized in the presence of a stabilizer – acetic acid (*101*). Herein, metal precursors exchanged sodium ions in laponite forming mesoporous solids with well dispersed metal oxides and a large surface area. In addition, bismuth and copper pillared montmorillonite was synthesized using mild reaction conditions that was utilized for the removal of organic contaminants from water (*102*). In that particular study, authors found that novel Claycunbic was able to simultaneously remove up to 95% of TOC and completely eliminate methylene blue, methyl orange, and dichlorophenol as well as entirely destroying microbial contamination in less than 2 hours. Furthermore, Claycunbic exhibited excellent stability up to 5 consecutive runs in terms of degradation of MB (100–88%) and regained nearly 100% of its catalytic activity after intermediate calcination at 400 °C for 2 h.

Conclusions

There is always a search for reliable and environmentally friendly processes to manufacture metal and metal oxide nanoparticles minimizing or even eliminating the use of hazardous chemicals. The only way to develop these "green" processes is to adapt benign synthesis approaches that use mild reaction conditions and non-toxic reaction precursors. It is likely that in the near future, synthesis of nanoparticles will emerge from the laboratory scale and will materialize to the synthesis of large quantities of structurally well-defined, stable, surface-functionalized, and effective nanoparticles that could be used in a great array of applications, environmental remediation included. However, to accomplish that goal, the efficient use of natural resources, such as biological systems, becomes necessary. The use of organisms (microorganisms and plants in particular) in the preparation of nanoparticles is usually performed close to ambient temperatures and pressures, at circum neutral pH, and with utilizing benign reaction precursors. According to the findings from numerous researchers, bacteria are relatively easy to control genetically, whereas plants and fungi are much easier to handle during downstream processing and during large scale production. However, the selection of the organisms is not vitally important when compared to understanding the biochemical and molecular mechanisms of nanoparticle formation.

This chapter focused on the use of various organisms (plants, bacteria, fungi), vitamins, and carbohydrates in environmentally-friendly preparation of metal and metal oxide nanoparticles under mild reaction conditions and with the limited use of hazardous chemicals. State-of-the-art in terms of the selection of synthesis precursors, shape control, and surface functionalization identifying drawbacks and challenges are described. Detailed examination of the case studies suggest that despite the vast amount of information available on the actual preparation of nanoparticles, more advancement is necessary to develop large scale continuous preparation of metal nanoparticles.

The key findings in the "green" synthesis of metal and metal oxide nanoparticles are as follows: 1) there is a need to eliminate the use of toxic reagents and organic solvents; therefore "greener" approaches that use benign reducing and capping agents are under continuous development, 2) "green" synthesis produces little or no hazardous by-products, 3) "green" synthesis provides better control of shapes, sizes and dispersivity of metal nanoparticles, and 4) "green synthesis" eliminates the need for purification of the as-prepared nanoparticles, which in turn eliminates the use of large amounts of organic solvents that are hazardous to the environment.

Despite the good advancement in the field, several issues still need to be resolved: 1) identification of the risks that are associated with exposure to nanomaterials, 2) development of even more "green" synthesis approaches, eliminating energy intensive processes and increasing the use of various wastes, and 3) development of various analytical techniques that would be able to analyze nanoparticles *in situ* for morphology, composition, structure, and purity regardless of the experimental matrix. Thus, despite the current innovations and encouraging results, further research is needed to improve current methods and processes that will provide opportunities and challenges for the scientific community and public for the years to come.

References

1. Eckelman, M. J.; Zimmerman, J. B.; Anastas, P. T. Toward green nano. *J. Ind. Ecol.* **2008**, *12* (3), 316–328.
2. Christian, P.; Von der Kammer, F.; Baalousha, M.; Hofmann, T. Nanoparticles: Structure, properties, preparation and behaviour in environmental media. *Ecotoxicology* **2008**, *17* (5), 326–343.
3. Wang, Z. L. Toward self-powered sensor networks. *Nanotoday* **2010**, *5*, 512–514.
4. Wu, W.; Wei, Y.; Wang, Z. L. Strain-gated piezotronic logic nanodevices. *Adv. Mater.* **2010**, *22*, 4711–4715.
5. Das, S. K.; Marsili, E. A green chemical approach for the synthesis of gold nanoparticles: Characterization and mechanistic aspect. *Rev. Environ. Sci. Bio/Technol.* **2009**, *9*, 199–204.

6. Iravani, S. Green synthesis of metal nanoparticles using plants. *Green Chem.* **2011**, *13* (10).
7. Huczko, A. Template-based synthesis of nanomaterials. *Appl. Phys. A: Mater. Sci. Process.* **2000**, *70* (4), 365–376.
8. Teng, X.; Black, D.; Watkins, N. J.; Gao, Y.; Yang, H. Platinum-maghemite core–shell nanoparticles using a sequential synthesis. *Nano Lett.* **2003**, *3* (2), 261–264.
9. Pomogailo, A. D.; Kestelman, V. N. *Metallopolymer Nanocomposites*; Springer: Berlin, Heidelberg, New York, 2005; Vol. 81, p 563.
10. Stubbs, D.; Gilman, P. *Nanotechnology Applications in Environmental Health: Big Plans for Little Particles*; Oak Ridge Center for Advanced Studies: Research Triangle Park, NC, 2007.
11. M, S.; A, A.; Islam, N. I.; K., R. Biosynthesis of metal nanoparticles using fungi and actinomycete. *Curr. Sci.* **2003**, *85*, 162–170.
12. Jebali, A.; Ramezani, F.; Kazemi, B. Biosynthesis of silver nanoparticles by Geotricum. sp. *J. Cluster Sci.* **2011**, *22* (2), 225–232.
13. Kirubaharan, C. J.; Kalpana, D.; Lee, Y. S.; Kim, A. R.; Yoo, D. J.; Nahm, K. S.; Kumar, G. G. Biomediated silver nanoparticles for the highly selective copper(II) ion sensor applications. *Ind. Eng. Chem. Res.* **2012**, *51* (21), 7441–7446.
14. Mandal, D.; Bolander, M.; Mukhopadhyay, D.; Sarkar, G.; Mukherjee, P. The use of microorganisms for the formation of metal nanoparticles and their application. *Appl. Microbiol. Biotechnol.* **2006**, *69* (5), 485–492.
15. Narayanan, K. B.; Sakthivel, N. Biological synthesis of metal nanoparticles by microbes. *Adv. Colloid Interface Sci.* **2010**, *156* (1–2), 1–13.
16. Mukherjee, P.; Ahmad, A.; Mandal, D.; Senapati, S.; Sainkar, S. R.; Khan, M. I.; Parishcha, R.; Ajaykumar, P. V.; Alam, M.; Kumar, R.; Sastry, M. Fungus-mediated synthesis of silver nanoparticles and their immobilization in the mycelial matrix: A novel biological approach to nanoparticle synthesis. *Nano Lett.* **2001**, *1* (10), 515–519.
17. Narayanan, K. B.; Sakthivel, N. Synthesis and characterization of nano-gold composite using *Cylindrocladium floridanum* and its heterogeneous catalysis in the degradation of 4-nitrophenol. *J. Hazard. Mater.* **2011**, *189* (1–2), 519–525.
18. Fayaz, A. M.; Balaji, K.; Girilal, M.; Yadav, R.; Kalaichelvan, P. T.; Venketesan, R. Biogenic synthesis of silver nanoparticles and their synergistic effect with antibiotics: A study against gram-positive and gram-negative bacteria. *Nanomed.: Nanotechnol., Biol. Med* **2010**, *6* (1), 103–109.
19. Korbekandi, H.; Abedi, D.; Jalali, M.; Fazeli, M.-R.; Heidari, M. Optimization of *Lactobacillus casei* growth and lactic acid production in batch culture. *J. Biotechnol.* **2007**, *131* (2) (Supplement), S182–S183.
20. Thakkar, K. N.; Mhatre, S. S.; Parikh, R. Y. Biological synthesis of metallic nanoparticles. *Nanomed.: Nanotechnol., Biol. Med* **2010**, *6* (2), 257–262.
21. Ahmad, A.; Senapati, S.; Khan, M. I.; Kumar, R.; Sastry, M. Extracellular biosynthesis of monodisperse gold nanoparticles by a novel extremophilic actinomycete, Thermomonospora sp. *Langmuir* **2003**, *19* (8), 3550–3553.

22. Kalishwaralal, K.; Deepak, V.; Ram Kumar Pandian, S.; Kottaisamy, M.; BarathManiKanth, S.; Kartikeyan, B.; Gurunathan, S. Biosynthesis of silver and gold nanoparticles using *Brevibacterium casei*. *Colloids Surf.,B* **2010**, *77* (2), 257–262.
23. Coker, V. S.; Bennett, J. A.; Telling, N. D.; Henkel, T.; Charnock, J. M.; van der Laan, G.; Pattrick, R. A. D.; Pearce, C. I.; Cutting, R. S.; Shannon, I. J.; Wood, J.; Arenholz, E.; Lyon, I. C.; Lloyd, J. R. Microbial engineering of nanoheterostructures: Biological synthesis of a magnetically recoverable palladium nanocatalyst. *ACS Nano* **2010**, *4* (5), 2577–2584.
24. Jha, A. K.; Prasad, K.; Kulkarni, A. R. Synthesis of TiO_2 nanoparticles using microorganisms. *Colloids Surf., B* **2009**, *71* (2), 226–229.
25. De Corte, S.; Hennebel, T.; Fitts, J. P.; Sabbe, T.; Bliznuk, V.; Verschuere, S.; van der Lelie, D.; Verstraete, W.; Boon, N. Biosupported bimetallic Pd–Au nanocatalysts for dechlorination of environmental contaminants. *Environ. Sci. Technol.* **2011**, *45* (19), 8506–8513.
26. Konishi, Y.; Ohno, K.; Saitoh, N.; Nomura, T.; Nagamine, S.; Hishida, H.; Takahashi, Y.; Uruga, T. Bioreductive deposition of platinum nanoparticles on the bacterium Shewanella algae. *J. Biotechnol.* **2007**, *128* (3), 648–653.
27. Ogi, T.; Saitoh, N.; Nomura, T.; Konishi, Y. Room-temperature synthesis of gold nanoparticles and nanoplates using Shewanella algae cell extract. *J. Nanoparticle Res.* **2010**, *12* (7), 2531–2539.
28. Kowshik, M.; Shriwas, A.; Sharmin, K.; Vogel, W.; Urban, J.; Kulkarni, S. K.; Paknikar, K. M. Extracellular synthesis of silver nanoparticles by a silver-tolerant yeast strain MKY3. *Nanotechnology* **2003**, *14* (1), 95.
29. Lim, H.-A.; Mishra, A.; Yun, S.-I. Effect of pH on the extra cellular synthesis of gold and silver nanoparticles by *Saccharomyces cerevisae*. *J. Nanosci. Nanotechnol.* **2011**, *11* (1), 518–522.
30. Kumar, S.; R, A.; Arumugam, P.; Berchmans, S. Synthesis of Gold Nanoparticles: An ecofriendly approach using *Hansenula anomala*. *ACS Appl. Mater. Interfaces* **2011**, *3* (5), 1418–1425.
31. Mishra, A.; Tripathy, S. K.; Yun, S.-I. Bio-Synthesis of gold and silver nanoparticles from *Candida guilliermondii* and their antimicrobial effect against pathogenic bacteria. *J. Nanosci. Nanotechnol.* **2011**, *11* (1), 243–248.
32. Subramanian, M.; Alikunhi, N. M.; Kandasamy, K. In vitro synthesis of silver nanoparticles by marine yeasts from coastal mangrove sediment. *Adv. Sci. Lett.* **2010**, *3* (4), 428–433.
33. Pimprikar, P. S.; Joshi, S. S.; Kumar, A. R.; Zinjarde, S. S.; Kulkarni, S. K. Influence of biomass and gold salt concentration on nanoparticle synthesis by the tropical marine yeast *Yarrowia lipolytica* NCIM 3589. *Colloids Surf., B* **2009**, *74* (1), 309–316.
34. Govindaraju, K.; Basha, S.; Kumar, V.; Singaravelu, G. Silver, gold and bimetallic nanoparticles production using single-cell protein (*Spirulina platensis*) Geitler. *J. Mater. Sci.* **2008**, *43* (15), 5115–5122.
35. Shakibaie, M.; Forootanfar, H.; Mollazadeh-Moghaddam, K.; Bagherzadeh, Z.; Nafissi-Varcheh, N.; Shahverdi, A. R.; Faramarzi, M. A.

Green synthesis of gold nanoparticles by the marine microalga *Tetraselmis suecica*. *Biotechnol. Appl. Biochem.* **2010**, *57* (2), 71–75.

36. Xie, J.; Lee, J.-Y.; Wang, D. I. C.; Ting, Y.-P. Identification of active biomolecules in the high-yield synthesis of single-crystalline gold nanoplates in algal solutions. *Small* **2007**, *3* (4), 672–682.
37. Dahoumane, S. A.; Djediat, C.; Yépremian, C.; Couté, A.; Fiévet, F.; Coradin, T.; Brayner, R. Recycling and adaptation of *Klebsormidium flaccidum* microalgae for the sustained production of gold nanoparticles. *Biotechnol. Bioeng.* **2012**, *109* (1), 284–288.
38. Xie, J.; Lee, J. Y.; Wang, D. I. C.; Ting, Y. P. Silver nanoplates: From biological to biomimetic synthesis. *ACS Nano* **2007**, *1* (5), 429–439.
39. Virkutyte, J.; Varma, R. S. Synthesis and visible light photoactivity of anatase Ag and garlic loaded TiO_2 nanocrystalline catalyst. *RSC Adv.* **2012**, *2* (6), 2399–2407.
40. Gardea-Torresdey, J. L.; Gomez, E.; Peralta-Videa, J. R.; Parsons, J. G.; Troiani, H.; Jose-Yacaman, M. Alfalfa sprouts: A natural source for the synthesis of silver nanoparticles. *Langmuir* **2003**, *19* (4), 1357–1361.
41. Gardea-Torresdey, J. L.; Parsons, J. G.; Gomez, E.; Peralta-Videa, J.; Troiani, H. E.; Santiago, P.; Yacaman, M. J. Formation and growth of Au nanoparticles inside live alfalfa plants. *Nano Lett.* **2002**, *2* (4), 397–401.
42. Gardea-Torresdey, J. L.; Tiemann, K. J.; Gamez, G.; Dokken, K.; Cano-Aguilera, I.; Furenlid, L. R.; Renner, M. W. Reduction and accumulation of gold(III) by *Medicago sativa* alfalfa biomass: X-ray absorption spectroscopy, pH, and temperature dependence. *Environ. Sci. Technol.* **2000**, *34* (20), 4392–4396.
43. Gardea-Torresdey, J. L.; Tiemann, K. J.; Gamez, G.; Dokken, K.; Tehuacanero, S.; José-Yacamán, M. Gold nanoparticles obtained by bio-precipitation from gold(III) solutions. *J. Nanopart. Res.* **1999**, *1* (3), 397–404.
44. Gardea-Torresdey, J. L.; Tiemann, K. J.; Parsons, J. G.; Gamez, G.; Yacaman, M. J. Characterization of trace level Au(III) binding to alfalfa biomass (*Medicago sativa*) by GFAAS. *Adv. Environ. Res.* **2002**, *6* (3), 313–323.
45. Montes, M.; Mayoral, A.; Deepak, F.; Parsons, J.; Jose-Yacamán, M.; Peralta-Videa, J.; Gardea-Torresdey, J. Anisotropic gold nanoparticles and gold plates biosynthesis using alfalfa extracts. *J. Nanopart. Res.* **2011**, *13* (8), 3113–3121.
46. Kaviya, S.; Santhanalakshmi, J.; Viswanathan, B.; Muthumary, J.; Srinivasan, K. Biosynthesis of silver nanoparticles using *Citrus sinensis* peel extract and its antibacterial activity. *Spectrochim. Acta, Part A* **2011**, *79* (3), 594–598.
47. Mondal, S.; Roy, N.; Laskar, R. A.; Sk, I.; Basu, S.; Mandal, D.; Begum, N. A. Biogenic synthesis of Ag, Au and bimetallic Au/Ag alloy nanoparticles using aqueous extract of mahogany (*Swietenia mahogani* JACQ.) leaves. *Colloids Surf., B* **2011**, *82* (2), 497–504.

48. Philip, D. Green synthesis of gold and silver nanoparticles using *Hibiscus rosa sinensis*. *Phys. E: Low-Dimens. Syst. Nanostruct.* **2010**, *42* (5), 1417–1424.
49. Roy, N.; Mondal, S.; Laskar, R. A.; Basu, S.; Mandal, D.; Begum, N. A. Biogenic synthesis of Au and Ag nanoparticles by *Indian propolis* and its constituents. *Colloids Surf., B* **2010**, *76* (1), 317–325.
50. Sheny, D. S.; Mathew, J.; Philip, D. Phytosynthesis of Au, Ag and Au-Ag bimetallic nanoparticles using aqueous extract and dried leaf of *Anacardium occidentale*. *Spectrochim. Acta, Part A* **2011**, *79* (1), 254–262.
51. Vilchis-Nestor, A. R.; Sánchez-Mendieta, V.; Camacho-López, M. A.; Gómez-Espinosa, R. M.; Camacho-López, M. A.; Arenas-Alatorre, J. A. Solventless synthesis and optical properties of Au and Ag nanoparticles using *Camellia sinensis* extract. *Mater. Lett.* **2008**, *62* (17-18), 3103–3105.
52. Singhal, G.; Bhavesh, R.; Kasariya, K.; Sharma, A.; Singh, R. Biosynthesis of silver nanoparticles using *Ocimum sanctum* (Tulsi) leaf extract and screening its antimicrobial activity. *J. Nanopart. Res.* **2011**, *13* (7), 2981–2988.
53. Valodkar, M.; Nagar, P. S.; Jadeja, R. N.; Thounaojam, M. C.; Devkar, R. V.; Thakore, S. Euphorbiaceae latex induced green synthesis of non-cytotoxic metallic nanoparticle solutions: A rational approach to antimicrobial applications. *Colloids Surf., A* **2011**, *384* (1−3), 337–344.
54. Vinod, V. T. P.; Saravanan, P.; Sreedhar, B.; Devi, D. K.; Sashidhar, R. B. A facile synthesis and characterization of Ag, Au and Pt nanoparticles using a natural hydrocolloid gum kondagogu (Cochlospermum gossypium). *Colloids Surf., B* **2011**, *83* (2), 291–298.
55. Karimi Zarchi, A.; Mokhtari, N.; Arfan, M.; Rehman, T.; Ali, M.; Amini, M.; Faridi Majidi, R.; Shahverdi, A. A sunlight-induced method for rapid biosynthesis of silver nanoparticles using an *Andrachnea chordifolia* ethanol extract. *Appl. Phys. A: Mater. Sci. Process.* **2011**, *103* (2), 349–353.
56. Prathna, T. C.; Chandrasekaran, N.; Raichur, A. M.; Mukherjee, A. Kinetic evolution studies of silver nanoparticles in a bio-based green synthesis process. *Colloids Surf., A* **2011**, *377* (1–3), 212–216.
57. Narayanan, K. B.; Sakthivel, N. Phytosynthesis of gold nanoparticles using leaf extract of *Coleus amboinicus Lour*. *Mater. Charact.* **2010**, *61* (11), 1232–1238.
58. Narayanan, K. B.; Sakthivel, N. Extracellular synthesis of silver nanoparticles using the leaf extract of *Coleus amboinicus Lour*. *Mater. Res. Bull.* **2011**, *46* (10), 1708–1713.
59. Leonard, K.; Ahmmad, B.; Okamura, H.; Kurawaki, J. In situ green synthesis of biocompatible ginseng capped gold nanoparticles with remarkable stability. *Colloids Surf., B* **2011**, *82* (2), 391–396.
60. Guidelli, E. J.; Ramos, A. P.; Zaniquelli, M. E. D.; Baffa, O. Green synthesis of colloidal silver nanoparticles using natural rubber latex extracted from *Hevea brasiliensis*. *Spectrochim. Acta, Part A* **2011**, *82* (1), 140–145.
61. Aromal, S. A.; Vidhu, V. K.; Philip, D. Green synthesis of well-dispersed gold nanoparticles using *Macrotyloma uniflorum*. *Spectrochim. Acta, Part A* **2012**, *85* (1), 99–104.

62. Vidhu, V. K.; Aromal, S. A.; Philip, D. Green synthesis of silver nanoparticles using *Macrotyloma uniflorum*. *Spectrochim. Acta, Part A* **2011**, *83* (1), 392–397.
63. Philip, D. *Mangifera indica* leaf-assisted biosynthesis of well-dispersed silver nanoparticles. *Spectrochim. Acta, Part A* **2011**, *78* (1), 327–331.
64. Das, R.; Gogoi, N.; Bora, U. Green synthesis of gold nanoparticles using *Nyctanthes arbortristis* flower extract. *Bioprocess Biosyst. Eng.* **2011**, *34* (5), 615–619.
65. Mohan Kumar, K.; Mandal, B. K.; Sinha, M.; Krishnakumar, V. *Terminalia chebula* mediated green and rapid synthesis of gold nanoparticles. *Spectrochim. Acta, Part A* **2012**, *86* (0), 490–494.
66. Njagi, E. C.; Huang, H.; Stafford, L.; Genuino, H.; Galindo, H. M.; Collins, J. B.; Hoag, G. E.; Suib, S. L. Biosynthesis of iron and silver nanoparticles at room temperature using aqueous sorghum bran extracts. *Langmuir* **2010**, *27* (1), 264–271.
67. Gao, X.; Wei, L.; Yan, H.; Xu, B. Green synthesis and characteristic of core-shell structure silver/starch nanoparticles. *Mater. Lett.* **2011**, *65* (19–20), 2963–2965.
68. El-Rafie, M. H.; El-Naggar, M. E.; Ramadan, M. A.; Fouda, M. M. G.; Al-Deyab, S. S.; Hebeish, A. Environmental synthesis of silver nanoparticles using hydroxypropyl starch and their characterization. *Carbohydr. Polym.* **2011**, *86* (2), 630–635.
69. Raveendran, P.; Fu, J.; Wallen, S. L. Completely "green" synthesis and stabilization of metal nanoparticles. *J. Am. Chem. Soc.* **2003**, *125* (46), 13940–13941.
70. Abdel-Halim, E. S.; Al-Deyab, S. S. Utilization of hydroxypropyl cellulose for green and efficient synthesis of silver nanoparticles. *Carbohydr. Polym.* **2011**, *86* (4), 1615–1622.
71. Virkutyte, J.; Jegatheesan, V.; Varma, R. S. Visible light activated TiO_2/microcrystalline cellulose nanocatalyst to destroy organic contaminants in water. *Bioresour. Technol.* **2012**, *113*, 288–293.
72. Baruwati, B.; Varma, R. S. High value products from waste: Grape pomace extract—A three-in-one package for the synthesis of metal nanoparticles. *ChemSusChem* **2009**, *2* (11), 1041–1044.
73. Panácek, A.; Kvítek, L.; Prucek, R.; Kolář, M.; Vecčeřová, R.; Pizúrová, N.; Sharma, V. K.; Nevěčná, T. j.; Zbořil, R. Silver colloid nanoparticles: Synthesis, characterization, and their antibacterial activity. *J. Phys. Chem. B* **2006**, *110* (33), 16248–16253.
74. Philip, D. Honey mediated green synthesis of silver nanoparticles. *Spectrochim. Acta, Part A* **2010**, *75* (3), 1078–1081.
75. Venu, R.; Ramulu, T. S.; Anandakumar, S.; Rani, V. S.; Kim, C. G. Bio-directed synthesis of platinum nanoparticles using aqueous honey solutions and their catalytic applications. *Colloids Surfac., A* **2011**, *384* (1–3), 733–738.
76. Hoag, G. E.; Collins, J. B.; Holcomb, J. L.; Hoag, J. R.; Nadagouda, M. N.; Varma, R. S. Degradation of bromothymol blue by 'greener' nano-scale

zero-valent iron synthesized using tea polyphenols. *J. Mater. Chem.* **2009**, *19* (45), 8671–8677.

77. Nadagouda, M. N.; Castle, A. B.; Murdock, R. C.; Hussain, S. M.; Varma, R. S. In vitro biocompatibility of nanoscale zerovalent iron particles (NZVI) synthesized using tea polyphenols. *Green Chem.* **2010**, *12* (1), 114–122.
78. Nadagouda, M. N.; Varma, R. S. Green synthesis of silver and palladium nanoparticles at room temperature using coffee and tea extract. *Green Chem.* **2008**, *10* (8), 859–862.
79. Moulton, M. C.; Braydich-Stolle, L. K.; Nadagouda, M. N.; Kunzelman, S.; Hussain, S. M.; Varma, R. S. Synthesis, characterization and biocompatibility of "green" synthesized silver nanoparticles using tea polyphenols. *Nanoscale* **2010**, *2* (5), 763–770.
80. Chandrasekharan, D. K.; Khanna, P. K.; Kagiya, T. V.; Nair, C. K. K. Synthesis of nanosilver using a vitamin C derivative and studies on radiation protection. *Cancer Biother. Radiopharm.* **2011**, *26*, 249–257.
81. Chien-Jung, H.; Pin-Hsiang, C.; Yeong-Her, W.; Teen-Hang, M.; Cheng-Fu, Y. Synthesis and characterization of gold nanodogbones by the seeded mediated growth method. *Nanotechnology* **2007**, *18* (39), 395603.
82. Zhang, L.; Shen, Y.; Xie, A.; Li, S.; Jin, B.; Zhang, Q. One-step synthesis of monodisperse silver nanoparticles beneath vitamin E Langmuir monolayers. *J. Phys. Chem. B* **2006**, *110* (13), 6615–6620.
83. Nadagouda, M. N.; Varma, R. S. Green and controlled synthesis of gold and platinum nanomaterials using vitamin B2: Density-assisted self-assembly of nanospheres, wires and rods. *Green Chem.* **2006**, *8* (6), 516–518.
84. Nadagouda, M. N.; Varma, R. S. Green synthesis of Ag and Pd nanospheres, nanowires, and nanorods using vitamin B2: Catalytic polymerisation of aniline and pyrrole. *J. Nanomater.* **2008**, *2008*, 1–8.
85. Nadagouda, M. N.; Polshettiwar, V.; Varma, R. S. Self-assembly of palladium nanoparticles: Synthesis of nanobelts, nanoplates and nanotrees using vitamin B1, and their application in carbon-carbon coupling reactions. *J. Mater. Chem.* **2009**, *19* (14), 2026–2031.
86. Virkutyte, J.; Varma, R. S. Green synthesis of metal nanoparticles: Biodegradable polymers and enzymes in stabilization and surface functionalization. *Chem. Sci.* **2011**, *2*, 837–846.
87. Abu-Reziq, R.; Wang, D.; Post, M.; Alper, H. Separable catalysts in one-pot syntheses for greener chemistry. *Chem. Mater.* **2008**, *20* (7), 2544–2550.
88. Balu, A. M.; Baruwati, B.; Serrano, E.; Cot, J.; Garcia-Martinez, J.; Varma, R. S.; Luque, R. Magnetically separable nanocomposites with photocatalytic activity under visible light for the selective transformation of biomass-derived platform molecules. *Green Chem.* **2011**, *13* (10), 2750–2758.
89. Baruwati, B.; Polshettiwar, V.; Varma, R. S. Magnetically recoverable supported ruthenium catalyst for hydrogenation of alkynes and transfer hydrogenation of carbonyl compounds. *Tetrahedron Lett.* **2009**, *50* (11), 1215–1218.

90. Sun, Y.; Guo, G.; Yang, B.; Zhou, X.; Liu, Y.; Zhao, G. One-step fabrication of Fe_2O_3/Ag core-shell composite nanoparticles at low temperature. *J. Non-Crys. Solids* **2011**, *357* (3), 1085–1089.
91. Adam, J. D.; Davis, L. E.; Dionne, G. F.; Schloemann, E. F.; Stitzer, S. N. Ferrite devices and materials. *IEEE Trans. Microwave Theory Tech.* **2002**, *50* (3), 721–737.
92. Li, C. C.; Li, Q. H.; Chen, L. B.; Wang, T. H. A facile titanium glycolate precursor route to mesoporous Au/$Li_4Ti_5O_{12}$ spheres for high-rate lithium-ion batteries. *ACS Appl. Mater. Interfaces* **2012**, *4* (3), 1233–1238.
93. Naidek, K. P.; Bianconi, F.; da Rocha, T. C. R.; Zanchet, D.; Bonacin, J. A.; Novak, M. A.; das Graças Fialho Vaz, M.; Winnischofer, H. Structure and morphology of spinel MFe_2O_4 (M = Fe, Co, Ni) nanoparticles chemically synthesized from heterometallic complexes. *J. Colloid Interface Sci.* **2011**, *358* (1), 39–46.
94. Özgür, Ü.; Alivov, Y.; Morkoç, H. Microwave ferrites, Part 1: Fundamental properties. *J. Mater. Sci.: Mater. Electron.* **2009**, *20* (9), 789–834.
95. Senapati, K. K.; Borgohain, C.; Phukan, P. Synthesis of highly stable $CoFe_2O_4$ nanoparticles and their use as magnetically separable catalyst for Knoevenagel reaction in aqueous medium. *J. Mol. Catal. A: Chem.* **2011**, *339* (1−2), 24–31.
96. Mangrulkar, P. A.; Polshettiwar, V.; Labhsetwar, N. K.; Varma, R. S.; Rayalu, S. S. Nano-ferrites for water splitting: Unprecedented high photocatalytic hydrogen production under visible light. *Nanoscale* **2012**.
97. Nasir Baig, R. B.; Varma, R. S. A highly active magnetically recoverable nano ferrite-glutathione-copper (nano-FGT-Cu) catalyst for Huisgen 1,3-dipolar cycloadditions. *Green Chem.* **2012**, *14* (3), 625–632.
98. Yang, Z.-p.; Gong, X.-y.; Zhang, C.-j. Recyclable Fe_3O_4/hydroxyapatite composite nanoparticles for photocatalytic applications. *Chem. Eng. J.* **2011**, *165* (1), 117–121.
99. Yallapu, M. M.; Othman, S. F.; Curtis, E. T.; Gupta, B. K.; Jaggi, M.; Chauhan, S. C. Multi-functional magnetic nanoparticles for magnetic resonance imaging and cancer therapy. *Biomaterials* **2011**, *32* (7), 1890–1905.
100. Garrido-Ramírez, E. G.; Theng, B. K. G.; Mora, M. L. Clays and oxide minerals as catalysts and nanocatalysts in Fenton-like reactions -- A review. *Appl. Clay Sci.* **2011**, *47* (3-4), 182–192.
101. Li, J. J.; Mu, Z.; Xu, X. Y.; Tian, H.; Duan, M. H.; Li, L. D.; Hao, Z. P.; Qiao, S. Z.; Lu, G. Q. A new and generic preparation method of mesoporous clay composites containing dispersed metal oxide nanoparticles. *Microporous Mesoporous Mater.* **2008**, *114* (1−3), 214–221.
102. Virkutyte, J.; Varma, R. S. Novel Claycunbic to eliminate micropollutants and *Vibrio fischeri* from water. *RSC Adv.* **2012**, *2* (8), 3416–3422.

Chapter 3

Greener and Other Approaches To Synthesize Fe and Pd Nanoparticles in Functionalized Membranes and Hydrogel

V. Smuleac, L. Xiao, and D. Bhattacharyya*

Department of Chemical and Materials Engineering, University of Kentucky, Lexington, Kentucky 40506
***E-mail: db@engr.uky.edu**

Nano-scale materials have applications in diverse fields, such as catalysis, electronics, and medical science. In the pollution remediation field Fe and Fe/Pd nanoparticles (NPs) are being used to detoxify chlorinated organics. The use of unsupported NPs often leads to aggregation and loss to environment. Our approach is based on direct synthesis of iron-based NPs in membranes and temperature responsive hydrogel supports by using "greener" approach. In our current work, we investigated a simple and fast method of Fe and Fe/Pd NPs synthesis in polyacrylic acid functionalized polyvinylidene fluoride (PAA/PVDF) membranes, using "green" reducing agents (green tea extract and epicatechin), as an alternative to the commonly used reducing agent, sodium borohydride. In addition, we have synthesized reactive Fe/Pd NPs in hydrogel, P(NIPAAm-AA (poly- N-isopropylacrylamide-acrylic acid). Both membrane and hydrogel immobilized NPs have been applied to successful dechlorination of TCE and various PCBs. For example, the surface area normalized reaction rates for TCE ranged between 0.008 to 0.04 L/m^2h with Fe/Pd systems. Membrane polymer support and hydrogel had insignificant loss of NPs to environment.

Introduction

In recent years, nanoscale materials and nanoparticles (NPs) in particular, received a great attention due to their unique physico-chemical and thermodynamic properties, different from those of bulk materials (*1*). As a consequence, these materials find applications in diverse fields, such as catalysis, electronics or medical science (*2–5*). NP synthesis with the desired properties is one of the most exciting and challenging aspects of modern nanotechnology.

Among various synthesis procedures, green chemistry techniques show a great potential by using economical, non-toxic and biodegradable materials, thus being an attractive alternative for conventional methods. A variety of materials originating from bio-renewable natural sources, such as plant surfactants (*6*), waste biomass (*7*), vitamins (*8–11*), aminoacids (*12*), pomace (wine waste) (*13*), polysaccharides (*14*), glucose (*15*), polyvinylpyrrolidone (*16*), oleic acid (*17*), as well as extracts of coffee (*18*), green tea (*19–23*), black tea (*24*) and plant leaves and seeds (*25–29*) have emerged as replacements for well established chemicals. These materials are nontoxic, biodegradable, and act as both dispersive and capping agents, thus minimizing the NPs oxidation and agglomeration (*30*). Various metallic NPs such as Au, Ru, Pd, Pt, Ag and Fe were synthesized using these novel techniques.

Our research group has been involved in synthesis of supported Fe and bimetallic Fe/Pd NPs in membranes and hydrogels, and their application toward remediation of toxic chloro-organics. Membrane's open structure and high internal surface area ensure a high NP loading and easy accessibility of the pollutant to the active site; in addition these can be operated in convective mode (permeation through the membrane). NPs incorporation in membranes can also be conducted in diffusive (soaking) mode that involved long (10-12 h) processing time.

In our current work, we investigated a simple and fast method of Fe and Fe/Pd NPs synthesis in polyacrylic acid functionalized polyvinylidene fluoride (PAA/PVDF) membranes, using "green" reducing agents (green tea extract and epicatechin), as an alternative to the commonly used reducing agent, sodium borohydride. All synthesis steps have been conducted in convective mode (mounting the membrane in a filtration cell) thus significantly reducing the synthesis time (from 12 h to less than 2 h) and promoting the NP formation inside membrane pores, rather than pore mout and external surface. In addition, we have synthesized reactive Fe/Pd NPs in hydrogel, P(NIPAAm-AA) (poly-N-isopropylacrylamide-acrylic acid). The membranes containing the Fe and bimetallic Fe/Pd NPs have been characterized by permeability studies and FTIR, SEM, TEM, EDX and selected area electron diffraction (SAED) pattern techniques. The applications of these immobilized NP systems are towards remediation of toxic chloro-organics (such as trichloroethylene and/or polychlorinated biphenyls) from water.

Materials and Methods

Chemicals

Potassium persulfate was purchased from EM Science. Deionized ultra-filtered water (DIUF) was purchased from Fisher Scientific. Acrylic acid (AA), potassium tetrachloropalladate (II), sodium borohydride, epicatechin, N-isopropylacrylamide (NIPAAm), poly (ethylene glycol) 600 dimethacrylate (PEG600DMA), 4-(4-dimethylaminophenylazo) aniline (DMPA), ethanol (>99.5%) were purchased from Sigma-Aldrich and ethylene glycol (EG) from Mallinckrodt. Hydrophilized PVDF microfiltration membranes, with a thickness of 125 μm and nominal pore size of 650 nm were obtained from Millipore Corporation.

Characterization of Nanoparticles (NPs)

Surface and cross-section of membrane/hydrogel and nanoparticles were examined by Hitachi S-4300 Scanning Electron Microscope (SEM). The samples were coated with gold for imaging purposes. A JEOL 2010F high-resolution Transmission Electron Microscopy (TEM) equipped with energy dispersive X-ray spectrometer (EDX) were used to observe the NPs morphology and analyze the elemental composition. A drop of nanoparticle solution was placed on a standard TEM copper (Cu) grid and then dried the samples in vacuum oven. Attenuated total reflectance Fourier transform infrared (ATR-FTIR) (Varian 7000e) was used to determine the presence of functional groups in membrane hydrogel. The samples were placed on the diamond crystal and the spectrum was obtained between 500 and 4000 cm-1 for 32 scans at a resolution of 8 cm^{-1}. The UV-Visible spectra were recorded in a Varian Cary Bio300 UV-visible absorption spectroscopy.

Results and Discussion

Aqueous Phase in Situ Polymerization of Acrylic Acid in PVDF Membranes

Our goal is to synthesize NPs inside a membrane domain and this approach involves metal cation exchange followed by reduction to metallic NPs. Therefore, the first step was to attach ion exchange groups on the membrane. Although various membrane materials can be used, PVDF was chosen for its high chemical and thermal stabilities with melting point in the range between 162-172 °C (*31*).

PVDF membranes were functionalized with poly(acrylic acid) by in situ polymerization of acrylic acid. A schematic was shown in Figure 1, the PVDF membrane (hydrophylized membranes were used, for proper wetting) was dipped in the polymerization solution for 2 min, sandwiched between two Teflon plates and placed in an oven at 90 °C for 4 hours. The polymerization was greatly influenced by several parameters, such as monomer and cross linker concentration, pH, initiator amount etc (*32*). The polymerization solution contained 30 wt% acrylic (cross-linker, added in a 1:10 molar ratio of EG to acrylic acid), and 1 wt% potassium persulfate (initiator). Raising the temperature is necessary for the formation of ester bonds between the ethylene glycol (crosslinker) and carboxylate on the formed polyacrylate. Ethylene glycol is a bidentate molecule (binds to two –COOH groups), and was used to prevent PAA leaching from the membrane. In order to maintain free –COOH groups for ion exchange, the amount of cross-linking agent had to be kept low. Under our experimental conditions at least 80% of the –COOH groups were free, this was established by quantifying the entrapment capacity of Ca^{2+} with –COOH groups (*33*, *34*). Nitrogen gas was continuously supplied to remove oxygen which acted as an inhibitor for the polymerization reaction.

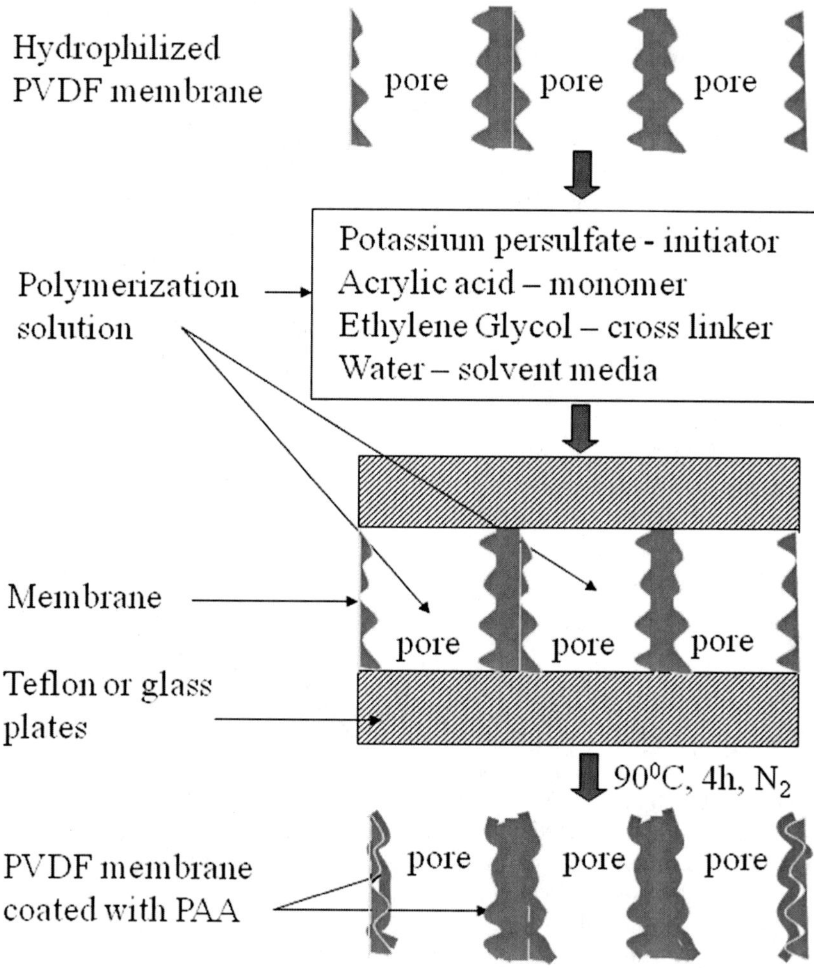

Figure 1. In situ polymerization of acrylic acid in PVDF membranes using green chemistry (no organic solvent used).

Fe and Bimetallic Fe/Pd NPs Synthesis in PAA-Functionalized PVDF Membranes

The flowchart for NPs synthesis in PAA/PVDF membranes as well as the properties of the base - PVDF membrane is shown in Figure 2. NPs synthesis in the PAA/PVDF membranes involved a two-step procedure and involves cation exchange (on -COO^- groups from PAA) followed by their reduction. Prior to Fe^{2+} ion exchange, PAA-functionalized PVDF membranes were immersed in NaCl (5 to 10 % wt) solution at pH 10 (adjusted with 0.1M NaOH) for at least 3 h to convert the -COOH to -COONa form. In the next step, the membrane was washed with DIUF until the pH of the washing solution became neutral.

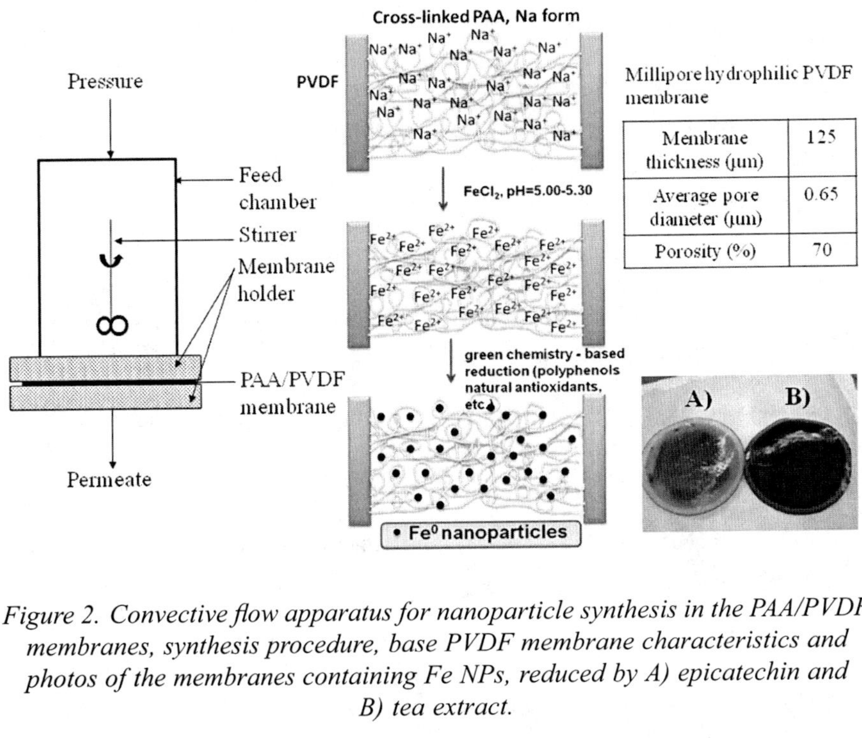

Membrane thickness (μm)	125
Average pore diameter (μm)	0.65
Porosity (%)	70

Figure 2. Convective flow apparatus for nanoparticle synthesis in the PAA/PVDF membranes, synthesis procedure, base PVDF membrane characteristics and photos of the membranes containing Fe NPs, reduced by A) epicatechin and B) tea extract.

The membrane was mounted in a filtration apparatus (also shown in Figure 2), and a solution of $FeCl_2$ solution at a pH of 5.5 (adjusted with 0.1M NaOH) was convectively permeated (flux of 12.6 x 10^{-4} cm^3/cm^2s) through the membrane. The concentration of Fe^{2+} can be varied, depending on the amount of Fe desired to be immobilized in the membrane. Typically the feed solution volume and concentration were 200 mL and 180 mg/L Fe^{2+}, respectively. Next a solution of green tea extract (50 ml, 20 g/L) or a solution of epicatechin (50 mL, 1 g/L), at pH 5 was permeated through the membrane, in order to reduce the Fe^{2+} to Fe NPs. The green tea extract was prepared by immersion of 1 tea bag (containing 1 g green tea, 100% purity) in boiling water for 15-20 minutes; the bag was removed and the liquid was filtered using a 450 nm syringe membrane (polytetrafluoroethylene, PTFE).

During the reduction process (1-2 h length), the membrane (initially white) changed color to grey or black, depending on the amount of Fe immobilized (darker with the increase of the amount of Fe). Photos in Figure 2 shows two membranes containing Fe NPs, reduced with epicatechin (Figure 2A) (3 mg Fe) and tea extract (Figure 2B) (8 mg Fe). For comparison purposes, Fe NPs immobilized on the membrane were also prepared with sodium borohydride (50 mL, 10 g/L) as a reducing agent.

Tea extract containing a number of polyphenols, has been well studied in nutrition science due to their numerous benefic effects and antioxidant properties in human body (*35*). The most common polyphenols are epicatechin and its ester forms with gallic acid. Polyphenols can directly complex with iron ions and then reduce to zero valent NPs. Cyclic voltammetry studies (*36*) showed that only dihydroxyl groups in ortho position form could complex with metal cations and one pair of ortho-dihydroxyl groups could exchange $2e^-$ during the complexation with Fe^{2+}. Therefore, in the case of epicatechin (with only 1 pair of ortho-dihydroxyl groups), the molar ratio of reductant to Fe^{2+} should be 1:1, for complete reduction.

The reduction potential for epicatechin is 0.57 V, sufficient for the reduction of Fe^{2+} to Fe^0 (-0.44 V). The standard reduction potentials for most polyphenols and flavonoids are in a range from 0.5 to 0.7 V (Table 1). However, it was reported that chelation of metals (such as Fe) to polyphenols could shift the redox potential toward either anodic or cathodic direction (making it more susceptible to either oxidation or reduction), as a function of the interacting polyphenol-metal species (*37*).

Although the exact overall mechanism for metal NPs formation is not known, several studies (supported by FTIR data) showed that hydroxyl groups used to form the complex with the metal cation were converted to carbonyl during the reduction process (*24*, *25*, *28*). It was also shown that among a multitude of compounds present in the tea extract, flavonoids and polyphenols played the most significant role in metal cation complexation with subsequent metal NPs formation (*24*).

In current study, the reduction of Fe was also confirmed by preparing NPs in homogeneous phase (mixing 10 mL $FeCl_2$ solution (0.1-0.4 M Fe^{2+}) with 10 mL tea extract) by UV spectra (shown in Figure 3). The blank tea extract has an absorption beginning at 500 nm, similar to $FeCl_2$ solutions. The reaction between $FeCl_2$ and tea extract was instantaneous and the color of the reaction mixture changed from yellow to black. After the reaction, the UV spectra had broad absorption at a higher wavelength, which increased as the concentration of $FeCl_2$ changed from 0.01 to 0.04 M.

In order to form Fe/Pd bimetallic NPs in the membrane, a solution containing a mixture of K_2PdCl_4 and tea extract was permeated through the membrane containing Fe NPs. In order to establish the appropriate reaction conditions to form Pd NPs, this reaction was firstly carried out in homogeneous phase (mixing 10 mL K_2PdCl_4 (0.03 M Pd) with10 mL tea extract) and analyzed by UV-Vis. It was shown previously (*28*) that Pd reduction with various plant extracts containing polyols proceeded slowly (12 h) when the reaction was carried out at 30 °C.

Table 1. Reduction Potential of Tea Polyphenols, Flavonoids, and Other Physiological Antioxidants. (adapted from reference (*20*, *38*))

Antioxidant	*Reduction potential (V)*
(-)-Epicatechin	0.57
(-)-Epicatechin gallate	0.55
(-)-Epigallocatechin	0.43
Catcchol	0.53
Theaflavin	0.51
Theaflavin digallate	0.54
Hesperidin	0.72
Rutin	0.60
Quercetin	0.33
3,5-Dihydroxy-anisol	0.84
Methyl gallate	0.56
Ascorbate	0.28
α-Tocopherol	0.48
Uric acid	0.59
Glutathion (Cysteine)	0.92

Here, the metal salt precursor (K_2PdCl_4) was reacted with tea extract both at room and elevated (80 °C) temperature (Figure 4); at room temperature (20°C), no color change or difference was observed in absorption spectra of the product (compared to the reactants) within 1 h reaction time. However, at elevated temperature (80 °C), in less than 30 min reaction time, the color of the reaction product turned to dark brown and the absorption spectra showed a broad band at a higher wavelength, indicating the formation of Pd NPs. This was further confirmed by TEM with SAED pattern, and it will be discussed later. Therefore, for bimetallic Fe/Pd NPs formation in the membrane, the mixture of K_2PdCl_4 and tea extract was permeated at 80 °C.

In order to quantify the amounts of Fe and Pd, membranes containing NPs were digested by nitric acid solution (0.02L, 35%) to release the metals into the solution phase. The concentrations of Fe and Pd in the digested solutions were analyzed by a Varian SpectrAA 220 Fast Sequential atomic absorption spectrometer equipped with a Fisher Scientific hollow cathode lamp that was operated at a wavelength of 386.0 nm for Fe and 246.6 nm for Pd. The calibration plot was created using 4 different concentrations of Fe ranging from 25 to 200 mg/L with $R^2 = 0.9995$ and average analytical error of 2%. In the case of Pd, the linear calibration range is between 0.2 and 28 mg/L Pd and the error of analysis was <2%.

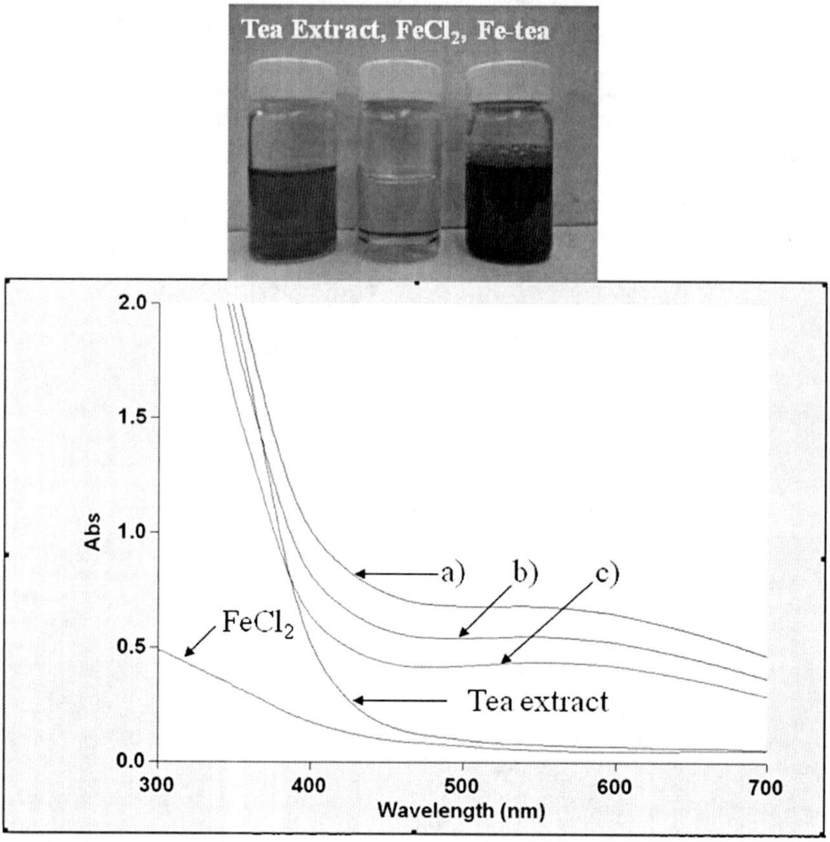

Figure 3. Photos and UV-Vis spectra for $FeCl_2$ (0.04 M), tea extract and solution phase Fe NPs, Fe concentration a) 0.04 M, b) 0.02 M, c) 0.01 M.

Characterization of the Metal NPs Immobilized in PAA/PVDF Membranes

The organic functionalities for the Fe/Pd-modified PAA/ PVDF membranes were recorded by ATR-FTIR. Membranes were dried for a day before testing and data were collected using 32 scans at a resolution of 8 cm^{-1}. The PAA-coated PVDF membrane showed a multitude of peaks (especially between 1100 and 1400 cm^{-1}), which are the characteristic peaks of CF_2. A comparison of the FTIR spectra for Fe NPs immobilized on PAA/PVDF membranes, reduced by borohydride and tea extract, respectively (Figure 5), shows a great similarity in a range of 1000 and 1800 cm^{-1}. However, in the case of Fe NPs reduced by polyphenols (present in tea extract), there is a pronounced peak at 1716 cm^{-1} which is characteristic to the stretch vibration of -C=O group (m1 and m2 in Figure 5 refers to two different membranes). It can be speculated that the peak was due to the presence of polyphenols (that have been oxidized to carbonyl, while reducing the cation to elemental metal) that remained adsorbed on the NPs surface, especially that this peak was missing for Fe-NPs reduced by borohydride. The adsorption of green reducing agents (containing polyols) on the NPs surface was also mentioned in the literature for Ag, Au and Pd NPs reduced by cinnamomum camphora leaf broth (*25*, *28*).

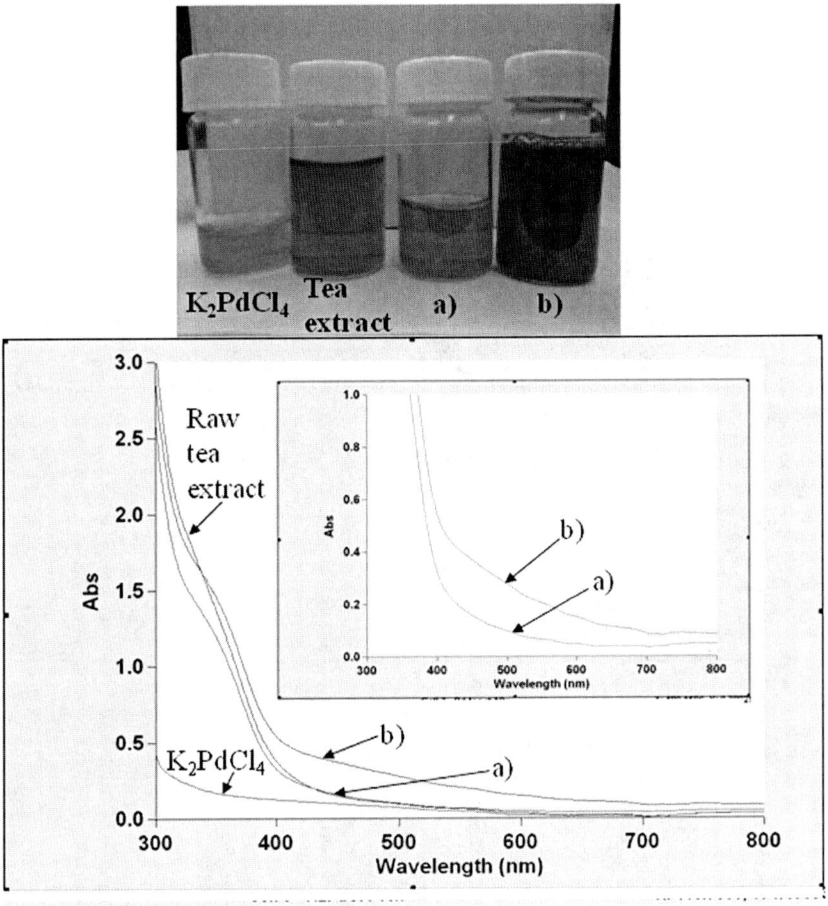

Figure 4. Photos and UV-Vis spectra for K_2PdCl_4, tea extract and solution phase Pd-tea extract prepared at a) room temperature, b) 80 °C.

With the PAA presence within the membrane pores, one would expect to see stimulus-responsive water permeability to pH. The effect of varying pH on water permeability through the PAA/PVDF and Fe-PAA/PVDF functionalized membranes is shown in Figure 6. For both cases, it can be observed that as the pH of the feed solution is increased, the permeability decreased. Since the pK_a of acrylic acid is approximately 4.25, the PAA network will maintain a relatively neutral charge at a low pH, leaving it in a compacted state. As the pH is increased, more -COOH groups would become ionized, resulting in an elongation of the PAA network due to the charge repulsion. While in the elongated state, the permeate flux and permeability were lower than those in the compacted state.

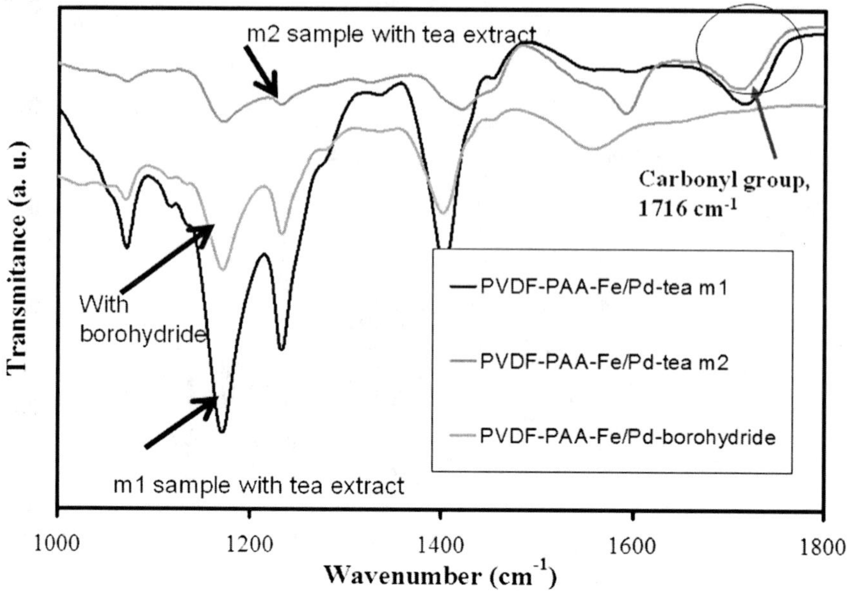

Figure 5. ATR-FTIR spectra for supported Fe/Pd NPs immobilized in PAA-PVDF membranes, synthesized by different reducing agents.

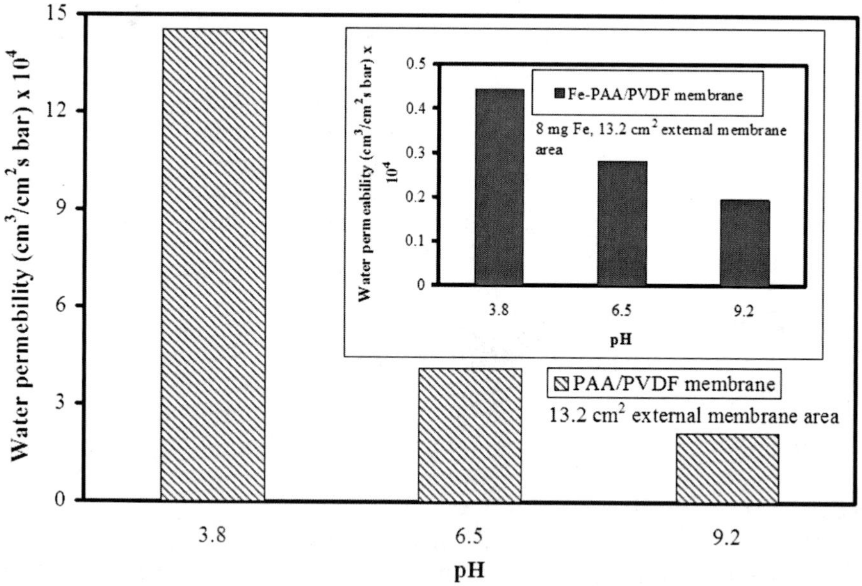

Figure 6. Water permeability in PAA-PVDF and Fe-PAA-PVDF membranes, as a function of the pH (water permeability of base PVDF membrane = 4824 x 10^{-4} cm^3/cm^2s bar).

There was a large discrepancy between the two cases, for PAA/PVDF membrane, the permeability decreased 3.5 and 7-fold as the pH increased from 3.8 to 6.5 and 9.2; while for Fe-NPs modified PAA/PVDF membrane, the increase was only 1.5 and 2-fold, respectively, for the samc pH values. This can be explained by NPs cross-linking of PAA chains as well as a severe reduction of the membrane pore void fraction. This also increased the resistance to water permeation at least one order of magnitude, as observed in Figure 6. At very high metal NPs loadings, one might expect little or no change in permeability as a function of pH in the feed solution.

Permeability measurements can be used to estimate the equivalent pore diameter, d_c, applying Hagen – Poisseuille's law of capillary flow:

$$d_c = 2\left(\frac{8J_v\mu L}{\pi N\Delta P}\right)^{\frac{1}{4}} \tag{1}$$

where N, ΔP, μ and L represent the number of pores, trans-membrane pressure, liquid viscosity, and membrane thickness, respectively. $J_v/\Delta P$ (the slope in J_v vs ΔP plot) represents the permeability A, (cm^3/cm^2 s bar). The equivalent diameter of the pores was calculated on the basis of permeability changes for the bare and functionalized membrane. Assuming no change in membrane thickness and no pores are completely blocked after functionalization,

$$A/A_0 = (d_c/d_{c0})^4 \tag{2}$$

where A_0 and d_{c0} are the permeability and the equivalent diameter for the bare membrane, and A and d_c are the pure water flux and the estimated equivalent diameter for the functionalized membrane. For the bare PVDF membrane, A_0 is 4824 x 10^{-4} cm^3/cm^2s bar from the manufacturer's data; for the functionalized membranes A is determined as the slope of the flux vs. pressure plot ($J_v/\Delta P$).

Assuming a cylindrical pore and uniform PAA and NPs distribution throughout the membrane pores, d_c can be obtained from Equation 2 and the values are included in Table 2. Just like for permeability, the d_c variation is more pronounced for PAA/PVDF membrane (from 152 nm to 94 nm) as compared to Fe-PAA/PVDF membrane (from 64 nm to 52 nm), as the pH is increased from 3.8 to 9.2. Again, NPs immobilized in the membrane increased its rigidity and did not allow the pores to expand as a response to pH change.

SEM imaging was used to examine the surface morphology for the cross-section and external surface (at the pore entrance) of the Fe/Pd NPs-modified PAA-PVDF membranes (Figure 7). It was observed that the base NPs has an average diameter less than 50 nm (Figure 7A, cross-section), and some aggregates in the range from 100-200 nm at the external surface (Figure 7B).

Table 2. Equivalent Pore Diameter (determined using Hagen-Poiseuille's equation (eq 1) as a Function of Water pH, Used for Permeability Measurements

pH	*PAA/PVDF membrane* $d_c(nm)$	Fe^0*-PAA/PVDF membrane* $d_c(nm)$
3.8	152	64
6.5	111	57
9.2	94	52

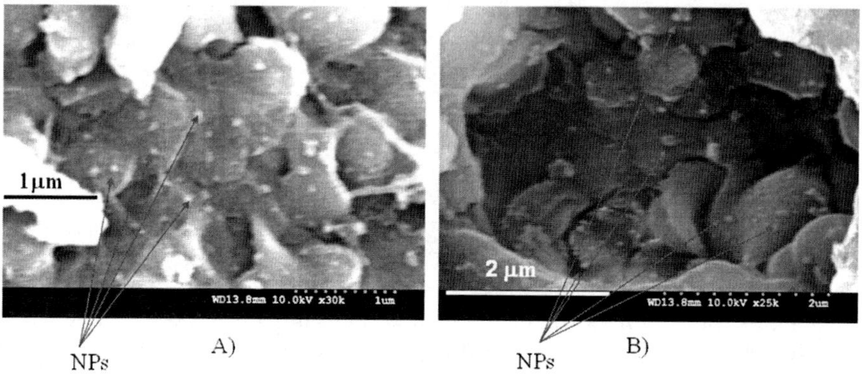

Figure 7. SEM image for the Fe NPs synthesized in the PAA-PVDF membrane, using tea extract as a reducing agent, A) cross-section, B) external surface, pore entrance.

For a more detailed observation of the NPs and for the ease of sample preparation, Fe and Pd NPs were synthesized in solution phase with epicatechin (0.02 M) as a reducing agent (the ratio of mmol metal : mmol epicatechin 1:1 or less, for complete reduction as mentioned above). A drop of mixed solution containing NPs was deposited on a carbon-coated copper grid, dried overnight and examined by TEM equipped with a scanning (STEM) unit. This instrument was coupled with an Oxford EDS detector (used for elemental analysis) and the selected area electron diffraction (SAED) pattern was also used to determine the form of the metallic NPs. The NPs has an average diameter 20-30 nm (Figure 8A) and the EDS spectrum shows the elemental composition (Fe, Pd and Cu-from the grid, Figure 8C). Additional STEM images are shown in Figure 9. In the case of Pd NPs reduced with epicatechin (at 80 °C), the average diameter is around 5 nm and the SAED pattern showed Pd to be in metallic (face centered cubic, fcc) structure, with (111), (200), (220) and (311) peaks (Figures 8B and 10). However, in the case of iron, a high degree of crystallinity was not observed, the rings being more diffuse (Figure 8B).

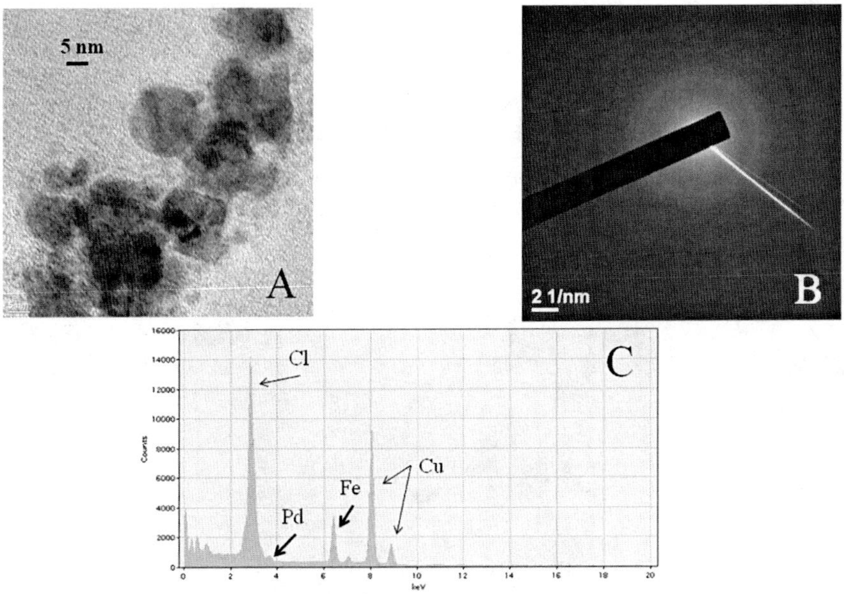

Figure 8. A) TEM B) the selected area electron diffraction (SAED) patterns and C) EDS for Fe/Pd nanoparticles reduced by epicatechin, (Cu is from TEM grid).

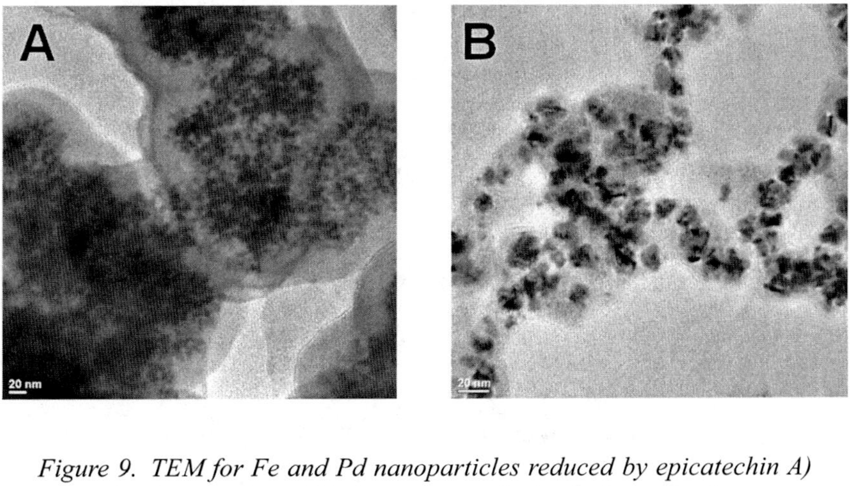

Figure 9. TEM for Fe and Pd nanoparticles reduced by epicatechin A) simultaneously and B) separately and mixed after reduction.

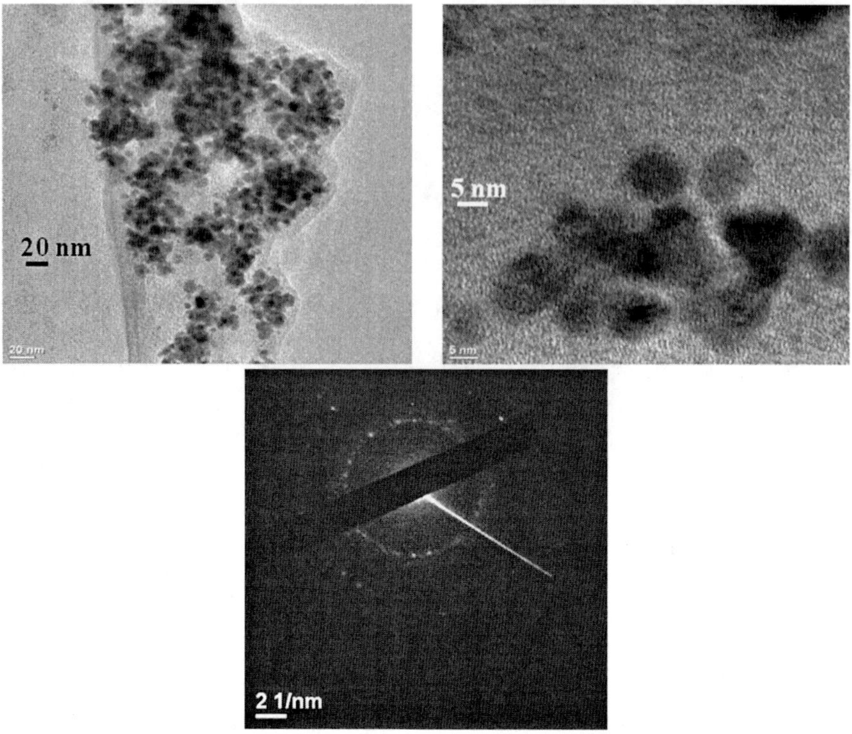

Figure 10. TEM images (at different magnifications) for Pd nanoparticles reduced by epicatechin (at 80 °C) and the selected area electron diffraction (SAED) pattern).

In summary, a fast and simple, "green" method for Fe and bimetallic Fe/Pd NPs synthesis in the membranes using the convective flow approach was presented. These membrane systems containing NPs were characterized by a variety of techniques with the NPs size about 20-30 nm for Fe and 5 nm for Pd from SEM and TEM images. SAED patterns revealed a crystalline (fcc) structure for Pd NPs. These membrane immobilized NPs reduced by green tea had been successfully demonstrated for the reductive degradation of a model compound, TCE, with surface area normalized reactivity for 0.005 L/m^2h (Fe) and 0.008 L/m^2h (Fe/Pd) (*23*).

Direct Nanoparticle Synthesis in Temperature Responsive Hydrogels

Another platform for nanoparticles synthesis is the utilization of temperature responsive hydrogel –NIPAAm with lower critical solution temperature (LCST) at about 32 °C (*39–41*), This hydrogel is non-toxic and widely used in drug delivery and tissue engineering (*42–45*), and has the advantage of both simple synthesis and the ease of functionalization. The systems consisting of nanoparticles immobilized in temperature-responsive hydrogel networks have the following key features: (i) the void volume can be utilized for in-situ formation and stabilization of uniformly-distributed NPs; (ii) NPs with tunable size and morphology can be obtained within the hydrogel network structure; (iii) the reaction rates (for toxic organics degradation) can be modulated by temperature-responsive hydrophobic/hydrophilic transitions within the hydrogel network.

The schematic of the synthesis process is shown in Figure 11, and the first step requires the preparation of PNIPAAm hydrogel co-polymerized with acrylic acid (AA). This provides ion exchange sites as NIPAAm has low affinity to metal ions, especially Fe^{2+}/Fe^{3+} (*46*, *47*). The P(NIPAAm-AA) hydrogel was prepared by Ultraviolet (UV) initiated free radical photopolymerization method (*48*). Monomer solution with initiator was pipetted into two 15x15 cm^2 clamped glass plates and a Teflon spacer was used to adjust the thickness to 0.5 mm. This was transferred to a UV source (LESCO) at wavelength of 365 nm and 14.8 mW/cm^2 intensity, to conduct the UV photo polymerization for 5 minutes. The most important variables in the polymerization are AA to NIPAAm ratios, and cross-linker amount which has a significant impact on the porous structure of the polymer network, and thus it can be used to control the nanoparticle size and morphology (*49*, *50*).

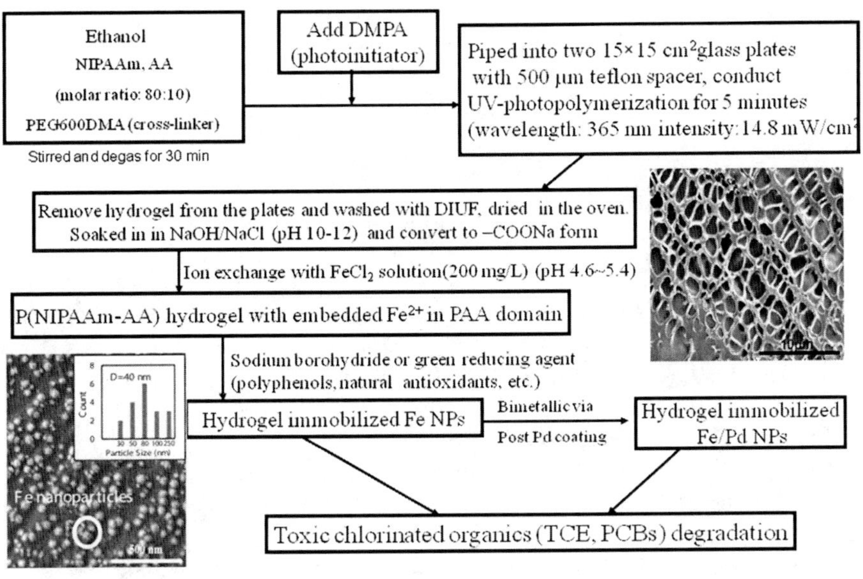

Figure 11. Flow chart for the synthesis of P(NIPAAm-AA) hydrogel immobilized NPs.

The nanoparticle (in situ) synthesis in hydrogels can be achieved using the same concept as presented in Figure 2, a two-step procedure (batch mode) consisting of ion exchange followed by sodium borohydride reduction (or other greener reducing agents). The hydrogel was characterized by ATR-FTIR, showing peaks at 1700 cm^{-1}, 1650 $cm^{-1}$and 1550 cm^{-1} corresponding to C=O group from acrylic acid and –NH groups from NIPAAm. The SEM images (Figure 11) shows a narrow nanoparticles size distribution (around 40 nm) and uniform dispersion in the hydrogel network.

The method presented here is applicable to synthesize other metal nanoparticles within the non-toxic hydrogel network, causing no environmental concerns. This approach has several advantages over other methods, such as: better stability, versatility (multiple catalytic reactions with one hydrogel system), increased catalyst life expectancy and regeneration. The P(NIPAAm-AA) hydrogel containing immobilized Fe or Fe/Pd nanoparticles was also used for the dechlorination of TCE and PCBs (as model compounds). Since the PNIPAAm is temperature responsive, the tunable reactivity was tested by changing the reaction temperature only by 4 °C (below and above LCST) and the surface-area normalized reaction rate for TCE increased from 0.0156 to 0.0411 L/m^{2}h (three fold) with Fe/Pd (1.5% Pd). In comparison, for the homogeneous phase nanoparticles, a 4 °C temperature increase enhanced the reaction rate only two fold (*51*).

Acknowledgments

This work was funded by NIEHS-SRP, DOE-KRCEE programs. We also thank the analytical support by the Environmental Research and Training Laboratory (ERTL) facility and the Electron Microscopy Center (EMC) at the University of Kentucky.

References

1. Nagarajan, R.; Hatton, A. T., Eds.; *Nanoparticles: Synthesis, Stabilization, Passivation, and Functionalization*; ACS Symposium Series 996; American Chemical Society: Washington, DC, 2008.
2. Salata, O. *J. Nanobiotechnol.* **2004**, *2*, 3.
3. Tratnyek, P. G.; Johnson, R. L. *Nano Today* **2006**, *1*, 44–48.
4. Clancy, P. *Nat. Nano* **2011**, *6*, 540–541.
5. Crane, R. A.; Scott, T. B. *J. Hazard. Mater.* **2012**, *211–212*, 112–125.
6. Nadagouda, M. N.; Hoag, G.; Collins, J.; Varma, R. S. *Cryst. Growth Des.* **2009**, *9*, 4979–4983.
7. Liu, Z.; Zhang, F.-S. *Bioresour. Technol.* **2010**, *101*, 2562–2564.
8. Nadagouda, M. N.; Varma, R. S. *Cryst. Growth Des.* **2007**, *7*, 2582–2587.
9. He, F.; Liu, J.; Roberts, C. B.; Zhao, D. *Ind. Eng. Chem. Res.* **2009**, *48*, 6550–6557.
10. Liu, J.; He, F.; Gunn, T. M.; Zhao, D.; Roberts, C. B. *Langmuir* **2009**, *25*, 7116–7128.
11. Smuleac, V.; Varma, R.; Baruwati, B.; Sikdar, S.; Bhattacharyya, D. *ChemSusChem* **2011**, *4*, 1773–1777.
12. Baruwati, B.; Polshettiwar, V.; Varma, R. S. *Green Chem.* **2009**, *11*, 926–930.
13. Baruwati, B.; Varma, R. S. *ChemSusChem* **2009**, *2*, 1041–1044.
14. Park, Y.; Hong, Y.; Weyers, A.; Kim, Y.; Linhardt, R. *IET Nanobiotechnol.* **2011**, *5*, 69–78.

15. Xu, L.; Wu, X. C.; Zhu, J. J. *Nanotechnology* **2008**, *19*, 305603.
16. Yan, X.; Liu, H.; Liew, K. Y. *J. Mater. Chem.* **2001**, *11*, 3387–3391.
17. Li, L.; Ding, J.; Xue, J. *J. Mater. Res.* **2010**, *25*, 810–813.
18. Nadagouda, M. N.; Varma, R. S. *Green Chem.* **2008**, *10*, 859–862.
19. Hoag, G. E.; Collins, J. B.; Holcomb, J. L.; Hoag, J. R.; Nadagouda, M. N.; Varma, R. S. *J. Mater. Chem.* **2009**, *19*, 8671–8677.
20. Nadagouda, M. N.; Castle, A. B.; Murdock, R. C.; Hussain, S. M.; Varma, R. S. *Green Chem.* **2010**, *12*, 114–122.
21. Moulton, M. C.; Braydich-Stolle, L. K.; Nadagouda, M. N.; Kunzelman, S.; Hussain, S. M.; Varma, R. S. *Nanoscale* **2010**, *2*, 763–770.
22. Vaseeharan, B.; Ramasamy, P.; Chen, J. *Lett. Appl. Microbiol.* **2010**, *50*, 352–356.
23. Smuleac, V.; Varma, R.; Sikdar, S.; Bhattacharyya, D. *J. Membr. Sci.* **2011**, *379*, 131–137.
24. Begum, N. A.; Mondal, S.; Basu, S.; Laskar, R. A.; Mandal, D. *Colloids Surf., B* **2009**, *71*, 113–118.
25. Huang, J.; Li, Q.; Sun, D.; Lu, Y.; Su, Y.; Yang, X.; Wang, H.; Wang, Y.; Shao, W.; He, N. *Nanotechnology* **2007**, *18*, 105104.
26. Kamal, S. S. K.; Sahoo, P. K.; Vimala, J.; Premkumar, M.; Ram, S.; Durai, L. *Acta Chim. Slov.* **2010**, *57*, 808–812.
27. Kumar, V.; Yadav, S. C.; Yadav, S. K. *J. Chem. Technol. Biotechnol.* **2010**, *85*, 1301–1309.
28. Yang, X.; Li, Q.; Wang, H.; Huang, J.; Lin, L.; Wang, W.; Sun, D.; Su, Y.; Opiyo, J. B.; Hong, L. *J. Nanopart. Res.* **2010**, *12*, 1589–1598.
29. Huang, X.; Wu, H.; Pu, S.; Zhang, W.; Liao, X.; Shi, B. *Green Chem.* **2011**, *13*, 950–957.
30. Virkutyte, J.; Varma, R. S. *Chem. Sci.* **2011**, *2*, 837–846.
31. Liu, F.; Hashim, N. A.; Liu, Y.; Abed, M. R. M.; Li, K. Progress in the production and modification of PVDF membranes. *J. Membr. Sci.* **2011**, *375*, 1–27.
32. Cutie, S.; Smith, P.; Henton, D.; Staples, T.; Powell, C. *J. Polym. Sci., Part B: Polym. Phys.* **1997**, *35*, 2029–2047.
33. Xu, J.; Dozier, A.; Bhattacharyya, D. *J. Nanopart. Res.* **2005**, *7*, 449–467.
34. Smuleac, V.; Bachas, L.; Bhattacharyya, D. *J. Membr. Sci.* **2010**, *346*, 310–317.
35. Ruzic, I.; Skerget, M.; Knez, Z. *Acta Chim. Slov.* **2010**, *57*, 263–271.
36. Sivaraman, S. K.; Elango, I.; Kumar, S.; Santhanam, V. *Curr. Sci.* **2009**, *97*, 1055–1059.
37. Chobot, V.; Hadacek, F. *Plant Signaling Behav.* **2010**, *5*, 4.
38. Frei, B.; Higdon, J. V. *J. Nutr.* **2003**, *133*, 3275S–3284S.
39. Schild, H. G. *Prog. Polym. Sci.* **1992**, *17*, 163–249.
40. Murakami, K.; Yu, X.; Kato, T.; Inoue, Y.; Sugawara, K. *J. Colloid Interface Sci.* **2012**, *376*, 189–195.
41. Makino, K.; Hiyoshi, J.; Ohshima, H. *Colloids Surf., B* **2000**, *19*, 197–204.
42. Alexander, C. *Nat. Mater.* **2008**, *7*, 767–768.

43. Fernández-Barbero, A.; Suárez, I. J.; Sierra-Martín, B.; Fernández-Nieves, A.; de las Nieves, F. J.; Marquez, M.; Rubio-Retama, J.; López-Cabarcos, E. *Adv. Colloid Interface Sci.* **2008**, *147-148*, 88–108.
44. Hamidi, M.; Azadi, A.; Rafiei, P. *Adv. Drug Delivery Rev.* **2008**, *60*, 1638–1649.
45. Stuart, M. A. C.; Huck, W. T. S.; Genzer, J.; Muller, M.; Ober, C.; Stamm, M.; Sukhorukov, G. B.; Szleifer, I.; Tsukruk, V. V.; Urban, M.; Winnik, F.; Zauscher, S.; Luzinov, I.; Minko, S. *Nat. Mater.* **2010**, *9*, 101–113.
46. Lehto, J.; Vaaramaa, K.; Vesterinen, E.; Tenhu, H. *J. Appl. Polym. Sci.* **1998**, *68*, 355–362.
47. Li, W.; Zhao, H.; Teasdale, P. R.; John, R.; Zhang, S. *React. Funct. Polym.* **2002**, *52*, 31–41.
48. Satarkar, N. S.; Zach Hilt, J. *Acta Biomater.* **2008**, *4*, 11–16.
49. Yin, X.; Hoffman, A. S.; Stayton, P. S. *Biomacromolecules* **2006**, *7*, 1381–1385.
50. da Silva, R.; de Oliveira, M. G. *Polymer* **2007**, *48*, 4114–4122.
51. Xiao, L.; Isner, A. B.; Hilt, J. Z.; Bhattacharyya, D. *J. Appl. Polym. Sci.* **2012**, *128*, 1804–1814.

Chapter 4

Nanostructured Materials for Environmentally Conscious Applications

Panagiotis Dallas,† Antonios Kelarakis,† and Emmanuel P. Giannelis*

Department of Materials Science and Engineering, Cornell University, Ithaca, New York, U.S.A.
†These authors contributed equally to this work.
***E-mail: epg2@cornell.edu**

We present and critically discuss a versatile range of smart and functional nanostructured materials that all share certain common characteristics: they are synthesized by green methods, they are non-toxic and are candidates for environmentally benign or environmentally conscious applications. In particular, we describe the synthesis, properties and applications of the following systems: photoluminescent carbogenic nanoparticles, eco-friendly clay nanocomposites, nanostructured coatings for advanced fouling release, nanocoated separators for lithium-ion batteries, nanoscale macroporous scaffolds for fuel cells, antibacterial silver based polymer nanocomposites, magnetic particles for oil removal and conductive polymer nanostructured sensors for the detection of toxic gases.

I. Introduction

In the recent years the synthesis and characterization of nanoparticles (NPs) are subject of intense investigation from both academia and industry. Interest lies on their use for optical (*1*), magnetic (*2*), electronic (*3*), renewable energy (*4*) and biomedical applications (*5*). In view of the increasing amount of Nps that is produced for numerous commercial formulations and unavoidably released to the environment, serious concerns have been raised about their impact to the ecosystem. Another concern is related with the use of toxic solvents or reactants for the preparation and purification of NPs. To that end, many green reactions have been studied and developed that involve mild and non-toxic reagents and aqueous solutions (*6*).

Also, there is evidence to suggest that certain types of NPs are toxic for both the ecosystem and the humans (*7*). At the same time, a substantial body of work suggests that the environmental impact of the preparation method and the NPs themselves can be controlled to a minimal level (*8*) since their toxicity depends on their size, shape, composition and surface chemistry. Green chemical routes that enable the synthesis of nanoparticles without the use of any harsh organic solvents and reducing or coating agents will help to reduce the negative impact of the nanoscience to the environment. Also, understanding the mechanism of the interaction between nanoparticles and cells will help to identify the characteristics that lead to the minimum or even no toxicity (*9*).

In this article we review a number of nanostructured materials that are synthesized by green methods and can open new pathways in addressing certain environmental concerns. Our focus is limited to systems and nanoparticles that have been developed or are currently under development in our laboratory and we do not intent to spherically cover the progress achieved in the field. Nevertheless, the systems and the approaches presented here adequately highlight the fact that nanomaterials due to their diverse and tunable properties can make the difference in various environmentally benign applications. In particular, we present the synthesis, morphology, properties and applications of the following systems: photoluminescent carbogenic nanoparticles, eco-friendly clay nanocomposites, nanostructured coatings for advanced fouling release, nanocoated separators for lithium-ion batteries, nanoscale macroporous scaffolds for fuel cells, antibacterial silver based polymer nanocomposites, magnetic particles for oil removal and conductive polymer nanostructured sensors for the detection of toxic gases. Particular emphasis is given to the chemical nature of the functional groups of the nanoparticles, their morphological and topological features and the interfacial structure and dynamics.

II. Synthesis, Properties, and Applications of Nanostructured Materials

1. Non-Toxic Photoluminescent Carbogenic Nanoparticles

The development of photoluminescent NPs is essential for biomedical applications (*10*), organic solar cells (*11*) and light emitting diodes (*12*). Photoluminescent quantum dots are typical representatives of this class of nanoscale materials, but are based on heavy metals (i.e. CdSe) and are toxic. In contrast, carbogenic nanoparticles (CNPs) are eco-friendly and biocompatible and their colloidal dispersions exhibit characteristic photoluminescent (PL) and electrogenerated luminescence properties (*13*). The CNPs can be prepared either by fragmentation of predominantly sp^2 hybridized, micron sized particles or by controlled pyrolysis of suitable carbon sources like 4-aminoantipyrine or citric acid (*14*). Their most important application that carbogenic nanoparticles have found is their use as biomarkers in cell imaging, both in vitro and in vivo

(*15*). To that end, the potential toxicity of carbogenic nanoparticles has been a major issue. Initial studies on the toxicity of the carbon dots have shown very low toxicity against mice cells presenting a significant advantage compared to fluorescent particles that are based on toxic metals (*16*). There are also other factors that may affect the toxicity of carbon dots, like for example the molecules being used for surface passivation and functionalization. These effects have been addressed in recent papers and found that indeed any observed toxicity is depending on the surface functionalization, however the materials can be considered of very low toxicity and safe for use as biomarkers (*17*). Wang et al performed toxicity studies on carbon dots with various surface passivation agents. They used O,O′-bis(3-aminopropyl) polyethylene glycol (PEG1500N), poly(propionylethyleneimine- co-ethyleneimine) (PPEI-EI), polyethyleneimine (PEI) and polyallyl amine (PAA). They concluded that the carbon dots are not intrinsically cytotoxic, and any observed cytotoxicity was due to the particle surface passivation molecules (*17*).

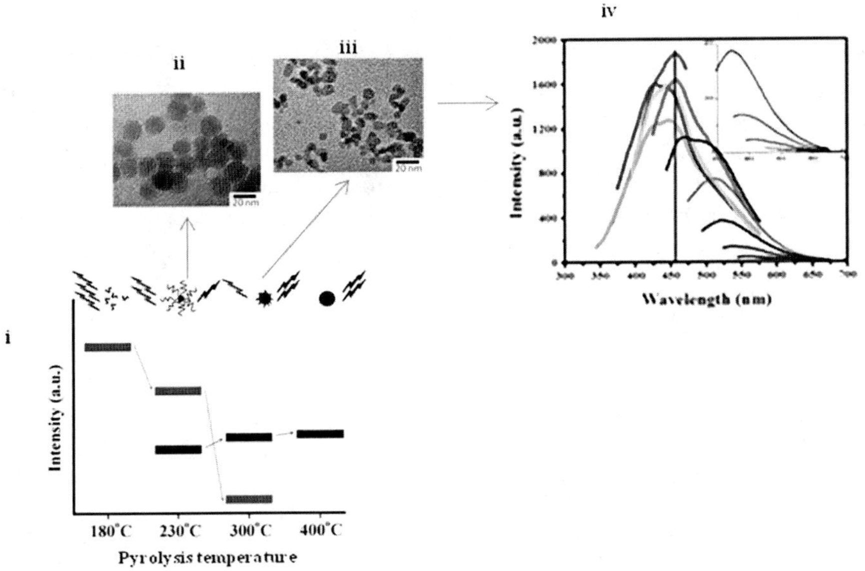

Figure 1. (i) Schematic representation of the PL properties of four photoactive species that are derived from the pyrolysis of a mixture of CA and EA. Upon increasing the pyrolysis temperature, the organic fluorophore groups (depicted as blue chains) are gradually decomposed to form the carbogenic core (black circles). Accordingly, the excitation independent PL component gets weaker (blue bars) and the excitation dependent component gets stronger (black bars). (ii and iii) TEM images of CNP230 and CNP300 respectively. (iv) PL spectra of CNP300. (Reproduced with permission from reference (18). Copyright American Chemical Society).

We have recently demonstrated that the thermal treatment of citric acid (CA) and ethanolamine (EA) generates a series of photoactive species that exhibit dual PL emission (*18*). At the initial stages of pyrolysis, the *in situ* formation of amide-containing fluorophores, via a simple intermolecular condensation reaction, gives rise to a strong excitation independent spectrum. At higher pyrolysis temperatures (230, 300, 400 °C) the organic components are partially consumed towards the formation of carbogenic cores (CNP230, CNP300 and CNP400, respectively) that impart excitation dependent PL properties (*19*) (Figure 1 i and iv). The CNP230 show spherical symmetry with an average diameter of 19 nm (Figure 1ii), while the CNP300 are considerably smaller in size (Figure 1iii) and are richer in carbon. By virtue of their high solubility in water and remarkable colloidal stability, high quantum yield and their non toxic character, the CNPs are ideal nanomarkers and nonoprobes in applications such as bio-imaging, oil tracking, environmental monitoring, water management.

2. Eco-Friendly Clay Nanocomposites with Enhanced Barrier Properties

Clay nanocomposites can be readily prepared by direct melt compounding that constitutes a solvent free and environmental benign method, thoroughly compatible with standard industrial processing (*20*). The presence of the clay modifiers during melt compounding is essential because they alter the thermodynamics of host/guest interactions, facilitating polymer intercalation within the clay galleries (*21*). To fulfill their role, the modifiers should withstand temperatures far above the melting point of the polymer matrix. In this respect, the incorporation of imidazolium and phosphonium, rather than ammonium, based surfactants has been proposed (*22*). However, there is now evidence to suggest that those modifiers are toxic and, therefore, their use should be restricted (*23*).

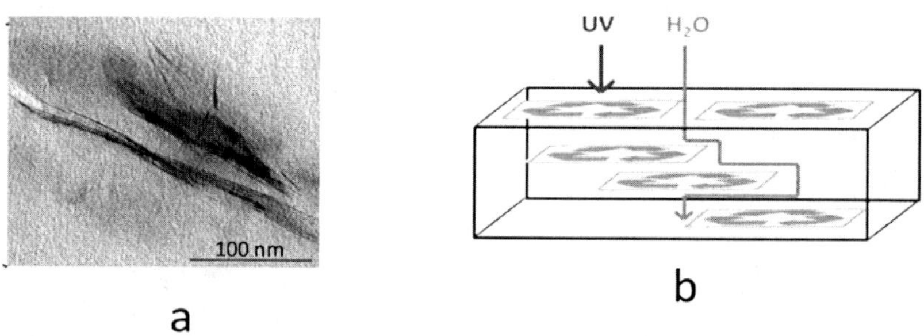

Figure 2. (a) TEM images of PET/MMT-MSPE nanocomposites and (b) a schematic depiction of the nanocomposites emphasizing the impermeable nature of the clay with respect to the enhanced water vapor and UV transmission properties. (Reproduced with permission from reference (25). Copyright 2012 Elsevier Ltd.).

To circumvent this problem we have recently introduced a new class of eco-friendly sulfonated polyester that can effectively function as thermally resistant clay compatibilizer (decomposition onset temperature higher than 350 °C). In particular, we have focused on a commercial polyester of diethylene glycol with isophthalic and sulfoisophthalic acid (SPE) that is an active ingredient in a variety of skin-care formulations (*24*). Addition of 5 wt% montmorillonite-sulfonated polyester (MMT-SPE) clay to poly(ethylene terephthalate) (PET) results in intercalated nanocomposites (Figure 2a) and reduces the water vapor and UV transmission by 22% and 67%, respectively without sacrificing the supreme transparency, clarity and mechanical strength of the neat matrix (*25*). The improved barrier properties can be attributed to the high level of dispersion of the impermeable clay platelets, (Figure 2b), and are highly desirable in food packaging applications.

3. Nanostructured Coatings for Improved Fouling Release Characteristics

The biofilm deposition, biofouling, to marine vessels is a major problem for the shipping industry because it increases friction resistance, reduces fuel efficiency and accelerates corrosion (*26*) thus having a severe economic impact which, as an example, for the U.S. Naval fleet, has been estimated to be $180–540 M annually (*27*).

Fine-tuning the surface topography, roughness and wettability can impart advanced fouling release characteristics to the ship's coatings (*28*). The underlying mechanism is simple; by introducing several levels of roughness to the surface the attached microorganisms cannot get a firm grasp and are easily removed, ideally by bringing the ship to speeding. Capitalizing on this concept, we have focused on the introduction of polyurea segments and fluorinated extenders to the polydimethysiloxane (PDMS) chains (a widespread ingredient of ships' coatings) in order to simultaneously enhance their mechanical strength and the phase-separation driven surface complexity. Incorporation of various types of nanoparticles (clays, polyhedral oligomeric silsesquioxane -POSS) to the segmented copolymers allows one more level of structural and morphological control (Figure 3) (*29*). The work needed for the removal of the attached microorganisms from the nanostructured surfaces is up to three times lower compared to the pristine PDMS surface (*29*).

The environmental benefits of our approach are multiple; it is based on a non-toxic material, it is a reliable alternative to the use of paintings containing biocides that have detrimental effects to the ecosystem, it results in improved fuel efficiency reducing the harmful gasses emissions and it dramatically reduces the otherwise substantial energy consumption for the surface cleaning (thus contributing to the sustainability).

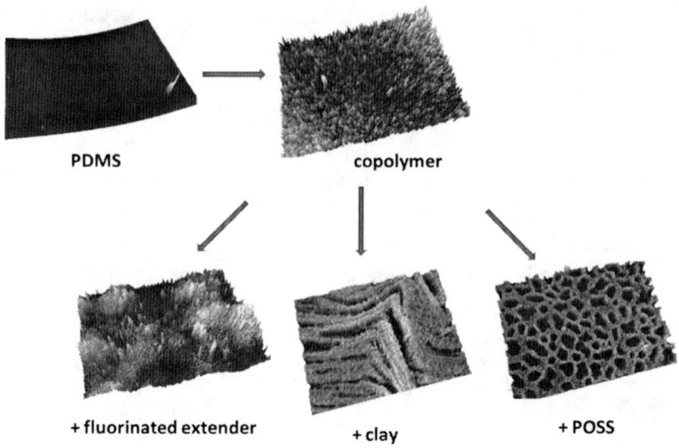

Figure 3. 3-D profilometer images of solvent-casted samples. Upper row: polydimethylsiloxane (PDMS) and polydimethylsiloxane/polyurea segmented copolymer (copolymer). Lower row: the derivates of the copolymer after incorporation of fluorinated chains extenders, clays, and POSS. (Reproduced with permission from reference (29). Copyright 2010 Elsevier Ltd.).

4. Functional Nanohybrids for Clean Energy

The drive towards sustainability requires the design of smart materials for the development of alternative, clean and renewable energy sources and technologies. For example, the advancement of lithium ion batteries will expand and promote the use of electric vehicles with zero-gas emissions.

In this respect, we have developed a nanocoated separator for lithium ion batteries with improved surface properties and wettability against common polar electrolytes (*30*). The coating approach relies on the electrostatic immobilization of cationically modified silica NPs on the surface of plasma treated macroporous polyolefin trilayer separator. The attached nanoparticles resist detachment even during harsh and prolonged treatments (*29*). The unique advantage of this strategy is that it allows the deposition of the NPs not only to the exterior of the surface, but also at a depth of several micrometers depending on the pH of the aqueous silica suspension (Figure 4).

The nanocoated separators have zero contact angle against propylene carbonate compared to 66° for the pristine separator, and their internal resistance is 5 times lower. The cells equipped with the modified separators exhibit superior cyclability and durability. In particular, the NPs-coated separator retains 92% of its charge capacity after 100 cycles compared to 77% for the pristine membrane.

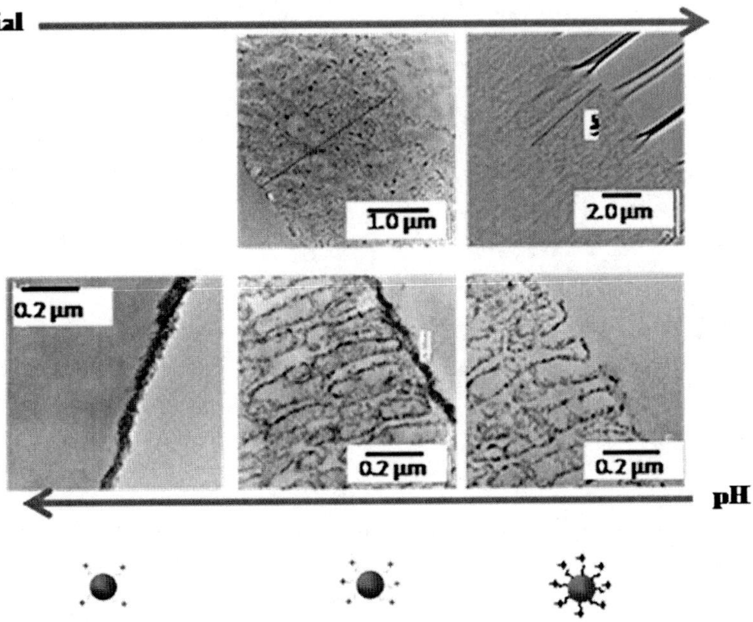

Figure 4. TEM images of the plasma treated separators coated with cationically modified silica NPs. The pH of the aqueous dispersion was controlled via acidification (to the values 4 to 7 and 7.5). The lower magnification images on the top reveal the depth of NPs coating. (Reproduced with permission from reference (31). Copyright Royal Society of Chemistry).

Hydrogen fuel cells are zero-emissive electrochemical converters that exploit the catalytic recombination of H_2 and O_2 to form water and generate electrical energy. We have synthesized macroporous multifunctional scaffolds that can improve the electrochemical performance of fuel cells by providing extensive triple phase boundaries (contact points of the gas reactants, the electronic conductor and the catalyst absorbed on an electronically conductive support) (*32*). The synthetic strategy is based on the cryostructuring (an environmentally benign method) of an aqueous dispersion of Nafion containing graphite oxide and a platinum salt. As the water rapidly freezes in a liquid nitrogen pool, the dispersed nanoparticles are confined in the area within the fleshly formed crystals so that the subsequent sublimation of the solvent gives rise to a macroporous scaffold. The monoliths are subjected to mild treatment with hydrazine to allow the *insitu* and simultaneous reduction of the graphite oxide towards graphene nanosheets and the platinum salt towards well-defined Pt NPs (Figure 5). The monoliths have large surface area and they combine high levels of electronic (graphene) and ionic conductivity (Nafion) with catalytic activity (Pt nanospheres).

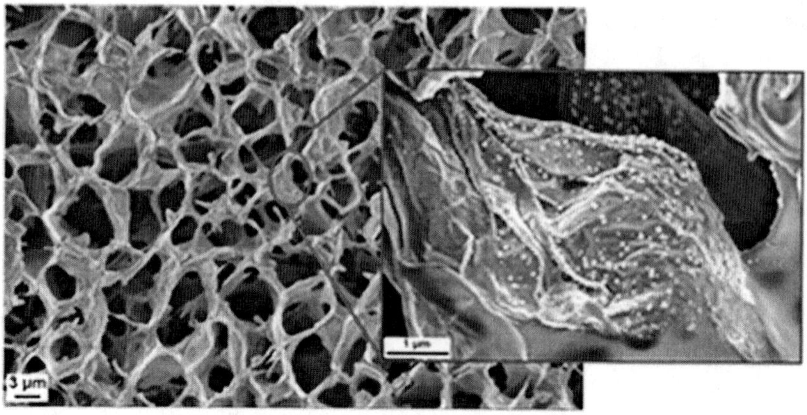

Figure 5. TEM images of the macroporous scaffold based on Nafion. The magnification focuses on an individual wrinkled graphene nanosheet that supports a number of platinum nanoparticles. (Reproduced with permission from reference (32). Copyright 2011 American Chemical Society).

5. Magnetic Nanoparticles for Oil Spill Removal

Oil leakage is a major environmental concern since it irreversibly affects the marine and coastal ecosystems in a significantly large scale, as an example, 1 ton of oil will form a film of 12 km^2 on the water surface. For example one of the most devastating oil spills, the Exxon Valdez incident released about 119.000 m^3 of crude oil and caused unpresentable environmental destruction (*33*). Furthermore, the oil spills take years to clean up while all their removal and recovery process is extremely difficult, time consuming and expensive. Various materials have been used for the effective removal of the oil spills but oftentimes they suffer for various problems and high cost. As an example, the activated carbons and porous materials that are used nowadays have high regeneration temperatures (e.g., in the range of 800-850 °C for activated carbon), and low separation efficiency due to the co adsorption of water (*34*), while other methods like the controlled burning or the use of detergents can cause air pollution and have toxic effects, while the polymeric membranes that can be used lack selectivity and stability and are unsuitable for the collection of a large oil spill. Currently, significant effort has been put to novel bioremediation technologies that are using microorganisms and bioremediation accelerators to break down the oil hydrocarbons. Constantly new materials are studied for oil absorption like for example the low cost and biocompatible beta-cyclodextrin which has been proven a very efficient absorbent that can also be reused for numerous times (*35*).

On the other hand, the magnetic iron oxides (maghemite and magnetite) have the advantage of being environmentally friendly, completely non-toxic (*36*), low cost materials that exhibit very high magnetization values. Numerous different reaction methods have been presented in the literature leading to a plethora of available sizes and shapes while its surface reactivity makes them versatile tools for the construction of functional nanomaterials since numerous organic molecules or polymers can be chemisorbed on their surface providing the particles with exciting properties and potential applications.

Recently, highly oleophilic functionalized iron oxide nanoparticles have been proven to be excellent removal agents for lubricating oils. The particles are introduced to siloxanes (*37*) or polyurethane foams to form composites that float on water (contact angles with water larger than 150 °C), absorb large amounts of oil (up to 3.8 times their weight) and can be easily directed and eventually removed by common magnets (*38*). Following their removal from the oil spill, the magnetic nanoparticles can be recovered from the foams simply by sonication in a suitable solvent. Furthermore, the materials proved to be recyclable and reusable: after heating at 600 °C the siloxane coating can be removed and after added again the materials possessed the same properties.

6. Silver Composites as Antibacterial Agents and Coatings

Silver nanoparticles exhibit remarkably low toxicity and excellent activity against a number of bacteria and fungi (*39*), while they leave beneficial enzymes intact. Furthermore, in contrast with a vast number of classical antibiotics, the bacteria have not yet developed a resistance towards silver.

The exact mechanism of the antibacterial activity of silver is still not yet well defined. However, there are strong indications that is based on either a penetration of the silver particles inside the membranes cell walls as proved by Sondi et al (*40*) (Figure 6) or from an ion flow from the surface of the silver nanoparticles towards the bacteria cell. The surface silver atoms can be oxidized and the ions can flow from the particle towards the cell. Consequently, the silver ions penetrate the membrane and form complexes with thiol, hydroxyl or amine groups of the proteins and the other functional organic biomolecules thus disrupting the functions of the cell and causing deformations in the cell walls (*41*).

Colloidal silver has been already used in medicine but specific applications like coatings and food packaging require the immobilization of the antibacterial agents inside suitably selected polymer matrices (*42*). Usually in order to synthesize silver-polymer nanocomposites, the silver cations are entrapped inside the polymer matrix and are reduced to metallic silver with the use of well known reducing agents. By firstly entrapping the silver ions inside the host matrix, the macromolecular chains prevent the formation of any undesired aggregates after the reduction of the ions to metallic silver. Up to now, numerous biocompatible matrices like poly(acrylamide-co-acrylic acid) hydrogels (*43*), biodegradable cellulose matrices (*44*) chitosan (*45*, *46*) and polysulfone fibers (*47*) have been already used for the synthesis of environmentally friendly antibacterial coatings.

Recently, many methods utilized the functional groups of specific polymers (*48*, *49*) and dendrimers in order the polymer chains themselves to reduce the silver cations and the synthesized nanoparticles to remain absorbed on its surface. With these methods any extra reducing agents are avoided. Since the particles remain absorbed on the surface of the polymer and still exhibit a high antimicrobial activity this is a strong indication that the mechanism of the activity can be an ion flow from the surface of the nanoparticles.

In a further development of the available nanocomposite materials, the introduction of both magnetic iron oxide NPs along with silver NPs to a polymeric matrix allows the preparation of magnetically controlled composites that can selectively disinfect a targeted area and then can be removed by a simple magnet for reuse or recycling (*48*, *49*). Some polymer linkers that have been used include phosphotriazine based polymers and biocompatible polyacrylate matrices and both showed excellent minimum inhibitory concentration (0.078-0.15 g/L) and magnetization values (more than 5 Am^2/kg at room temperature). These magnetization values are high enough to enable the magnetic control of the nanocomposites.

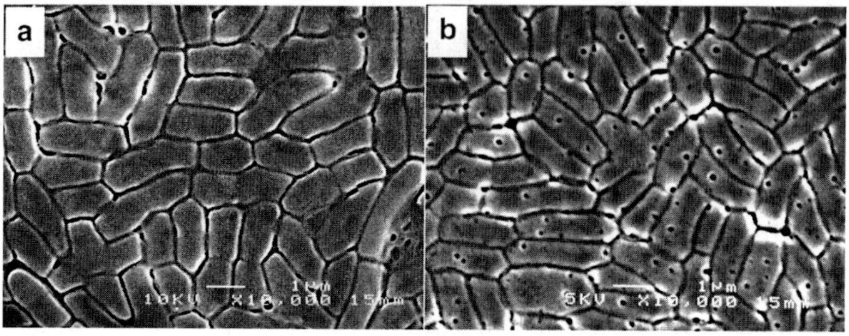

Figure 6. SEM images of E. coli cells (a) and the same cells treated with 50 μg cm^{-3} of silver nanoparticles (b). The images are indicating the presence of silver nanoparticles inside the cell itself. The cell is affected from the penetration of the silver particles. (Reproduced with permission from reference (40). Copyright 2004 Elsevier Ltd.).

7. Conductive Polymer Nanostructures for Sensor Applications

Typically the organic polymers are insulators with high resistivity values. To that end, the discovery of the intrinsically conductive polymers with their extraordinary and novel electrical properties lead to a revolution in the field of polymers physics providing new perspectives and applications for the polymer science.

Most efforts in the field of conductive polymers were concentrating in synthesizing either conductive polymers with high values of conductivity or even polymers with a metallic-like behavior, i.e. the conductivity is increasing as the temperature is decreasing. Usually the conductivity values are reaching a maximum in a specific temperature. In this temperature a metal-insulator transition takes place something that is typical for 1-D conductors due to their intrinsic Peierls instability. In a previous work by Dallas et al (*50*) demonstrating the interfacial polymerization of aniline, the needle shaped polyaniline (PANI) nanoparticles with good ordering and ultra small dimensions were exhibiting a very interesting electrical and magnetic behavior (Figure 7). Temperature dependent conductivity measurements indicated that the resistance from 330 to

230 K was following a metallic behavior, with the maximum conductivity value to be 3.5 σ•cm^{-1} (*50*). The fitting of the resistance values in the semiconducting region according to the Mott relation: $\ln(R_T/R_{330})$ vs $1/T^{1/2}$ defined the polymer as an 1-D conductor (x~ 1/2) with the Mott temperature, a characteristic value for conductive materials that is reflecting the energy required for charge hopping, calculated to be 1.5•10^4 K. Furthermore the electrical properties were correlated with a transition from a room temperature (RT) diamagnetism to a low temperature paramagnetism and a phase diagram reflecting these changes was constructed (Figure 7c). In this point, the work by Lee et al should be highlighted since it presents a significant contribution in the field of conductive polymers. They synthesized PANI particles that even in the lower temperatures the electrical behavior was metallic (*51*).

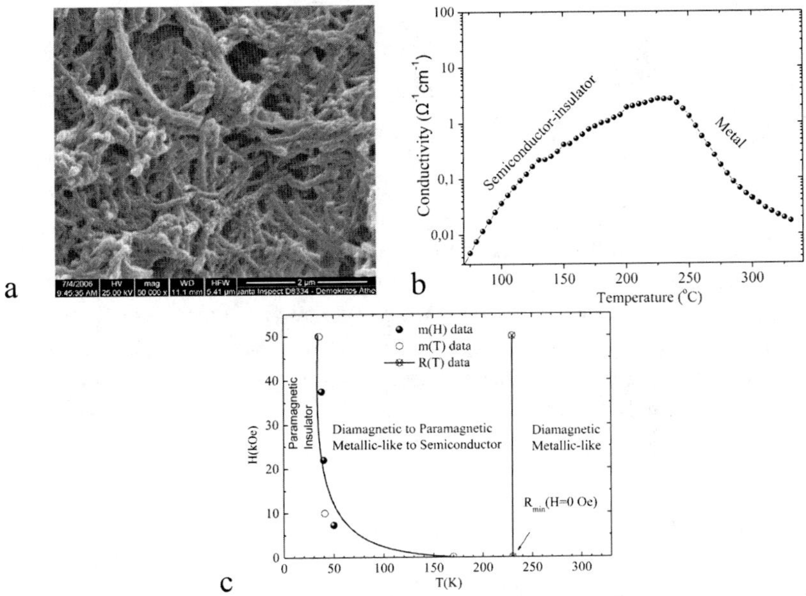

Figure 7. SEM image (a) and electronic transport (conductivity vs temperature) behavior (b) of polyaniline nanoneedles synthesized through an interfacial polymerization. The nanostructured polymer is exhibiting a metal-insulator transition at 230 K. (c) The diagram is representing how the changes of the electrical and magnetic behavior are correlated. (Reproduced with permission from reference (50). Copyright 2007 Elsevier Ltd.).

In general the mechanism of the electron transport of these polymers is more complex than in conventional inorganic crystals or organic polymers and depends on both their ability to transport charge carriers along the macromolecular aromatic backbone and on the ability of the carriers to hop between adjacent polymer chains.

Furthermore, the most commonly used conductive polymers like polyaniline and polypyrrole can exist in different oxidation states each one exhibiting different conductivity values as also optical properties (i.e. different UV-Visible absorbance). Consequently, their acid-base and redox reactions can be exploited for the construction of gas sensors with numerous environmental applications. As an example, Palaniappan, et al synthesized efficient sensors (*52*) based on PANI for toxic gases like hydrogen sulfide, while Kaner et al used PANI nanofibers usually obtained through interfacial polymerization, as sensing materials for the detection of acids, bases, reducing agents, organic vapors, and alcohols (*53*). Thin films (~2.5 μm) of both nanofibers and of standard PANI with non-fibrilar morphology have been compared in order to evaluate the response time and sensitivity of the nanofibers in various gases. The films were deposited on an array of gold electrodes consisting of pairs of electrodes on a glass substrate. The conductivity is changing as a function of time as the materials were exposed to gases like hydrochloric acid, (HCl), ammonia, (NH_3), hydrazine (N_2H_4), chloroform ($CHCl_3$), and methanol (CH_3OH).

Each of the above mentioned gases induces different changes in the polyaniline chains and in their oxidation state and leads to a different electrical response. These changes were protonation, deprotonation, reduction, swelling, and conformational changes depending on the gas that was used. The importance of the formation of ultra-small nanofibers was demonstrated by the fact that the response time was significantly better for polyaniline nanofiber thin films than that observed for larger polyaniline particles. This can be assigned to their higher surface area ratio and also in the fact that these films have a very porous structure consisting from a three-dimensional, interconnected network of fibers. In conclusion, the easy preparation method, the exciting physical properties, the reactivity and the various oxidation states is making conductive polymers excellent candidates for the construction of chemical sensors for a wide range of gases with extremely high sensitivity as also selectivity.

Conclusions

In this review we present a range of new platform materials that can play a key role in environmentally conscious applications in the near future. Those versatile multifunctional materials can be fine tuned to meet specific requirements and performance characteristics. Major challenges include the synthesis of carbogenic particles with tunable photoluminescence emission, the development of recyclable clay nanocomposites with improved macrosopic properties, the design of anti-adhesive surfaces that can withstand harsh treatment over a long period of time and the advancement of nanostructured electrolytes and electrodes for electrochemical converters (fuel cells, solar cells and batteries). In addition, considerable efforts are directed towards the development of stable oil absorbents from inexpensive sources, the coupling of silver nanoparticles with conventional antimicrobial agents, and the accurate structural control of the polyalinilie nanofibers in order to enhance their sensitivity and selectivity as toxic gas sensors.

References

1. Liu, Q.; Cui, Y.; Gardner, D.; Li, X.; He, S.; Smalyukh, I. I. *Nano Lett.* **2010**, *10*, 1347–1353.
2. Sun, S.; Murray, C. B.; Weller, D.; Folks, L.; Moser, A. *Science* **2000**, *287*, 1989.
3. Homberger, M.; Simon, U. *Phil. Trans. R. Soc., A* **2010**, *368*, 1405–1453.
4. Leonard, B. M.; Zhou, Q.; Wu, D.; DiSalvo, F. J. *Chem.Mater.* **2011**, *23*, 1136–1146.
5. Huang, X.; Neretina, S.; El-Sayed, M. *Adv. Mater.* **2009**, *21*, 4880–4910.
6. Nadagouda, M. N.; Varma, R. S. *Green Chem.* **2008**, *10*, 859–862.
7. Hansen, T.; Clermont, G.; Alves, A.; Eloy, R.; Brochhausen, C.; Boutrand, J. P.; Gatti, A. M.; Kirkpatrick, C. J. *J. R. Soc., Interface* **2006**, *3*, 767–775.
8. Suresh, A. K.; Pelletier, D. A.; Wang, W.; Morrell-Falvey, J. L.; Gu, B.; Doktycz, J. *Langmuir* **2012**, *28* (5), 2727–2734.
9. Bar-Illan, O.; Albrecht, R. M.; Fako, V. E.; Furgeson, D. Y. *Small* **2009**, *5*, 1897.
10. Pellegrino, T.; Kudera, S.; Liedl, T.; Javier, A. M.; Manna, L.; Parak, W. J. *Small* **2005**, *1*, 48.
11. Huynh, W. U.; Dittmer, J. J.; Alivisatos, A. P. *Science* **2002**, *295*, 2425.
12. Zorn, M.; Bae, W. K.; Kwak, J.; Lee, H.; Lee, Ch.; Zentel, R.; Char, K. *ACS Nano* **2009**, *3*, 1063.
13. Wang, F.; Chen, Y.; Liu, C.; Ma, D. *Chem. Commun.* **2011**, *47*, 3502–3504.
14. Baker, S. N.; Baker, G. A. *Angew. Chem. Int. Ed.* **2010**, *49*, 6726.
15. Tao, H. Q.; Yang, K.; Ma, Z.; Wan, J. M.; Zhang, Y. Z.; Kang, Z. H.; Liu, Z. *Small* **2012**, *8* (2), 281–290.
16. Yang, S. T.; Wang, X.; Wang, H. F.; Lu, F.; Luo, P. G.; Cao, L.; Meziani, M. J.; Liu, J. H.; Liu, Y.; Chen, M.; Huang, Y.; Sun, Y. P. *J. Phys. Chem. C* **2009**, *113*, 18110–4.
17. Wang, Y.; Anilkumar, P.; Cao, L.; Liu, J. H.; Luo, P. G.; Tackett, K. N.; Sahu, S.; Wang, P.; Wang, X.; Sun, Y. P. *Exp. Biol. Med.* **2011**, *236*, 1231–1238.
18. Krysmann, M. J.; Kelarakis, A.; Dallas, P.; Giannelis, E. P. *J. Am. Chem. Soc.* **2012**, *134*, 747.
19. Eda, G.; Lin, Y.-Y.; Mattevi, C.; Yamaguchi, H.; Hsin-An Chen, H.-A.; Chen, I.-S.; Chen, C. W.; Chhowalla, M. *Adv. Mater.* **2010**, *22*, 505.
20. Giannelis, E. P. *Adv. Mater.* **1996**, *8*, 29.
21. Pavlidou, S.; Papaspyrides, C. D. *Prog. Pol. Sci.* **2008**, *33*, 1119.
22. Calderon, J. U.; Lennox, B.; Kamal, M. R. *Appl. Clay Sci.* **2008**, *40*, 90.
23. Docherty, K. M.; Kulpa, C. F. *Green Chem.* **2005**, *7*, 185.
24. Ash, M.; Ash, I. *Handbook of Green Chemicals*, 2nd ed.; Synapse Information Resources: New York, 2004; p 184.
25. Hayrapetyan, S.; Kelarakis, A.; Estevez, L.; Lin, Q.; Dana, K.; Chung, YL.; Giannelis, E. P. *Polymer* **2012**, *53*, 422.
26. Omal, I. *Chem. Rev.* **2003**, *103*, 3431.
27. Schultz, M. P.; Bendick, J. A.; Holm, E. R.; Hertel, W. M. *Biofouling* **2011**, *27* (1), 87–98.

28. Schumacher, J. F.; Carman, M. L.; Callow, M. E.; Callow, J. A.; Finlay, J. A.; Brennan, AB. *Biofouling* **2007**, *23*, 55.
29. Fang, J.; Kelarakis, A.; Wang, D.; Giannelis, E. P.; Finlay, J. A.; Callow, M. E.; Callow, J. A. *Polymer* **2010**, *51*, 2636.
30. Fang, J.; Kelarakis, A.; Lin, Y. W.; Kang, C. Y.; Yang, M. H.; Cheng, C. L.; Wang, Y.; Giannelis, E. P.; Tsai, L. D. *Phys. Chem. Chem. Phys.* **2011**, *13*, 14457.
31. Fang, J.; Kelarakis, A.; Estevez, L.; Wang, Y.; Rodriguez, R.; Giannelis, E. P. *J. Mater .Chem.* **2010**, *20*, 1651.
32. Estevez, L.; Kelarakis, A.; Gong, Q.; Da'as, E. H.; Giannelis, E. P. *J. Am. Chem. Soc.* **2011**, *133*, 6122.
33. Maki, A. W. *Environ. Sci. Technol.* **1991**, *25*, 24–29.
34. Sabio, E.; González, E.; González, J. F.; González-García, C. M.; Ramiro, A.; Ganan, J. *Carbon* **2004**, *42*, 2285–2293.
35. Ding, L.; Li, Y.; Jia, D.; Deng, J. P.; Yang, W. T. *Carbohydr. Polym.* **2011**, *83* (4), 1990–1996.
36. Pankhurst, Q. A.; Connoly, J.; Jones, S. K.; Dobson, J. *J. Phys. D: Appl. Phys.* **2003**, *13*, R167.
37. Zhu, Q.; Tao, F.; Pan, Q. *ACS Appl. Mater. Interfaces* **2010**, *2*, 3141–4146.
38. Calcagnile, P.; Fragouli, D.; Bayer, IS.; Anyfantis, GC.; Martiradonna, L.; Cozzoli, P.; Cingolani, R.; Athanassiou, A. *ACS Nano* **2012**, *6*, 5413–5419.
39. Chaloupka, K.; Malam, Y.; Seifalian, AM. *Trends Biotechnol.* **2010**, *28*, 580.
40. Sondi, I.; Sondi, S. B. *J. Colloid Interface Sci.* **2004**, *275*, 177.
41. Zhang, Y.; Peng, H.; Huang, W.; Zhou, Y.; Zhang, X.; Yan, D. *J. Phys. Chem. C* . **2008**, *112*, 2330.
42. Dallas, P.; Sharma, V.; Zboril, R. *Adv. Colloid Interface Sci.* **2011**, *166*, 199–135.
43. Kidambi, S.; Bruening, M. L. *Chem. Mater* **2005**, *17*, 301.
44. Pinto, R. J. P.; Marques, P. A. A. P.; Neto, C. P.; Trindade, T.; Daina, S.; Sadocco, P. *Acta Biomater.* **2009**, *5*, 2279.
45. Kong, H.; Jang, J. *Langmuir* **2008**, *24*, 2051.
46. Rhim, J. W.; Hong, S. I.; Park, M.; Ng, P. K. W. *J. Agric. Food. Chem.* **2006**, *54*, 5814.
47. Schiffman, J. D.; Wang, Y.; Giannelis, E. P.; Elimelech, M. *Langmuir* **2011**, *27*, 13159–13164.
48. Dallas, P.; Tucek, J.; Jancik, D.; Kolar, M.; Panacek, A.; Zboril, R. *Adv. Funct. Mater.* **2010**, *20*, 2347–2354.
49. Prucek, R.; Tucek, J.; Kilianova, M.; Panacek, A.; Kviket, L.; Filip, J.; Kolar, M.; Tomankova, K.; Zboril, R. *Biomaterials* **2011**, *32* (21), 4704–4713.
50. Dallas, P.; Stamopoulos, D.; Boukos, N.; Tzitzios, V.; Niarchos, D.; Petridis, D. *Polymer* **2007**, *48*, 3162.
51. Lee, K.; Cho, K. S.; Sung, H. P.; Heeger, A. J.; Lee, C. W.; Lee, S. H. *Nature* **2006**, *441* (1), 65–68.
52. Palaniappan, S.; Saravanan, C. *J. Appl. Polym. Sci.* **2010**, *118*, 518–524.
53. Huang, J.; Virji, S.; Weiller, B. H.; Kaner, R. B. *J. Am. Chem. Soc.* **2003**, *125*, 314–315.

Chapter 5

Nanomaterials in Agricultural Production: Benefits and Possible Threats?

Jie Hong,[1] Jose R. Peralta-Videa,[1] and Jorge L. Gardea-Torresdey[*,1,2]

[1]Environmental Science and Engineering PhD Program and The University of Texas at El Paso, 500 West University Ave., El Paso, Texas 79968
[2]Chemistry Department, The University of Texas at El Paso, 500 West University Ave., El Paso, Texas 79968
[*]E-mail: jgardea@utep.edu. Phone: 915-747-5359. Fax: 915-747-5748.

Food and agricultural production are among the most sensitive fields of application of nanotechnology. It is predicted that nanotechnology will revolutionize agriculture through new tools for disease treatment and detection, smart delivery systems, sensors, and better management devices. However, the rapid development of nanotechnology could release a huge amount of engineered nanoparticles, which may cause adverse effects on edible plants. In this review article the authors analyzed the most recent literature on the application of nanotechnology in agriculture, including nanofertilizers, nanosensors, crop protection, pollution control, waste management, and pesticide detection. The negative effects caused by nanoparticles on edible plants are also discussed.

Keywords: Nanotechnology; Agriculture; Nanoparticles; Applications; Negative effects

Introduction

Nanotechnology, the fabrication and utilization of nanosized materials, is increasingly attracting attention not only for its variety of applications in modern life, but for the potential negative effects that nanomaterials (NMs) or engineered nanomaterials (ENMs) can cause in the environment and human beings. According to the U.S. National Nanotechnology Initiative (NNI), nanotechnology is defined as "the understanding and control of matter at dimensions of roughly 1 to 100 nanometers (a nanometer is one billionth of a meter), where unique phenomena enable novel applications" (*1*). The building blocks of nanotechnology are nanoparticles (NPs), where particles with at least two dimensions between 1 and 100 nm are included (*2–4*). Their small size and large surface area make them different from bulk materials. Differences between NPs and bulk materials include physical strength, chemical reactivity, electrical conductivity, magnetism, and optical effects (*4*, *5*). These properties enable the use of NMs in electronic engineering, energy, material applications, pharmaceutics, biomedics, cosmetics, textiles, food industry, and agriculture (*6–8*).

This variety of applications has made a fast investment in nanotechnology worldwide. In the United States, the annual budget of the NNI was \$464 million in 2001; however, this amount was increased to \$1.9 billion in 2010 with a request of 2.1 billion for 2012. During this period, the NNI received nearly \$12 billion in nanotechnology research and the investment of the U.S. Department of Agriculture (USDA) and the National Institute of Food and Agriculture (NIFA) increased from \$0 to \$13.2 million (*9*). The above numbers clearly demonstrate that nanotechnology will play an increasingly important role in agriculture.

Studies over the past decade showed that nanotechnology has the potential to revolutionize agriculture with the use of biosensors (*10*), plant growth regulators (*11*), food additives (*12*), genetic improvement of plants and animals (*13*, *14*), smart delivery systems for drugs, pesticides and fertilizers (*15*, *16*) and nanopesticides (*16*, *17*). Currently, there is a new agricultural methodology widely used in the U.S, Europe, and Japan, which is called Controlled Environment Agriculture (CEA). With this methodology, plants grown in controlled environment are maintained in good conditions by computerized system monitors. CEA also brings a perfect platform for the utilization of nanotechnology for food production.

Developing countries will soon start using nanotechnology in agricultural activities. According to Salamanca-Buentello et al. (*18*), within the next ten years, several nanotechnology applications will be available for agriculture in developing countries. Applications such as the use of nanoporous zeolites for slow-release and efficient dosage of water and fertilizers, slow release of nutrients and drugs for livestock, nanocapsules for herbicide delivery, nanosensors for soil quality and plant health monitoring, and nanomagnets for removal of contaminants from soil will be in the market soon.

Although nanotechnology can benefit agriculture, it could represent potential negative impacts to ecosystems that cannot be ignored (*19*). Current studies have shown that some engineered NPs have negative effects on plants growth. For example, single walled carbon nanotubes (SWCNTs) have been found to affect root elongation of tomato, cabbage, carrot and lettuce (*20*). ZnO NPs show toxicity to *Arabidopsis* (*21*), *Prosopis* (*22*, *23*), and *Glycine max* (*24–26*). ZnO and CeO_2 cause genotoxic effects on soybean (*27*). Soybean seedlings treated with either ZnO or CeO_2 NPs at 2000 and 4000 mg/l showed new bands in the genomic DNA, which suggest these NPs could be mutagenic agents (*27*). TiO_2 NPs inhibit leaf growth and transpiration of maize seedlings due to the reduction of hydraulic conductivities (*28*), while Ag NPs disrupt the cell division process in *Allium cepa* (*29*).

In a recent review, Rico et al. (*26*) analyzed the most recent literature related to the interaction of NPs with edible plants and their possible implications in the food chain. The authors analyzed the positive and negative impacts of NPs in edible plants and the potential risk of the bioaccumulation of NPs in the next plant generation. In a more recent study, Zhao et al. (*30*) have shown that corn plants grown in a sandy loam soil treated with ZnO NPs at 200 mg/kg can concentrate about 200 mg Zn/kg dry shoot biomass. This could indicate that an unintentional or deliberate release of ZnO NPs in the ecosystem could put an excess of Zn in the food chain.

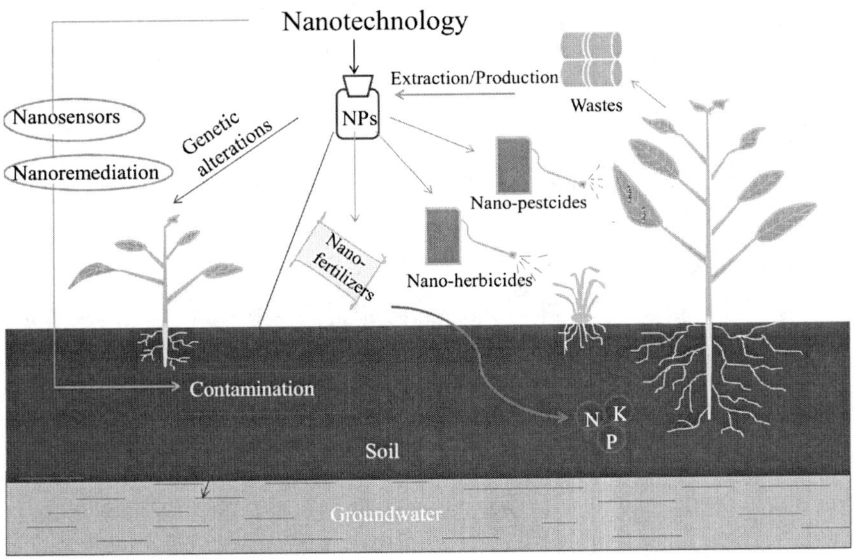

Figure 1. Interactions of nanomaterials with agricultural activities.

The present review is focused on the applications of nanotechnology as a tool in agricultural production. More than 80 articles published in the last five years were analyzed. A flux diagram depicting the interactions of nanotechnology with agricultural activities is presented in Figure 1. As shown in this figure, nanofertilizers, nanopesticides and nanoherbicides can help to protect plants and animals for better productivity. However, the applications of nanoproducts can be a source of soil and groundwater contamination (*31*). On the other hand, there is evidence that NPs such as nano-TiO_2, nano-ZnO, and nano-Fe_3O_4 can be used to remediate contaminated bodies of water (*32*, *33*). Nanosensors can be used to monitor the growth condition of crops in order to improve the use of water, fertilizers, and pesticides and to examine the presence of unwanted substances; thus, preventing contamination (*34*). Agricultural wastes have been associated with pollution (*35*). However, recent reports indicate that NPs such as SiO_2 NPs can be obtained from agricultural wastes (*36*, *37*), which can reduce contamination and at the same time, generate added value to agricultural production. The analyzed literature has shown that the application of nanotechnology in agricultural activities could have positive and negative effects.

Positive Effects of the Application of Nanotechnology in Agriculture

Traditional agricultural activities such as fertilization, weed control, and pesticide applications are labor-consuming, expensive, and in most cases, risky for agricultural workers (*34*, *38*). There is a wide range of products and devices currently used to simplify product applications and cost reductions in commercial crop production. However, new pests, or pesticide-resistant pests, challenge agricultural production worldwide. In developed countries, farmers have turned to techniques such as transgenic varieties or several nano-based goods such as nanofertilizers (*39*, *40*, *92*), nanopesticides (*41*, *42*, *92*), and nanosensors (*43*, *44*, *92*) to reduce production costs (*34*). Obviously, the use of these new products and devices has benefited agricultural production; however, the collateral environmental impacts of some of these nanoproducts are still unknown.

Enhancement of Plant Growth by Nanofertilizers

It is well known that nitrogen is a key element in structural and functional proteins that determine plant growth and the utilization of raw or processed plant products. However, the nitrogen use efficiency (NUE) by crops is very low. About 50 to 70% of the nitrogen applied to soil by conventional fertilizers is not timely utilized by crop plants because of leaching as water soluble nitrates, emission of gaseous ammonia and nitrogen oxides, or incorporated into soil organic matter by soil microorganisms (*45*). The use of nanofertilizers can help to reduce nitrogen losses in soils. Nutrients, such as N, P, and K can be encapsulated in nanotubes or nanoporous materials, coated by thin protective polymer films, or delivered as NPs or emulsions (*39*).

Kottegoda et al. (*40*) studied nitrogen release from a nanofertilizer composition based on urea-modified hydroxyapatite (HA) NPs encapsulated into the cavities of the soft wood of *Gliricidia sepium*. The research was performed in soils of different degrees of acidity (pH 4.2, 5.2 and 7) in Sri Lanka. A commercial fertilizer was applied and used as a control. Compared to the control, the urea-modified *G. sepium*-encapsulated nano HA-fertilizer showed a slow and sustained release of nitrogen at the three different pH values. Even 60 days after application, the nanofertilizer was still releasing more than10 mg nitrogen to the soil. On the other hand, the release from the commercial fertilizer was heavy at the beginning, and low and non-uniform around day 30th after the application. In another study, a nano-composite superabsorbent polymer (NC-SAP) was used to encapsulate plant nutrients for slow-release. The NC-SAP is a material that has the ability to absorb and retain large volume of water and aqueous solution that can be used for slow-release fertilizers. The release rate of nutrients (ammonium & N^+) from the NC-SAP was 56.85% after ten days, which could suggest practical applications (*46*).

In addition to nanofertilizers, some NPs can directly improve plants' growth. Pandey et al. (*47*) found that ZnO NPs can enhance seed germination and root growth of common gram *Cicer arietinum* (Chick pea). They found that ZnO NPs increased the level of indole acetic acid (IAA) in roots, a phytohormone that regulates plant growth. Also, it has been proven that carbon nanotubes (CNTs) can penetrate seed coats enhancing the germination and plant growth (*48*). These researchers treated tomato seeds on standard agar-based Murashige and Skoog medium with different concentrations of CNTs (0, 10, 20, 40 μg/mL). The results showed that the seeds in the control experiment (without CNTs) did not germinate by the third day, but seeds treated with CNTs did germinate. Furthermore, as shown in Figure 2, tomato seedlings treated with CNTs had longer stems and better aspect compared with control seedlings. Tripathi et al. (*49*) treated common gram (*Cicer arietinum*) plants with 6 μg/mL water soluble CNTs for 10 days. They reported that CNT treated plants showed an increase in roots, shoots and branching growth rate. These authors concluded that the water soluble CNTs attached to the cell walls can help plants to uptake more water. This can be used in agriculture to prevent water losses, optimizing water uptake and transport.

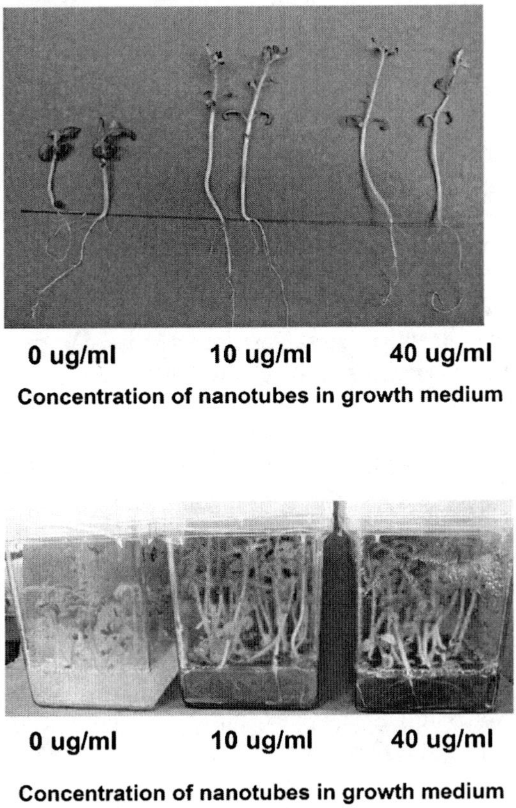

Figure 2. Seedlings grown condition of CNT-exposed and control tomato plants. (Reproduced with permission from reference (48). Copyright © 2009 American Chemical Society).

Crop Protection with Nanopesticides

Nanotechnology-based pesticides (NBPs) are defined by Stone et al. (*42*) as "deliberately engineered particles having a size of 1 to 300 nm in any dimension that display novel characteristics and are used for preventing, destroying, repelling, or mitigating any pest." Some NPs can be used directly as pesticides. For example, cuticular lipids are necessary for insect pests to protect their water barrier in order to prevent death from desiccation. Nanosilica can be absorbed into the cuticular lipids and cause death of insects merely by physical action (*41*). This makes nanosilica a kind of safe and effective pesticide. Silver NPs have also been used as pesticide. Marimuthu et al. (*50*) reported that silver NPs can cause effective mortality on the larvae of *Anopheles subpictus* and *Culex quinquefasciatus* with LC_{50} of 13.90 mg/l and 11.73 mg/l, respectively. Novaluron is a new pesticide (insecticide) belonging to the family of insect growth regulators. Insect growth regulators slowly kill immature insects in a short period by disturbing the normal growth and development. Very recently, novalucron was produced in the form of NPs. This formulation was tested in Egyptian cotton leaf worm (*Spodoptera littoralis*) larvae with similar results as the commercial novalucron formulation (*51*). Stadler et al. (*52*) reported the insecticidal effect of nanostructured alumina (NSA) on two primary grain pests, the rice weevil (*Sithophilus oryzae*) (L.) and lesser grain borer (*Rhyzopertha dominica* (Fabricius). These two pests were treated with NSA at 0, 125, 250, 500 and 1000 mg/kg for three and nine days. There was a significant mortality after three days for both species, which increased as NSA concentration and exposure time increased. Compared to commercial pesticides, the NSA may provide an inexpensive, reliable and safe alternative to control pests. Goswami et al. (*94*) studied the application of different kinds of nanoparticles (silica, aluminum oxide, zinc oxide, and titanium dioxide) against insect pests and pathogens, and they found that nanosilica caused nearly 100% mortality.

NPs can also be used to encapsulate pesticides to improve delivery systems. For instance, aluminosilicate filled nanotubes have proven to stick to plant surfaces delivering NPs that bind the insect pests' bodies affecting certain physiological functions (*53*). Yang et al. (*54*) observed that polyethylene glycol (PEG) coated NPs loaded with garlic essential oil are effective against adult red flour beetles (*Tribolium castaneum*). The pest was treated with 160, 320, 480, and 640 mg kg^{-1} of free garlic essential oil with and without NPs. At the optimal ratio of essential oil/PEG (10%), the control efficacy against adult *T. castaneum* remained over 80% after five months with NPs, but was only 11% without NPs. The authors attributed the difference to the "slow and persistent release of the active components from the NPs" (*54*). Porous hollow silica NPs can also be used as controlled delivery system for water-soluble pesticide (*95*).

Nanotechnology can also be used to improve herbicide application through penetration in cuticles and tissues, and slow release of the active substance. In an interesting review, Perez and Rubiales (*55*) analyzed the literature with reference to utilization of nanotechnology for parasitic plant control. The authors concluded that the potential use of NP application in agricultural production is just starting and not affordable for low income crops, but it will be cost-effective in the near future. However, collateral effects such as genotoxicity needs more research. Compared to conventional pesticides, NBPs have advantages in bioavailability, sensitivity, dosimetry, and pharmacokinetics. However, as stated by Stone et al. (*42*), a complete evaluation of exposure and risks is needed previous to the use of NBPs for agricultural and urban pest control. It is very likely that the U.S. EPA will set sufficient restrictions in the applications for registration of pesticide-containing nanomaterials (*56*).

Control and Management of Agriculture-Based Pollution and Wastes

For several decades, agricultural production worldwide has increased with the use of chemical fertilizers and pesticides. However, this increase has had collateral environmental impacts. Chemical fertilizers and pesticides have been found to cause groundwater and soil pollution, resulting in diseases and adverse effects to humans and animals. Several techniques, including the *ex situ*, pump and treat and *in situ* procedures such as thermal treatment, chemical oxidation, surfactant cosolvent flushing, and bioremediation have been used to reduce the impacts derived from agricultural activities (*57*). Though effective, some of these techniques are expensive and time consuming. Nanotechnology appears as a cost-effective way to degrade agricultural pollutants (*58*). Nanoscaled materials like zeolites, metal oxides, carbon nanotubes and nanofibers, enzymes, and bimetallic NPs have been used for remediation purposes (*57*). Nanotechnology-based remediation reduces costs of cleaning up large scale contaminated places in a short time, avoiding ground water pumping for above-ground treatment, or to mobilize soil for treatment/disposal (*59*).

Agricultural organic pollutants in groundwater can be degraded through photocatalysis by metal oxide semiconductor nanostructures (e.g, nano TiO_2, nano ZnO, nano Fe_3O_4). These nanocomposites absorb photons and initiate redox reactions, which can break up organic molecules into simpler fragments (*32*, *33*). Mahmoodi et al. (*60*) studied the degradation, mineralization, and toxicity reduction of the organic compounds Alachlor and Fenitrothion by nanophotocatalysis using immobilized titanium. Results showed that after 240 min of irradiation with two UV-C lamps (15 W, Philips), carboxylic acids of these compounds were mineralized into CO_2. The residual and acute toxicity of these compounds to *Daphnia magna* was reduced during the photocatalytic degradation.

Agriculture not only causes contamination, it also generates wastes. Agricultural wastes have traditionally been used as feedstock. However, very recently, they are being used in combination with NPs to remove pollutants. Panneerselvam et al. (*61*) studied the ability of magnetic NPs mixed with tea waste biomass to remove Ni(II) from an aqueous solution. Results showed that Fe_3O_4-tea waste had the highest adsorption capacities (38.30 mg/g) for the removal of Ni(II) from water, amongst 17 different adsorbents. It is expected that other NP-biomass mixtures will soon demonstrate their capabilities for heavy metal removal from aqueous matrices. Agricultural wastes have also attracted attention as sources for novel green alternatives such as biofuels, and very recently, as sources for nanomaterial production. Espindola-Gonzalez et al. (*36*) reported the synthesis of silicon dioxide (SiO_2) NPs from waste rice husk, coffee husk and sugarcane bagasse by vermicomposting the wastes using the annelid *Eisenia foetida*. The processes include stabilization, elimination of organic matter, acid treatment, and neutralization. Compared with standard chemical methods, this is an inexpensive and relatively eco-friendly approach to synthesize SiO_2 NPs. Wang et al. (*93*) also successfully obtained biogenic silica nanoparticles (25–30 nm in diameter) from rice husks (Figure 3). As shown in this figure, the biogenic porous silica nanoparticles can be further converted to semicrystalline meso/macroporous frameworks with tunable pore size, which may lead to new applications in the near future. Ghasemi et al. (*62*), synthesized nanozeolite sodalite using extracted silica from rice husk ash, using a hydrothermal method without organic additives. Approximately 60 kg of nanozeolite sodalite per ton of rice husk were produced. Another nanoproduct obtained from rice husk is nanosized silicon carbide (SiC) or carborundum. Producing SiC can help to solve some ecological problems in rice-productive agricultural districts, giving added value to rice farmers (*63*). Wheat straw has also been used to produce nano-silica materials through combustion and acid leaching (*37*). Other nanoproducts such as carbon nanofibers and biocellulose nanofibers have been produced from agricultural waste kernel shell, coconut, and wheat straw (*64*, *65*).

Figure 3. Silica nanoparticles extracted from rice husks. (Reproduced with permission from reference (93). Copyright © 2011 American Chemical Society).

Other Applications of Nanotechnology in Agriculture

Nanosensors

Sensors are devices that collect physical, chemical, or biological information and convert them into signals that can be read by humans. Nanoenabled sensors are high sensitivity information collecting devices that allow earlier responses to environmental changes (*38*, *66*). According to Gruere et al. (*34*), nanosensors can be used in agriculture for detection of pollutants, nutrient content of soil, and plant stress produced by drought, temperature, insect and pathogen pressures. By using nanosensors, farmers can improve the use of water, fertilizers, and pesticides, saving resources and reducing the release of contaminants to the environment.

Hu et al. (*44*) detected multiple residues of organophosphorus pesticides by using nanomagnetic particles in an enzyme-linked immune sorbent assay (ELISA) test. The results demonstrated that ELISA with magnetic NPs had both significantly higher sensitivity and better class specificity than conventional ELISA. Loh et al. (*43*) used Fe_3O_4 NPs to improve an electrochemical biosensor based on immobilized alkaline phosphatase. The biosensor measured the reaction of the alkaline phosphatase with ascorbic acid 2-phosphate used as substrate. The Fe_3O_4 NPs improved twice the sensitivity towards 2,4-D and demonstrated sensitivity for other contaminants such as Hg^{2+}, Cu^{2+}, Ag^{2+} and Pb^{2+}. Other nanosensors such as carbon nanotubes or nano-cantilevers are small enough to trap and measure individual proteins or smaller molecules. Some NPs and nanosurfaces can be engineered to produce electrical or chemical signals in the presence of new contaminants such as bacteria (*38*). It is predicted that nanosensors will be widespread in agriculture-related activities in the near future.

Animal Production

Nanotechnology has the potential to revolutionize the livestock sector due to its potential to solve puzzles in animal production, reproduction, and disease prevention and treatment. Satishkumar and Vertegel (*67*) attached polystyrene latex NPs to lysozyme, an antimicrobial enzyme. These researchers found that the activity of lysozyme conjugated to positively charged NPs was almost twice as large as that of free lysozyme, on the Gram-positive bacteria *Micrococcus luteus,* an inhabitant of mammalian skin and contaminant in sick patients. In Canada and Denmark, researchers developed photocatalyst coatings used in poultry houses to reduce the concentration of food-borne pathogen (*68*, *69*). These coatings incorporate natural mineral titanium dioxide (TiO_2) and nano-silver. Titanium dioxide oxidizes in the presence of fluorescence or UV light and humid air, converting complex organic molecules into water and carbon dioxide. The photocatalyst power of TiO_2 is able to destroy bacterial cell walls. On the other hand, the nano-silver does not need UV light. The silver ions released from the nano-silver can kill bacteria preventing the development of biofilms. Kuzma (*70*) reviewed the applications of nanotechnology in animal production including pathogen detection and removal, veterinary medicine, feed improving, waste remediation, and animal breeding. The author highlights that "dialogue and deliberation about specific cases and their associated issues prior to market entry can help to ensure the safe, responsible, and equitable deployment of nanotechnology to livestock production."

Pesticide Detection

The literature from the last lustrum has shown that multi-walled carbon nanotubes (MWCNTs) can be used to detect pesticides in agricultural samples. For instance, Fang et al. (*71*) used a MWCNTs-based solid-phase dispersion extraction method to analyze pesticide residues in market produces with comparable results to those obtained from accelerated solvent extraction. Compared to C_{18} column and diatomite extractions, the extracts were cleaner, with limits of detection varying from 0.1 to 3.1 μg/kg for an apple matrix and 0.1 to 4.0 μg/kg for a potato matrix. Asensio et al. (*72*) also used MWCNTs as solid-phase extraction stationary phase to determine organophosphorus pesticides in soil samples. The results proved that MWCNTs is an efficient extracting agent for the analysis of pesticides in soils. Siswana et al. (*73*) reported that nanostructured nickel(II) phthalocyanine/MWCNTs composite supported on a basal plane pyrolytic electrode (NiPcNP/MWCNT-BPPGE) can be used as viable platform for the detection of asulam pesticide in phosphate-buffered solution (pH 7.0 conditions). The NiPcNP/MWCNT-BPPGE presented fast electron transport and good analytical performance compared to the electrode without MWCNTs or bare BPPGE, with a detection limit of 0.285 μM, and a sensitivity of 44.6 μA/mM.

Adverse Effects of Nanotechnology in Agriculture

Although nanotechnology can bring many benefits to the agriculture, it also can cause numerous adverse effects. Like traditional fertilizers, nanofertilizers also can have some environmental problems, such as eutrophication. What is more, because of the slowly release of the nutrients, nanofertilizers contamination may take longer time to resolve. More studies are needed to assess the collateral impacts of nanofertilizers and nanopaeticides.

Agricultural soils contain naturally occurring nanoscaled materials that surely play important roles in ecosystem dynamics (*5*). However, in recent times carbon-based and metal-based engineered nanomaterials (ENMs) have been produced and used in goods and consumer products in such amounts that have raised concerns about their accumulation in excess in soils (*5*, *74*).

In the event of intentional or unintentional environmental release of ENMs, these could cause physical or chemical changes in edible plants with unknown consequences (*26*). Researchers concerned with the interactions of NPs with plants have shown different degrees of phenotoxicity and genotoxicity. Table 1 summarizes the negative effects produced by ENMs in edible plants. As shown in this table, most of the studied nanomaterials affect plants at different growth stages. For instance, Si NPs completely inhibit zucchini (*Cucurbita pepo*) seed germination (*75*); CeO_2 NPs reduce seed germination in corn, tomato, cucumber, and spinach (*24*, *76*). Root development is also affected by ENMs. For example, TiO_2 NPs reduce root growth in onions and tobacco (*77*) and Cu NPs reduce root growth in zucchini (*75*). Other NPs like La_2O_3, Gd_2O_3, and Yb_2O_3 inhibit root growth in cabbage, cucumber, lettuce, radish, rape, tomato, and wheat (*78*); Co and Zn NPs inhibit root growth in onions (*79*) and ZnO NPs completely block root growth in garlic (*80*). Other ENMs like ZnO NPs reduce biomass production in zucchini (*75*), Ag and Cu NPs reduce biomass production in squash (*81*), while TiO_2 and ZnO NPs reduce biomass production in wheat (*82*). Reports also indicate that NPs can exhibit genotoxic effects. TiO_2 NPs have shown to produce DNA damage in onions, tobacco, maize and Narbon beans (*77*, *83*). In addition,

Lopez-Moreno et al. (*27*) reported that CeO_2 and ZnO NPs cause the appearance of new bands in the genomic DNA of soybean. A review by Rico et al. (*26*) highlighted that the uptake, bioaccumulation, biotransformation, and risks of NPs for food crops are still not well understood. Currently, several researchers worldwide are performing research to answer questions about the interaction of ENMs with agricultural plants.

To prevent the possible adverse effects from the use of nanotechnology in agriculture, studies on issues such as the fate and transport of the nanomaterials in the ecosystem, uptake and accumulation in plants and animals, and evaluation of the toxicity of nanomaterials are yet to be performed. Risk assessment studies should also be performed before the application of nanoproducts to the agricultural field, and the effects should be monitored.

Concluding Remarks

Nanotechnology is gaining a relevant place in agriculture due to its use in devices like biosensors or for the application of new pesticides, plant grow regulators, genetic improvement of plants and animals, fertilizers, and food additives. It is foreseen that nano-based devices or product delivery will improve the production and protection of crops with high efficiency and reduced costs. Also, new reports suggest nanotechnology could also help to protect the environment through the waste management and pollution control. However, there are many aspects of agricultural nanotechnology in need of immediate action including: (1) the fate and transportation of NPs in soil and bodies of water, (2) the possible eutrophication due to the NPs or ions released by the nanoparticles, and (3) the potential genotoxicity of ENMs and their accumulation in reproductive organs, which could threaten the next plant generations and the food chain. Other issues foreseen for the near future of agricultural nanotechnology are: (1) the lack of specificity of nanopesticides on useful microorganisms and plants, (2) the lack of biodegradable natural nanomaterials which can be obtained from natural fibers, (3) the use of nanotechnology for the increase in shelf life and quality of perishable fruits like strawberries, bananas, and avocados, (4) the production of nanomaterials with low or no impact in soil microflora, (5) the determination of possible biomagnification of very stable metal nanomaterials used in agricultural production, and (6) improvement of policy issues for the production and use of nanomaterials. In summary, although nanotechnology has shown to be useful, and appears as very promising for agriculture and food technology, the above list clearly shows that there is still a long way to go before finding a balance between the benefits of the application of ENMs in agricultural production and their possible threats to the food chain.

Table 1. Negative Effects of Nanomaterials in Agricultural Plants

Nanoparticles	*Concentration*	*Crops*	*Adverse effects*	*References*
Ag	20-100 mg/L	Onion	Decrease mitotic index and promotes\ cell wall disintegration	(*29*)
Ag & nZVI (Zero-valent iron)	Ag 0-100 mg/L nZVI 0-5000 mg/L	Flax, Barley, Ryegrass	Inhibition of shoot growth and germination	(*86*)
Ag, Cu	100 mg/L & 500 mg/L	Squash	Ag NPs reduce biomass by 74-83%;61% & 67% reduction in transpiration by Ag and Cu NPs	(*81*)
Ag, Cu, ZnO, Si, MWCNTs	1000 mg/L	Zucchini	Si NPs completely inhabit germination; Cu NPs decrease root length after 24h; ZnO NPs reduce biomass by 78-90%	(*75*)
Au	3.5 nM	Tobacco	Leaf necrosis after 14 days	(*91*)
CdSe/ZnS QD	100 nM	Alfalfa	Increase of ROS by 50% after 72 hours	(*89*)
CeO_2	2000 mg/L	Alfalfa, Tomato, Cucumber, Corn.	Reduction in corn, tomato and cucumber germination. Decrease in alfalfa and tomato root growth	(*24*)
CeO_2	0.64 mg/mL	Lettuce, Cucumber, Tomato, Spinach.	Reduce seed germination	(*76*)
CeO_2, La_2O_3 , Gd_2O_3, Yb_2O_3	2000 mg/L	Radish, Rape, Tomato, Lettuce, Wheat, Cucumber and Cabbage.	La_2O_3 , Gd_2O_3, Yb_2O_3 inhibit root elongation at different growth stages	(*78*)
CeO_2, ZnO	2000 mg/L	Soybean	Appearance of new DNA bands in genomic DNA	(*27*)

Nanoparticles	*Concentration*	*Crops*	*Adverse effects*	*References*
CNTs	20 mg/L	Rice	Increase reactive oxygen species (ROS) in cells	(*84*)
CNTs	40 μg/L	Mustard, Bean	Sever reduction in the number of root hairs and their length	(*88*)
Co, Zn	5-20 μg/mL	Onion	Inhibition of root elongation.	(*79*)
Pd	0.4 mg/L	Kiwi fruit	Sever inhibition of pollen tube emergence and elongation	(*87*)
Si, Pd, Cu, Au.	0.013% & 0.066%(w/w)	Lettuce	Increase shoot/root ration after 15 days of incubation	(*85*)
TiO_2	1000 mg/L	Maize	Reduce root hydraulic conductivity and leaf growth	(*28*)
TiO_2	0-2 mM	Onion & Tobacco	Reduction in root growth and DNA damage	(*77*)
TiO_2	0-4.0‰	Maize, Narbon bean	Germination delay and genotoxic effect	(*83*)
TiO_2, ZnO	Ti: 4307.5 mg/kg , Zn: 214.5 mg/kg	Wheat	Decrease biomass	(*82*)
ZnO	500-4000 mg/L	Velve tmesquite	Increased the specific activity of stress enzymes CAT in root, stem and leaves	(*22*)
ZnO	50 mg/L	Garlic	Completely block root growth	(*80*)
ZnO	0-2000 mg/L	Mung, Gram	Retardation of root seedlings growth (above 20 mg/L for mung and above 1 mg/L for gram)	(*90*)

MWCNTs = Multiwalled carbon nanotubes CNTs = Carbon nanotubes QD = Quantum dots

Acknowledgments

This material is based upon work supported by the National Science Foundation and the Environmental Protection Agency under Cooperative Agreement Number DBI-0830117. Any opinions, findings, and conclusions or recommendations expressed in this material are those of the author(s) and do not necessarily reflect the views of the National Science Foundation or the Environmental Protection Agency. This work has not been subjected to EPA review and no official endorsement should be inferred. The authors also acknowledge USDA grant numbers 2008-38422-19138 and 2011-38422-30835, and the NSF Grant # CHE-0840525. J. L. Gardea-Torresdey acknowledges the Dudley family for the Endowed Research Professorship in Chemistry.

References

1. What It Is and How It Works? National Nanotechnology Initiative. http://www.nano.gov/nanotech-101/what (accessed August 24, 2012).
2. Roco, M. C. *Curr. Opin. Biotechnol.* **2003**, *14*, 337–346.
3. Moore, M. N. *Environ. Int.* **2006**, *32*, 967–976.
4. Klaine, S. J.; Alvarez, P. J. J; Batley, G. E.; Fernandes, T. F.; Handy, R. D.; Lyon, D. Y.; Mahendra, S.; McLaughlin, M. J.; Lead, J. R. *Environ. Toxicol. Chem.* **2008**, *27*, 1825–1851.
5. Peralta-Videa, J. R.; Zhao, L.; Lopez-Moreno, M. L.; de la Rosa, G.; Hong, J.; Gardea-Torresdey, J. L. *J. Hazard. Mater.* **2011**, *186*, 1–15.
6. Biswas, P.; Wu, C. *J. Air Waste Manage. Assoc.* **2005**, *55*, 708–746.
7. Chen, H. D.; Weiss, J. C.; Shahidi, F. *Food Technol.* **2006**, *60*, 30–36.
8. Maynare, A. D.; Maynard, A. D.; Aitken, R. J.; Butz, T.; Colvin, V.; Donaldson, K.; Oberdörster, G.; Philbert, M. A.; Ryan, J.; Seaton, A.; Stone, V.; Tinkle, S. S.; Tran, L.; Walker, N. J.; Warheit, D. B. *Nature* **2006**, *444*, 267–269.
9. NNI Budget. National Nanotechnology Initiative. http://www.nano.gov/about-nni/what/funding (accessed August 24, 2012).
10. Day, W. *Ann. Appl. Biol.* **2005**, *146*, 155–162.
11. Choy, J. H.; Choi, S. J.; Oh, J. M.; Park, T *Appl. Clay Sci.* **2006**, *36*, 122–132.
12. Shi, Y. H.; Xu, Z. R.; Feng, J. L.; Wang, C. Z. *Anim. Feed. Sci. Technol.* **2006**, *129*, 138–148.
13. Kuzma, J. *J. Nano. Res.* **2007**, *9*, 165–182.
14. Scott, N. R. *Vet. Res. Commun.* **2007**, *31* (Suppl.), 139–144.
15. Maysinger, D. *Org. Biomol. Chem.* **2007**, *5*, 2335–2342.
16. Das, M.; Sexena, N.; Dwivedi, P. D. *Nanotoxicology* **2009**, *3*, 10–18.
17. Bouwmeester, H.; Dekkers, S.; Noordam, M.; Hagens, W.; Bulder, A.; De-Heer, C.; Voorde, S.; Wijnhoven, S.; Marvin, H.; Sips, A. *Regul. Toxicol. Pharm.* **2009**, *53*, 52–62.
18. Salamanca-Buentello, F.; Persad, D. L.; Court, E. B.; Martin, D. K.; Daar, A. S.; Singer, P. A. *PLoS Med.* **2005**, *2*, 383–386.
19. Knauer, K.; Bucheli, T. *Agrarforschung* **2009**, *16*, 390–395.

20. Canas, J. E.; Long, M.; Nations, S.; Vadan, R.; Dai, L.; Luo, M. *Environ. Toxicol. Chem.* **2008**, *27*, 1922–1931.
21. Lee, W.; An, Y.; Yoon, H.; Kweon, H. *Toxicol. Chem.* **2008**, *27*, 1915–1921.
22. Hernandez-Viezcas, J. A.; Castillo-Michel, H.; Servin, A. D.; Peralta-Videa, J. R.; Gardea-Torresdey, J. L. *Chem. Eng. J.* **2011**, *170*, 346–352.
23. De la Rosa, G.; López-Moreno, M. L.; Hernández-Viescaz, J.; Peralta-Videa, J. R.; Gardea-Torresdey, J. L. *Int. J. Nanotechnol.* **2011**, *8*, 492–506.
24. Lopez-Moreno, M. L.; de la Rosa, G.; Hernandez-Viezcas, J. A.; Peralta-Videa, J. R.; Gardea-Torresdey, J. L. *J. Agric. Food Chem.* **2010**, *58*, 3689–3693.
25. Thomas, C. R.; George, S.; Horst, A. M.; Ji, Z.; Miller, R. J.; Peralta-Videa, J. R.; Xia, T.; Mädler, L.; Gardea-Torresdey, J. L.; Holden, P. A.; Keller, A.; Lenihan, H. S.; Nel, A. E.; Zink, J. I. *ACS Nano* **2011**, *5*, 3–20.
26. Rico, C. M.; Majumdar, S.; Duarte-Gardea, M.; Peralta-Videa, J. R.; Gardea-Torresdey, J. L. *J. Agric. Food Chem.* **2011**, *59*, 3485–3498.
27. López-Moreno, M. L.; de la Rosa, G.; Hernández-Viezcas, J. A.; Castillo-Michel, H.; Botez, C.; Peralta-Videa, J. R.; Gardea-Torresdey, J. L. *Environ. Sci. Technol.* **2010**, *44*, 7315–7320.
28. Asli, S.; Neumann, P. M. *Plant Cell Environ.* **2009**, *32*, 577–584.
29. Kumari, M.; Mukherjee, A.; Chandrasekaran, N. *Sci. Total Environ.* **2009**, *407*, 5243–5246.
30. Zhao, L. J.; Peralta-Videa, J. R.; Ren, M. H.; Varela-Ramirez, A.; Li, C. Q.; Hernandez-Viezcas, J. A.; Aguilera, R. J.; Gardea-Torresdey, J. L. *Chem. Eng. J.* **2012**, *184*, 1–8.
31. Handy, R. D.; Kammer, F.; Lead, J. R.; Hassellov, M.; Owen, R.; Crane, M. *Ecotoxicology* **2008**, *17*, 287–314.
32. Ulah, R.; Dutta, J. *J. Hazard. Mater.* **2008**, *156*, 194–200.
33. Baruah, S.; Rafique, R. F.; Dutta, J. *Nano: Brief. Rep. Rev.* **2008**, *3*, 399–407.
34. Gruere, G.; Narrod, C.; Abbott, L. IFPRI Discussion Paper 01064; February, 2011.
35. Li, P.; Lee, C.; Chen, W. *Adv. Mat. Res.* **2011**, *343-344*, 278–282.
36. Espindola-Gonzalez, A.; Martinez-Hernandez, A. L.; Angeles-Chavez, C.; Castano, V. M. *Nanoscale Res. Lett.* **2010**, *5*, 1408–1417.
37. Chen, H.; Wang, F.; Zhang, C.; Shi, Y.; Jin, G.; Yuan, S. *J. Non-Cryst. Solids* **2010**, *356*, 2781–2785.
38. Joseph, T.; Morrison, M. *Nanotechnology in Agriculture and Food*; Nanoforum Report; Institute of Nanotechnology, European Nanotechnology Gateway, May 2006. http://www.nanoforum.org/dateien/temp/nanotechnology%20in%20agriculture%20and%20food.pdf (accessed August 24, 2012).
39. De Rosa, M. C.; Monreal, C.; Schnitzer, M.; Walsh, R.; Sultan, Y. *Nat. Nanotechnol.* **2010**, *5*, 91.
40. Kottegoda, N.; Munaweere, I.; Madusanka, N.; Karunaratne, V. *Current. Sci.* **2011**, *101*, 73–78.
41. Barik, T. K.; Sahu, B; Swain, V. *Parasitol. Res.* **2008**, *103*, 253–258.
42. Stone, D.; Harper, B. J.; Lynch, I.; Dawson, K.; Harper, S. L. *Int. J. Occup. Environ. Health* **2010**, *16*, 467–474.

43. Loh, K.; Lee, Y. H.; Musa, A.; Salmah, A. A.; Zamri, I. *Sensors* **2008**, *8*, 5775–5791.
44. Hu, Y; Shen, G.; Zhu, H.; Jiang, G. *J. Agric. Food Chem.* **2010**, *58*, 2801–2806.
45. Monreal, C. M.; McGill, W. B.; Nyborg, M. *Can. J. Soil Sci.* **1986**, *66*, 499–511.
46. Barati, A. U.S. Patent 2010/0139347 A1, 2010.
47. Pandey, A. C.; Sanjay, S. S.; Yadav, R. S. *J. Exp. Nanosci.* **2010**, *5*, 488–497.
48. Khodakovskaya, M.; Dervishi, E.; Mahmood, M.; Xu, Y.; Li, Z.; Watanable, F.; Biris, A. S. *ACS Nano* **2009**, *3*, 3221–3227.
49. Tripathi, S.; Sonkar, S. K.; Sarkar, S. *Nanoscale* **2011**, *3*, 1176–1181.
50. Marimuthu, S.; Rahuman, A. A; Rajakumar, G.; Santhoshkuma, T.; Kirthi, A. V.; Jayaseelan, C.; Bagavan, A.; Zahir, A. A.; Elango, G.; Kamaraj, C. *Parasitol. Res.* **2011**, *108*, 1541–1549.
51. Elek, N.; Hoffman, R.; Raviv, U.; Resh, R.; Ishaaya, I.; Magdassi, S. *Colloid Surf., A* **2010**, *372*, 66–72.
52. Stadler, T.; Buteler, M.; Weaver, D. K. *Pest. Manage. Sci.* **2010**, *66*, 577–579.
53. Bhattacharyya, A.; Bhaumik, A.; Rani, P. U.; Mandal, S.; Epidi, T. T. *Afr. J. Biotechnol.* **2010**, *9*, 3489–3493.
54. Yang, F. L.; Li, X. G.; Zhu, F.; Lei, G. L. *J. Agric. Food Chem.* **2009**, *57*, 10156–10162.
55. Perez-de-Luque, A.; Rubiales, D. *Pest Manage. Sci.* **2009**, *65*, 540–545.
56. Mittal, A.; Erdmann, E.; Bennett, D.; Capaccio, A.; Crothers, N.; Gilbert, C.; Guggolz, G.; Harkin, N.; Raheb, K.; Tran, H. *Nanotechnology: Nanomaterials Are Widely Used in Commerce, But EPA Faces Challenges in Regulating Risk*; Report to the Chairman, Committee on Environment and Public Works, U.S. Senate; Report No.GAO-10-549; Government Accountability Office, Washington, DC, 2010.
57. Karn, B.; Kuiken, T.; Otto, M. *Environ. Health Perspect.* **2009**, *117*, 1823–1831.
58. Saruah, S.; Duta, J. *Environ. Chem. Lett.* **2009**, *7*, 191–204.
59. Otto, M.; Floyd, M.; Bajpai, S. *Remediation* **2008**, *19*, 100–108.
60. Mahmoodi, N. M.; Arami, M. *J. Alloy Compd.* **2010**, *506*, 155–159.
61. Panneerselvam, P.; Morad, N.; Tan, K. A. *J. Hazard. Mater.* **2011**, *186*, 160–168.
62. Ghasemi, Z.; Younesi, H.; Kazemian, H. *Can. J. Chem. Eng.* **2011**, *89*, 601–608.
63. Radev, D.; Uzunov, I. *Solid State Phenom.* **2010**, *159*, 153–156.
64. Chen, X.; Timpe, O.; Hamid, S. B. A; Schlogl, R.; Su, D. S. *Carbon* **2008**, *47*, 340–343.
65. Dahman, Y; Jayasuriya, K. E.; Kalis, M. *Appl. Biochem. Biotechnol.* **2010**, *162*, 1647–1659.
66. Van Zee, R. D.; Pomrenke, G. S.; Evans, H. M. *Nanotechnology-Enabled Sensing*; Report of the National Nanotechnology Initiative Workshop; Arlington, VA, May, 2009. http://www.dtic.mil/cgi-bin/

GetTRDoc?Location=U2&doc=GetTRDoc.pdf&AD=ADA523650 (accessed August 24, 2012).

67. Satishkumar, R.; Vertegel, A. *Biotechnol. Bioeng.* **2008**, *100*, 403–412.
68. CENS-NANO-Self-Sanitizing Coating for Walls & Ceilings. Green Earth Nano Science. http://www.canadianpoultrymag.com/content/view/1653/67 (accessed August 24, 2012).
69. Clements, M. Pullet Production Gets Silver Lining. Poultry International, April, 2009. http://www.wattagnet.com/Poultry_International/4166.html (accessed August 24, 2012).
70. Kuzma, J. *Livest. Sci.* **2010**, *130*, 14–24.
71. Fang, G.; Min, G.; He, J.; Zhang, C.; Qian, K.; Wang, S. *J. Agric. Food Chem.* **2009**, *57*, 3040–3045.
72. Asensio-Ramos, M.; Hernandez-Borges, J; Borges-Miquel, T. M; Rodriguez-Delgado, M. A. *Anal. Chim. Acta.* **2009**, *647*, 167–176.
73. Siswana, M. P.; Ozoemena, K. I.; Geraldo, D. A.; Nyokong, T. *J. Solid State Electrochem.* **2010**, *14*, 1351–1358.
74. Ma, X; Geiser-Lee, J.; Deng, Y.; Kolmakov, A. *Sci. Total. Environ.* **2010**, *408*, 3053–3061.
75. Stampoulis, D.; Sinha, S. K.; White, J. C. *Environ. Sci. Technol.* **2009**, *43*, 9473–9479.
76. Garcia, A.; Espinosa, R.; Delgodo, L.; Casals, E.; Gonzalez, E.; Puntes, V; Barata, C.; Font, X.; Sanchez, A. *Desalination* **2011**, *269*, 136–141.
77. Ghosh, M.; Bandyopadhyay, M.; Mukherjee, A. *Chemosphere* **2010**, *81*, 1253–1262.
78. Ma, Y; Kuang, L.; He, X.; Bai, W.; Ding, Y; Zhang, Z.; Zhao, Y; Chai, Z. *Chemosphere* **2010**, *78*, 273–279.
79. Ghodake, G.; Seo, Y. D.; Lee, D. S. *J. Hazard. Mater.* **2011**, *186*, 952–955.
80. Shaymurat, T.; Gu, J.; Xu, C; Yang, Z.; Zhao, Q.; liu, Y.; Liu, Y. *Nanotoxicology* **2012**, *6*, 241–248.
81. Musante, C.; White, J. C. *Environ. Toxicol.* **2012**, *27*, 510–517.
82. Du, W.; Sun, Y.; Ji, R.; Zhu, J.; Wu, J.; Guo, H. *J. Environ. Monit.* **2011**, *13*, 822–828.
83. Castiglione, M. R.; Giorgetti, L.; Geri, C.; Cremonini, R. *J. Nanopart. Res.* **2011**, *13*, 2443–2449.
84. Tan, X.; Lin, C.; Fugetsu, B. *Carbon* **2009**, *47*, 3479–3487.
85. Shah, V.; Belozerova, I. *Water, Air, Soil Pollut.* **2009**, *197*, 143–148.
86. Ei-Temsah, Y. S.; Joner, E. J. *Environ. Toxicol.* **2012**, *27*, 42–49.
87. Speranza, A.; Leopold, K.; Maier, M.; Taddei, A. R.; Scoccianti, V. *Environ. Pollut.* **2010**, *158*, 873–882.
88. Ghodake, G.; Seo, Y. D; Park, D.; Lee, D. S. *J. Nanoelectron. Optoelectron.* **2010**, *5*, 157–160.
89. Santos, A. R.; Miguel, A.; Tomaz, L.; Malho, R.; Maycock, C.; Patto, M. V.; Fevereiro, P.; Oliva, A. *J. Nanobiotechnol.* **2010**, *8*, 24–37.
90. Mahajan, P.; Dhoke, S. K.; Khanna, A. S. *J. Nanotechnol.* **2011**, *2011*, DOI: 10.1155/2011/696535.
91. Sabo-Attwood, T.; Unrine, J. M.; Stone, J. W.; Murphy, C. J.; Ghoshroy, S.; Blom, D.; Bertsch, P. M.; Newman, L. A. *Nanotoxicology* **2012**, *6*, 353–360.

92. Ghormadea, V.; Deshpandeb, M. V.; Paknikar, K. M. *Biotechnol. Adv.* **2011**, *29*, 792–803.
93. Wang, w.; Martin, J. C.; Fan, X.; Han, A.; Luo, Z.; Sun, L. *ACS Appl. Mater. Interfaces* **2012**, *4*, 977–981.
94. Goswami, A.; Roy, I.; Sengupta, S.; Debnath, N. *Thin Solid Films* **2010**, *519*, 1252–1257.
95. Liu, F.; Wen, L. X.; Li, Z. Z.; Yu, W.; Sun, H. Y.; Chen, J. F. *Mat. Res. Bull.* **2006**, *41*, 2268–2275.

Chapter 6

Nanotechnology for More Sustainable Manufacturing: Opportunities and Risks

David G. Rickerby*

European Commission Joint Research Centre, Institute for Environment and Sustainability, 21027 Ispra VA, Italy
***E-mail: david.rickerby@jrc.ec.europa.eu**

It is anticipated that nanotechnology will have an influence on virtually every area of industrial activity. Potential advantages include increased efficiency for renewable energy generation, resource saving due to reduced consumption of raw materials and the substitution of more abundant or less toxic materials than the ones presently used. This should contribute to more sustainable manufacturing through the development of cleaner and less wasteful production processes, leading to reductions in energy consumption, emissions, effluents and the demand for rare or strategic materials, while protecting and improving the environment. Applications in various industrial sectors are discussed together with the measures needed to minimize the possible health and environmental risks due to the release of and exposure to nanomaterials at each stage of the product life cycle from the production phase, during use, to final disposal or recycling.

Introduction

The oft-cited Bruntland report (*1*) defines sustainability as "development that meets the needs of the present without compromising the ability of future generations to meet their needs". Reducing the impact of industry and transport on the environment is crucial to sustainability. Product design must therefore be modified and processes optimized, with increased attention to the disposal of waste and recycling. There are opportunities for nanotechnology to contribute by supporting industrial innovation and competitiveness, reducing consumption of raw materials and energy and promoting a cleaner environment.

Applications of nanotechnology can be expected to have significant impact on all industrial sectors including healthcare, agrifood, transport, energy, water and electronic materials (*2*). The present dependence on fossil fuels for energy generation and transport, and the waste streams created by industry, are in the long term unsustainable. Nanotechnology can enable efficient use of resources through advanced renewable energy systems, decreased materials requirements and the substitution of rare materials with more abundant ones. It shows promise for improving the environment by reducing emissions and the demand for non-renewable natural resources.

Nanotechnological manufacturing methods comprise both top-down and bottom-up processes. Examples of top-down techniques include lithography, etching, electrospinning and milling, while bottom-up techniques employ vapor phase, liquid phase, and self assembly methods (*3*). There are already several hundred nanotechnology based products on the market (*4*) and this number is destined to increase rapidly. Some of the areas in which innovative applications of nanotechnology could provide tangible benefits in terms of energy savings and materials efficiency will be discussed.

Any technological innovation carries with it associated risks. In the case of traditional manufacturing processes the health and environmental risks can be assessed by means of already existing frameworks, methodologies and tools. However, in the case of nanotechnologies these might no longer be applicable due to the inherent uncertainties in quantifying hazard and exposure (*5*). Release of nanomaterials may occur during production, transport, storage, distribution, use, final disposal or recycling (*6*) and there hence exist multiple pathways for dispersion in the environment. Since comparatively little is known at present regarding the environmental fate of nanomaterials, it is urgently necessary to identify those properties that determine their transport and degradation in air, soil and water (*7*).

Potential applications of nanotechnology need to be reassessed in the light of sustainability, taking into account the societal, economic and environmental factors and interdependencies (*8*, *9*). Life cycle analysis is expected to provide valuable insights in this respect (*10*, *11*). Because the standard methods for risk assessment may be inadequate to identify the particular hazards associated with nanomaterials, nano-specific risk assessment tools are being developed. In the absence of reliable data, however, worst case scenarios must be assumed for risk management (*12*).

Resource and Energy Efficiency

There are various ways in which nanotechnology can make possible a more sustainable use of raw materials and energy. Significant improvements can be foreseen in the areas of materials efficiency, reduction of waste and emissions, catalysis, substitution of hazardous chemicals and the conservation of water and energy resources.

Materials Resources

Increasing pressure on natural resources due to industrial development and population growth has led to increased emphasis on eco-innovative and resource efficient products (*13*). The concept of resource efficiency entails the design of products that are more material and energy efficient over their entire life cycle, thus reducing their environmental footprint. Eco-innovation through resource efficiency could lead to significant progress towards sustainable development. Nanotechnology, since it requires minimal amounts of raw materials to achieve functionality, can contribute to increased resource efficiency. Already over sixty elements have been commercialized in nanomaterial form (*14*) for applications in semiconductor devices, magnetic memories, display technologies, renewable energy systems and biomedicine.

Of especial interest are the critical rare earth elements that have applications in components of clean energy technologies such as magnets for wind turbines, electric and hybrid vehicles, or the phosphors used in energy efficient solid state lighting. Concern regarding the restricted supplies of neodymium has stimulated research on magnetic materials utilizing smaller amounts of this scarce element and nanocomposites that combine two types of magnetic materials: one with a high coercive field and the other with a high saturation magnetization (*15*).

While nanotechnology offers remarkable possibilities for resource saving, products containing nanomaterials represent a challenge for recycling because the materials are so finely dispersed. Present recycling rates for rare earths are typically less than 1%, compared with greater than 50% for the transition metals (*16*). Increased recycling would contribute to a low carbon, resource efficient economy by decreasing the need for mining and its consequent environmental impact but to realize this aim will require improved design of products in order to facilitate disassembly and materials recovery.

Waste minimization is another objective through which nanotechnology can potentially help contribute to a more sustainable economy (*17*, *18*). This could be achieved through source reduction, avoiding generation of waste by adopting more efficient manufacturing processes requiring smaller quantities of materials, and creating less waste and fewer by-products. In addition to the environmental problems it causes, hazardous waste also represents a significant economic loss because of the waste of raw materials. It is essential, however, to consider the total environmental impact of nanotechnology based manufacturing processes in order to determine whether they are genuinely sustainable or merely replace one environmental problem with another.

Catalysis

Catalysts play a key role in the petrochemical industries, energy generation and environmental protection. Mixed oxide catalysts represent a total worldwide market of about $4 billion but their economic impact is far greater due to their use in the chemical and petrochemical industries (*19*). Catalysts typically consist of nanosized particles dispersed on a large surface area support. Perovskite-type oxides are the most widely studied on account of their high activity and stability.

Advances in characterization techniques have provided molecular level insights into the relationships between nanoscale properties and catalyst behavior (*20*). This knowledge has aided in the development of more efficient catalysts using innovative nanomaterials synthesis methods.

Nanostructured mixed oxide catalysts are currently under development for stationary and automotive applications. Their catalytic activity is determined not only by the constituent atoms but also by the local crystal or surface structures as well as by the particle size. High selectivity is essential in catalyst applications to attain high efficiency in raw materials and energy consumption and reduce the costs of product separation (*21*). Increasing concern about the health effects of automotive and industrial emissions has moreover led to increasingly stringent regulatory standards and a demand for more efficient catalysts.

Nanostructured spinel type oxides are one of the more promising materials for abatement of NO_x and soot particles in diesel engine applications (*22*). There are numerous potential applications in the petrochemical industry: zirconia based composite oxide catalysts have been shown to have exceptional performance for dehydrogenation of ethylbenzene to styrene (*23*); Cr_2O_3-ZrO_2 nanocomposites have been developed for application in oxidative dehydrogenation of ethane to ethylene (*24*); and monolithic $LaNiO_x/CeO_2$–ZrO_2/α-Al_2O_3 catalysts have been investigated for synthetic gas production by partial oxidation of methane (*25*).

Chemical Substitution

It has been postulated that nanotechnology could be applied to reduce or replace the use of certain hazardous chemical substances (*26*). The candidate substances include, among others, heavy metals, brominated flame retardants and chlorinated hydrocarbons. Although the meaning of chemical substitution is rather narrowly defined, i.e., the replacement of a substance with another one, substitution in reference to hazardous substances is a much broader concept. In this context it can relate to not only to the replacement of a hazardous substance by a less hazardous one but also to the modification of a process or the use of an alternative technology to reduce the risk.

One example is the replacement of the potent biocides, such as tributyltin (TBT), cybutryne, copper oxide, copper thiocyanate and zinc pyrthione, used in marine antifouling coatings, by innovative nanostructured anti-adhesive coatings based on molecular chemistry, self-assembled nanostructures, and low surface energy polymers (*27*). The polymers may be used alone, or in combination with carbon nanotubes or nanoclays in nanocomposites, to form the antifouling layer. Rather than employing a toxic chemical to destroy marine micro-organisms, this approach employs a coating designed to prevent their attachment to the surface. The implementation of non-biocidal solutions to aquatic biofouling reduces the environmental impact of antifouling treatments. Furthermore, maintaining ship hulls cleaner helps reduce fuel consumption and consequently carbon dioxide emissions.

Energy and Water

Water and energy are inextricably linked to form the so-called water-energy nexus (*28*). Water is needed in power generation for cooling and energy is used for pumping, purification of drinking water and wastewater treatment. Without energy there would be limited water distribution and without water there would be limited energy production. Renewable energy technologies, such as wind and solar, have a smaller water footprint compared with traditional nuclear, coal and oil power generation (*29*). Nanomaterials find extensive application in advanced energy systems, for the purpose of increasing efficiency, while nanotechnology can also save energy and reduce the use of chemicals in water treatment (*30*).

Renewable energy technologies in which nanotechnology already plays or might in future play a significant role have been evaluated (*31*). These include fuel cells, biofuels, and wave power, in addition to the previously mentioned wind power and photovoltaics. Nanotechnology could make alternative energy sources more attractive by reducing costs and increasing their efficiency. This would enable the transition from a nuclear and fossil fuel based economy to a greener solar and hydrogen powered one. One of the major drawbacks of these sustainable energy sources is that the electricity they produce needs to be stored; the output cannot be increased at will to meet demand as in conventional power plants. Rechargeable batteries and supercapacitors are promising devices for this purpose but their energy densities are still comparatively low. To overcome this limitation, improved batteries and capacitors are being developed exploiting the unique properties of nanostructured materials (*32*).

Advances in understanding nanoscale processes have led to improvements in nanostructured membranes, oxidants and absorbents for water treatment (*33*). The ability to functionalize and modify the structure of the membrane surface is utilized to increase selectivity. Because of the high reactivity of nanomaterials, faster redox reaction rates can be achieved with photocatalytic titanium dioxide and zerovalent iron, resulting in more efficient degradation of pollutants. There are hence outstanding opportunities for the application of nanotechnology in the treatment of drinking water (*34*) and wastewater (*35*), as well as in groundwater remediation (*36*, *37*).

Industrial Applications of Nanotechnology and Sustainability

Some selected applications of nanotechnology in products and processes in various industrial sectors will now be outlined with focus on the sustainability aspects. It will be shown how the adoption of greener manufacturing methods based on nanotechnology can reduce the environmental impacts of products and processes by allowing more efficient use of raw materials and reducing energy consumption, emissions and waste. Depending on the category of product these savings may be direct or indirect and can be achieved in the production phase, the use phase, or both.

Healthcare

The range of applications of nanotechnology in the biomedical field is vast, comprising in-vitro diagnostic systems incorporating biochips and nanoarrays, nanoparticle contrast agents for in-vivo imaging, biocompatible and biomimetic implant materials, pharmaceuticals, targeted drug delivery and gene therapies (*38*, *39*). Innovations in these areas can be expected to increase the efficacy and cost effectiveness of medical treatment. Nanomedicine can improve the quality of healthcare by making possible therapies with fewer side-effects. The precise targeting of nanoparticle based pharmaceuticals at the site of disease serves to increase drug uptake and bioavailability, while lowering the dose that needs to be administered (*40*).

Nanotechnology is expected to exert a significant influence by improving our knowledge of the physiological causes of disease and revolutionizing drug discovery within the pharmaceutical industry (*41*). High throughput analytical techniques based on microarrays and biochips, needing a smaller sample volume than conventional methods, are driving the development of minimally invasive diagnostic systems of increased reliability. Their integration with informatics systems enables real-time monitoring of patients at the point of care. Current advances in genomics and proteomics can ultimately be expected to allow the development of personalized medicine employing molecular biomarkers for the prediction, detection, diagnosis, prognosis, and treatment of disease (*42*). This will lead to an evolution in healthcare from treatment only when symptoms of illness are present to much earlier detection and prevention before the onset of clinical disease.

Nanotechnology offers evident advantages to patients in the form of more efficacious, less invasive diagnosis and therapies. It could increase sustainability of healthcare systems by reducing costs and shifting the point of care from the hospital to the home. Potential environmental benefits would accrue from the reduction of pharmaceutical residues as a result of more precise drug targeting and decreased use of reagents because of smaller analytical volumes. However, to estimate the global environmental impact is complex – the increased use of inexpensive, disposable diagnostic devices, for example, might negate to some extent the resource savings achieved by miniaturization (*43*).

Agriculture and Food

The main application areas of nanotechnology in the agrifood sector are in food processing, food packaging and agricultural production. A recent review summarizes how nanotechnology can assist in the development of more efficient and environmentally friendly farming methods (*44*). Nanoscale carriers can be used for controlled delivery and release of fertilizers, pesticides and herbicides in a similar way as in drug delivery, thereby preventing overuse and reducing pollution due to run-off. Networks of nanosensors could be deployed to enable real-time monitoring of crops and soil and climatic conditions for optimization of the use of resources and increased crop yields. Application of the results of genome research has potential for increasing the resistance of crops to drought and disease.

Nanoparticles can be used in the food industry to improve the properties, like solubility and absorption, of bioactive compounds in functional foods, while nanoencapsulation can help to preserve flavor enhancers, additives and nutrients (*45*, *46*). Nanosensors may be employed to monitor and control food quality and safety in the processing chain or embedded in the packaging (*45*, *47*). They are capable of detecting spoilage due to excess moisture or oxygen and the presence of microorganisms. Additional applications of nanotechnology can be found in antimicrobial and biodegradable packaging materials consisting of multilayer barriers of polymer nanocomposites that incorporate silver or titanium dioxide nanoparticles to give antibacterial protection (*48*).

Nanotechnology represents an opportunity for addressing the ever growing challenges of agricultural sustainability and food security (*49*). At the same time it presents risks due to our incomplete knowledge regarding the toxicity of the nanoparticles themselves. Thus the use of nanotechnology in the food sector is a potentially controversial topic, for which reason the health and safety concerns need to be properly dealt with. Critical to this discussion are a clear definition of nanoparticles for regulatory purposes, the availability of analytical instruments for the characterization of nanoparticles in complex biological matrices such as food, the establishment of relevant dose metrics, studies of nanoparticle toxicity and estimation of consumer exposure (*50*).

Solid State Lighting

Lighting consumes approximately 19% of the electricity that is produced in developed countries, equivalent to 1,900 million tonnes of CO_2 per year (*51*). It has been estimated that the widespread introduction of solid state lighting could reduce the electricity consumed by lighting by up to 50% (*52*). Light emitting diodes (LEDs) and organic light emitting diodes (OLEDs) have a much greater energy efficiency compared to halogen and incandescent lamps. Solid state light sources have a smaller environmental impact due to their longer lifetime, which means that fewer resources are consumed in manufacturing them. They contain no mercury and less phosphorous in comparison to fluorescent lamps.

Evolution of the cost-efficiency of LEDs with time is governed by Haitz's law (*53*). On the basis of historical data, it predicts that the cost per lumen will decrease by a factor of 10 each decade, while the light emission will increase by a factor of 20. Technical problems remaining to be solved are the requirements for high intensity emission across the entire visible spectrum and good efficiency at the current densities and temperatures attained during high power operation (*54*). Present day LEDs are manufactured using conventional microelectronics technology employing planar architectures (*55*) but there is scope for additional improvement by the application of innovative phosphors and low dimensional nanostructures.

Vertically aligned arrays of gallium arsenide nanorod emitters are currently being developed as a novel way of producing high efficiency LEDs at low cost (*56*). Their emission wavelength can be tailored by varying the nanorod diameter and spacing on the substrate (*57*). It is thus possible to fabricate devices which are able to emit white light by red-green-blue color mixing. A core shell metal organic chemical vapor deposition fabrication technique is used; the inner core of the nanorod is n-type semiconductor material while the outer shell consists of p-type material. Large increases in the active area of the device and the internal quantum efficiency can be achieved in this way. However, while these nanorod LEDs have been successfully tested in the laboratory, quantitative evaluation of their performance in comparison with planar devices still needs to be carried out.

Lithium Ion Batteries

The performance characteristics of batteries can be described in terms of two parameters (*58*): energy density (Wh/kg) and power density (W/kg). The energy density reflects the total storage capacity, while the power density relates to the speed at which it can be delivered. Despite considerable research efforts to improve performance, battery energy densities have increased only relatively slowly, with a doubling time of the order of a decade (*59*). While a great deal of research has been dedicated to studying the electrode material composition and the choice of the electrode-electrolyte, rather less attention has been paid to the possibility of employing nanostructured electrodes in lithium batteries.

The use of nanomaterials provides shorter diffusion paths for lithium ions, leading to higher charge/discharge rates and much greater energy densities (*60*). Transition metal nanoparticles embedded in a Li_2O matrix can be formed in-situ from transition metal oxides by a process known as conversion (*61*). Due to the more favorable chemistry, a higher number of electrons are transferred for each transition metal atom than in the conventional insertion reaction. Lithium ions are also better accommodated in the electrode lattice, extending the life of the battery and making it intrinsically safer.

Hydrogen Storage

Hydrogen is a highly efficient energy carrier for transport applications and can be cleanly produced by photocatalytic water splitting with solar power (*62*). Fuel cell vehicles require on board hydrogen storage capability, which must be lightweight and compact but has also to be safe. The main storage options are compressed hydrogen gas, liquid hydrogen, or solid state absorption systems (*63*). Both high pressure tanks and liquefied gas storage systems have practical limitations for mobile applications (*64*). For this reason, solid hydrogen storage materials, and in particular the metal hydrides, have generated a great deal of interest (*65*). Metal hydrides can store higher volumetric densities of hydrogen (120 g/l compared with around 70g/l for liquid hydrogen).

Magnesium hydride has an excellent hydrogen storage capacity but suffers from the disadvantage of a relatively high desorption temperature, in excess of 300°C. Nanocrystalline magnesium hydride shows faster sorption kinetics due to the high fraction of material in grain boundaries, which provide rapid diffusion paths for the hydrogen as well as suitable nucleation sites for the formation and decomposition of the hydride phase (*66*). Further improvements in the sorption rates can be effected by the addition of metal oxide catalysts. Attempts have also been made to reduce the reaction enthalpy by alloying with 3d transition metals or by using reactive hydride composites.

Photovoltaics

Although costs have decreased as a result of more economic materials use for 2nd generation thin film solar cells, the corresponding increases in efficiency compared to crystalline silicon (c-Si) technology have, in the majority of cases, been only marginal (*67*). Amorphous silicon (a-Si), cadmium tellurium (CdTe), copper indium diselenide (CIS) and copper indium gallium selenide (CIGS) thin film cells have somewhat lower efficiencies than the best c-Si solar cells, while gallium arsenide (GaAs) cells have slightly higher efficiencies. Organic polymer and dye sensitized cells rank lower, having at most about half the efficiency of typical thin film devices.

Improved efficiencies can be expected to be achieved with 3rd generation inorganic solar cells based on ultra thin nanostructured absorbers and quantum nanostructures (*68*). Devices based on nanostructured $CuInS_2$ (*69*) and CdTe (*70*) ultra thin absorbers have been demonstrated experimentally. However, the relative scarcity of tellurium and indium restricts the potential for large scale application of solar cells fabricated using these elements. In addition there is concern regarding the toxicity of CdTe, CIS and CIGS (*71*), which could affect long term sustainability of solar energy unless adequate recycling procedures are in place to reduce the environmental impact (*72*). The search is therefore on for cheaper, and possibly less toxic, semiconductor materials with sufficiently good photovoltaic conversion efficiencies that the running costs over the life of the solar cell make them economically viable (*73*).

Water Treatment

Photocatalysis is increasingly employed for purification of drinking water and treatment of wastewater. The commonest and most efficient photocatalytic material is anatase titanium dioxide; the nanocrystalline form is used to increase the surface area of exposed reactive facets. Irradiation of the TiO_2 with UV light causes electrochemical reactions with adsorbed acceptor and donor molecules, leading to formation of reactive hydroxyl radicals and oxygen species that break down toxic organic and inorganic compounds in water (*74*). The operating costs depend to a large extent on the power consumed by the UV light source and this has motivated research activities in the field of solar photocatalysis.

Various solar reactor designs have been developed for the construction of pilot plants for recycling industrial wastewater (*75*). These include the parabolic trough reactor (PTR), the thin film fixed bed reactor (TFFBR), the compound parabolic collecting reactor (CPCR) and the double skin sheet reactor (DSSR). Field tests have been carried out at several different sites on water contaminated with organic chemicals, such as trichloroethylene (TCE) and benzene, toluene, ethylbenzene and xylene (BTEX), with varying degrees of success. Treatment and reuse of wastewater is clearly beneficial in reducing the amount of effluents discharged into the environment. However, widespread industrial application of photocatalytic wastewater treatment will be determined by its cost effectiveness in comparison with established processes already on the market.

Photocatalysis can also be used for chlorine-free disinfection of drinking water (*76*). Inactivation of bacteria occurs within a few hours and this method has the advantage of also being effective against chlorine-resistant organisms. Significantly increased disinfection rates are achieved compared to irradiation with UV light alone (*77*). Solar photocatalytic reactors facilitate decentralized treatment of drinking water and are suitable for applications in remote areas, in developing countries, and in emergency situations. The use of an immobilized catalyst avoids the presence of nanoparticles in the water following treatment but additional research is required on preventing the formation of toxic by-products. Investigation is also needed of the potential long term health and environmental risks of using nanomaterials for water purification.

Life Cycle Analysis

Two questions arise with respect to the sustainability of nanotechnology: (i) will it lessen the environmental impact of industrial processes and products? (ii) do the potential benefits outweigh any risks associated with the release of nanomaterials into the environment? Life cycle analysis can provide a suitable framework for assessment of the overall impacts of nanotechnology but so far relatively little work has been performed in this area (*78*). To determine whether or not nanotechnology is truly sustainable it is necessary to consider all energy and material inputs over the entire life cycle, as well as any outputs in the form of emissions and waste.

Life cycle analysis is a systematic approach for collecting data on material and energy flows for products and processes from extraction of raw materials, through refinement and manufacturing, use and disposal or recycling at end of life (*79*). In assessing the sustainability of nanotechnology, factors to be taken into account, as already mentioned, are more efficient use of rare and strategic materials, substitution of hazardous materials, and reduction in the consumption of raw materials and energy. Certain nanomaterials such as carbon nanotubes, while offering energy savings due to lightweight construction are very energy intensive to produce and this additional energy input during production needs to be correctly incorporated in the life cycle analysis (*80*). A three-tiered approach to assessing the impact of nanotechnologies has been suggested (*81*), integrating technology characterization, ecological evaluation, and incorporation of health, safety and environmental aspects into product development.

Risk Assessment

Estimation of exposure due to release into the environment represents a key challenge in determining nanomaterials safety. It is therefore necessary to know to what extent existing risk assessment approaches and tools can be applied or will require modification to account for the specific properties of nanomaterials. The main steps of classical risk assessment involve problem formulation, hazard identification, release assessment, exposure assessment and risk estimation. Risk management requires exposure limits based on dose-response relationships and evaluation of the exposure. Unfortunately, in the case of nanomaterials, neither the hazards nor the exposure can be properly quantified and this leads to major uncertainties in carrying out a risk assessment.

There is still insufficient information regarding the possible release routes for nanomaterials during production, use and final disposal or recycling to carry out accurate life cycle and risk assessments (*82*). For this reason, it is essential to gather more information regarding the release of nanomaterials from products during all phases of the life cycle. A comprehensive survey of the experimental and theoretical knowledge on the release of engineered nanomaterials into the environment has been published (*83*). The findings indicate that the majority of nanomaterials released from products are embedded in a matrix but a significant fraction is released as individual nanoparticles. To arrive at a correct estimate of the exposure it is essential to distinguish between inherently dispersive and non-dispersive applications (*84*).

A major problem in risk assessment of nanomaterials is the lack of accurate information about fate and behavior in the environment and the wide variability of the available data. For this reason the life cycle assessment must consider the most likely and the worst case scenarios (*85*), taking into account the estimated nanoparticle release from products and the flows in the various environmental compartments. A probabilistic flow model has been developed to calculate the most likely concentrations in the absence of reliable data (*86*). Comparison of the predicted environmental concentrations (PEC) with the predicted no effect concentrations (PNEC) allows a first estimate of the potential risk.

Conclusion

Nanotechnology offers remarkable opportunities for increasing the efficient use of natural resources and energy. It can contribute to sustainability through the development of cleaner, less wasteful production processes and transport systems. Potential industrial applications include healthcare, agrifood, lighting, renewable energies, energy storage, catalysis and water treatment. Reductions in materials use, energy consumption, emissions and effluents can be achieved, thereby protecting the environment by avoiding pollution. The sustainability of nanotechnology based products needs to be considered over the entire life cycle, not just during the manufacturing phase. Technology assessment, taking into account the whole product life cycle, is therefore required at an early stage of development to achieve the correct balance between cost effectiveness and the environmental impacts. Appropriate recycling and recovery strategies are also needed to limit dispersion of nanomaterials in the environment. Measures must also be adopted to minimize the health and environmental risks due to release of nanomaterials at each stage of the product life cycle from the production phase, during use, through to final disposal or recycling.

Acknowledgments

This work leading to this publication received funding from the European Union Seventh Framework Programme (FP7/2007-2013) under grant agreement n° 247989.

References

1. Bruntland G., Ed.; *World Commission on Environment and Development. Our Common Future*; Oxford University Press: Oxford, 1987.
2. Rickerby, D. G.; Morrison, M. *Sci. Technol. Adv. Mater.* **2007**, *8*, 19–24.
3. Şengül, H.; Theis, T. L.; Ghosh, S. *J. Ind. Ecol.* **2008**, *12*, 329–359.
4. Maynard, A. D. *Ann. Occup. Hyg.* **2007**, *51*, 1–12.
5. Ostiguy, C.; Roberge, B.; Ménard, L.; Endo, C. A. *J. Phys.: Conf. Ser.* **2009**, *151*, 012037.
6. Friedrichs, S.; Schulte, J. *Sci. Technol. Adv. Mater.* **2007**, *8*, 12–18.
7. Scheringer, M. *Nature Nanotechnol.* **2008**, *3*, 322–323.
8. Helland, A.; Kastenholz, H. *J. Cleaner Prod.* **2008**, *16*, 885–888.
9. Wiek, A.; Lang, D. J.; Siegrist, M. *J. Cleaner Prod.* **2008**, *16*, 988–999.
10. Fleischer, T.; Grunwald, A. *J. Cleaner Prod.* **2008**, *16*, 889–898.
11. Karn, B. *J. Ind. Ecol.* **2008**, *12*, 263–266.
12. Hutchison, J. E. *ACS Nano* **2008**, *2*, 395–402.
13. Lang-Koetz, C.; Pastewski, N.; Rohn, H. *Chem. Eng. Technol.* **2010**, *33*, 559–566.
14. Karn, B. *J. Environ. Monit.* **2011**, *13*, 1184–1189.
15. Hadjipanayis, G.; Gabay, A. *IEEE Spectrum* **2011**August, 36–41.
16. Graedel, T. E.; Allwood, J.; Birat, J.-P.; Buchert, M.; Hagelüken, C.; Reck, B. K.; Sibley, S. F.; Sonnemann, G. *J. Ind. Ecol.* **2011**, *15*, 355–366.
17. Lewinski, N. *Nanotechnology for Waste Minimization and Pollution Prevention*; Rice University: Houston ,TX, 2008.
18. Kassim, T. A. In *Handbook of Environmental Chemistry*; Hutzinger, O., Ed.; Springer: Berlin, 2005; Vol. 5, Part F, Sub Vol. 3, pp 191−229.
19. Cortés Corberán, V.; Conesa Cegarra, J. C. *The CONCORDE NSOCRA White Paper. A Technology Roadmap on NanoStructured Oxide Catalysts for Redox Applications*; SOCINTEC Grupo AZERTIA - CONCORDE Consortium: Granads, 2006.
20. Bell, A. T. *Science* **2003**, *299*, 1688–1691.
21. Cortés Corberán, V. *Catal. Today* **2005**, *99*, 33–41.
22. Fino, B. *Sci. Technol. Adv. Mater.* **2007**, *8*, 93–100.
23. Reddy, B. M; Han, D. S.; Jiang, N. Z.; Park, S. E. *Catal. Surv. Asia* **2008**, *12*, 56–69.
24. Deng, S.; Li, H. Q.; Li, S. G.; Zhang, Y. *J. Mol. Catal. A: Chem* **2007**, *268*, 169–175.
25. Pavlova, S.; Sazonova, N.; Sadykov, V.; Pokrovskaya, S.; Kuzmin, V.; Alikina, G.; Lukashevich, A.; Gubanova, E. *Catal. Today* **2005**, *105*, 367–371.
26. Fiedeler, U. *J. Ind. Ecol.* **2008**, *12*, 307–315.

27. Rosenhahn, A.; Ederth, T.; Pettitt, M. E. *Biointerphases* **2008**, *3*, IR1–IR5.
28. Lazarova, V.; Choo, K.-H.; Cornel, P. *Water* **2012**, *21*, 12–16.
29. Gipe, P. *Wind Energy Comes of Age*; Wiley: New York, 1995.
30. Rickerby, D. G.; Morrison, M. *Nanotechnol. Perceptions* **2007**, *3*, 193–207.
31. Guo, K. W. *Int. J. Energy Res.* **2012**, *36*, 1–17.
32. Serrano, E.; Rus, G.; García-Martínez, J. *Renewable Sustainable Energy Rev.* **2009**, *13*, 2373–2384.
33. Bottero, J.-Y.; Rose, J.; Wiesner, M. R. *Integr. Environ. Assess. Manage.* **2006**, *2*, 391–395.
34. Brame, J.; Li, Q.; Alvarez, P. J. J. *Trends Food Sci. Technol.* **2011**, *22*, 618–624.
35. El Saliby, I. J.; Shon, H. K.; Kandasamy, J.; Vigneswaran, S. In *Water and Wastewater Treatment Technologies*; Vigneswaran, S., Ed.; Encyclopedia of Life Support Systems, EOLSS: Oxford, 2008; Vol. 3, pp 1–22.
36. Mueller, N. C.; Braun, J.; Bruns, J.; Černic, M.; Rissing, P.; Rickerby, D.; Nowack, B. *Environ. Sci. Pollut. Res.* **2012**, *19*, 550–558.
37. Karn, B.; Kuiken, T.; Otto, M. *Environ. Health Perspect.* **2009**, *117*, 1823–1831.
38. Rickerby, D. G. *Int. J. Healthcare Technol. Manage.* **2006**, *7*, 463–473.
39. Wagner, V.; Dullaart, A.; Bock, A.-K; Zweck, A. *Nature Biotechnol.* **2006**, *24*, 1211–1217.
40. Suri, S. S.; Fennir, H.; Singh, B. *J. Occup. Med. Toxicol.* **2007**, *2*, 16.
41. Jain, K. K. In *Pharmaceutical Biotechnology*; Guzmán, C. A.; Feuerstein G. Z., Eds.; Advances in Experimental Medicine and Biology; Springer: Berlin, 2009; Vol. 655, pp 37–-43.
42. Marchant, G. E. *Curr. Pharmacogenomics Pers. Med.* **2009**, *7*, 231–237.
43. Rickerby, D. G. *J. Nanosci. Nanotechnol.* **2007**, *7*, 1–8.
44. Chen, H.; Yada, R. *Trends Food Sci. Technol.* **2011**, *22*, 585–594.
45. Sozer, N.; Kokini, J. L. *Trends Biotechnol.* **2009**, *27*, 82–89.
46. Sekhon, B. S. *Nanotechnol. Sci. Appl.* **2010**, *3*, 1–15.
47. Neetirajan, S.; Jayas, D. S. *Food Bioprocess Technol.* **2011**, *4*, 39–47.
48. Duncan, T. V. *J. Colloid Interface Sci.* **2011**, *363*, 1–24.
49. Datta, P. S. *Curr. Sci.* **2008**, *94*, 22–23.
50. Bouwmeester, H.; Dekkers, S.; Noordam, M. Y.; Hagens, W. I.; Bulder, A. S.; de Heer, C.; ten Voorde, S. E. C. G.; Wijnhoven, S. W. P.; Marvin, H. J. P.; Sips, A. J. A. M. *Regul. Toxicol. Pharmacol.* **2009**, *53*, 52–62.
51. International Energy Agency. *Light's Labour's Lost - Policies for Energy Efficient Lighting*; IEA Publications: Paris, 2006.
52. Haitz, R.; Kish, F.; Tsao, J.; Nelson, J. *The Case for a National Research Program on Semiconductor Lighting*; Optoelectronics Industry Development Association: Washington, DC, 1999.
53. Haitz, R.; Tsao, J. Y. *Phys. Status Solidi A* **2011**, *208*, 17–29.
54. Crawford, M. H. *IEEE J. Sel. Top. Quantum Electron.* **2009**, *15*, 1028–1040.
55. Schubert, E. F.; Kim, J. K.; Luo, H.; Xi, J.-Q. *Rep. Prog. Phys.* **2006**, *69*, 3069–3099.

56. Li, S.; Fündling, S.; Sökmen, U.; Merzsch, S.; Neumann, R.; Hinze, P.; Weimann, T.; Jahn, U.; Trampert, A.; Riechert, H.; Peiner, E.; Wehmann, H.-H.; Waag, A. *Phys. Status Solidi C* **2010**, *7*, 84–87.
57. Waag, A.; Wang, X.; Fündling, S.; Ledig, J.; Erenburg, M.; Neumann, R.; Al Suleiman, M.; Merzsch, S.; Wei, J.; Li, S.; Wehmann, H.-H.; Bergbauer, W.; Strassburg, M.; Trampert, A.; Jahn, U.; Riechert, H. *Phys. Status Solidi C* **2011**, *8*, 2296–2301.
58. Ragone, D. *Review of Battery Systems for Electrically Powered Vehicles*; SAE Technical Paper 680453; SAE International: Warrendale, PA, 1968.
59. Tarascon, J.-M.; Armand, M. *Nature* **2001**, *414*, 359–367.
60. Tarascon, J.-M.; Armand, M. *Nature* **2008**, *451*, 652–657.
61. Amatucci, G. G.; Pereira, N. *J. Fluorine Chem.* **2007**, *128*, 243–262.
62. Matsuoka, M.; Kitano, M.; Takeuchi, M.; Tsujimaru, K.; Anpo, M.; Thomas, J. M. *Catal. Today* **2007**, *122*, 51–61.
63. Von Helmolt, R.; Eberle, U. *J. Power Sources* **2007**, *165*, 833–843.
64. Schlapbach, L.; Züttel, A. *Nature* **2001**, *414*, 353–358.
65. Sakintuna, B.; Lamari-Darkrim, F.; Hirscher, M. *Int. J. Hydrogen Energy* **2007**, *32*, 1121–1140.
66. Dornheim, M.; Eigen, M.; Barkhordarian, G.; Klassen, T.; Bormann, R. *Adv. Eng. Mater.* **2006**, *8*, 377–385.
67. Green, M. A.; Emery, K.; Hishikawa, Y.; Warta, W. *Prog. Photovolt: Res. Appl.* **2011**, *19*, 84–92.
68. Dittrich, T.; Belaidi, A.; Ennaoui, A. *Sol. Energy Mater. Sol. Cells* **2011**, *95*, 1527–1536.
69. Kaiser, I.; Ernst, K.; Fischer, Ch.-H.; Könenkamp, R.; Rost, C.; Sieber, I.; Lux-Steiner, M. Ch. *Sol. Energy Mater. Sol. Cells* **2001**, *67*, 89–96.
70. Ernst, K.; Belaidi, A.; Könenkamp, R. *Semicond. Sci. Technol.* **2003**, *18*, 475–479.
71. Fthenakis, V. M.; Morris, S. C.; Moskowitz, P. D.; Morgan, D. L. *Prog. Photovolt. Res. Appl.* **1999**, *7*, 489–497.
72. Fthenakis, V. M. *Renewable Sustainable Energy Rev.* **2009**, *13*, 2746–2750.
73. Wadia, C.; Alivisatos, A. P.; Kammen, D. M. *Environ. Sci. Technol.* **2009**, *43*, 2072–2077.
74. Hoffmann, M. R.; Martin, S. T.; Choi, W.; Bahnemann, D. W. *Chem. Rev.* **1995**, *95*, 69–96.
75. Bahneman, D. *Sol. Energy* **2004**, *77*, 445–459.
76. Malato, S.; Blanco, J.; Alarcón, D. C.; Maldonado, M. I.; Fernández-Ibáñez, P.; Gernjak, W. *Catal. Today* **2007**, *122*, 137–149.
77. Sichel, C.; Blanco, J.; Malato, S.; Fernández-Ibáñez, P. *J. Photochem. Photobiol., A* **2007**, *189*, 239–246.
78. Bauer, C.; Buchgeister, J.; Hischier, R.; Poganietz, W. R.; Schebek, L.; Warsen, J. *J. Photochem. Photobiol., A* **2007**, *189*, 239–246.
79. Dhingra, R.; Naidu, S.; Upreti, G.; Sawhney, R. *Sustainability* **2010**, *2*, 3323–3338.
80. Theis, T. L.; Bakshi, B. R.; Durham, D.; Fthenakis, V. F.; Gutowski, T. G.; Isaacs, J. A.; Seager, T.; Wiesner, M. R. *Phys. Status Solidi RRL* **2011**, *5*, 1–6.

81. von Gleich, A.; Steinfeldt, M.; Petschow, U. *J. Cleaner Prod.* **2008**, *16*, 899–909.
82. Shatkin, J. A. *J. Ind. Ecol.* **2008**, *12*, 278–281.
83. Gottschalk, F.; Nowack, B. *J. Environ. Monit.* **2011**, *13*, 1145–1155.
84. Reijnders, L. *J. Ind. Ecol.* **2008**, *12*, 297–306.
85. Mueller, N. C.; Nowack, B. *Environ. Sci. Technol.* **2008**, *42*, 4447–4453.
86. Gottschalk, F.; Scholz, R. W.; Nowack, B. *Environ. Modell. Software* **2010**, *25*, 320–332.

Chapter 7

Biological Synthesis of Silver Nanoparticles and Assessment of Their Bactericidal Activity

M. Sathishkumar,[1] A. Mahadevan,[1] S. Pavagadhi,[1,2] R. Kaushik,[1,2] V. K. Sharma,[3] and R. Balasubramanian*,[2,1]

[1]Singapore-Delft Water Alliance, National University of Singapore, Singapore 117577, Singapore
[2]Division of Environmental Science & Engineering, National University of Singapore, Singapore 117576, Singapore
[3]Department of Chemistry, Florida Institute of Technology, Melbourne, Florida 32901, United States
***E-mail: ceerbala@nus.edu.sg**

Synthesis of silver nanoparticles (AgNPs) was carried out by a simple green method using extracts of two brown algal species (*Sargassum* sp. and *Turbinaria* sp.) as reducing agents for the first time. The effect of duration of the extraction of the algal species from the respective biomass was studied in water at 110 °C over a range of 5 s to 5 min to achieve a maximum yield of bioactive components that can reduce silver ions. In addition, the effects of the concentration of algal extracts and pH (2.0 - 9.0) were investigated to identify optimum experimental conditions for the effective synthesis of AgNPs. The synthesized AgNPs were characterized using UV–Vis, TEM, and XRD techniques. The optimal experimental conditions for the effective biological synthesis of AgNPs included an extraction time of 30 s at pH 9.0. This finding was supported by the results of radical scavenging of the aqueous extracts, which were also high at the same extraction time. Phenolic contents of the extracts appeared to be the main reductants for the conversion of Ag(I) to Ag(0). The antibacterial activity of synthesized AgNPs was examined using four opportunistic pathogens (*Klebsiella pneumonia*, *Aeromonas hydrophila, Pseudomonas aeruginosa*, and *Escherichia coli*). In the present study, as low as 50 mg/L of AgNPs was found to inhibit all the four bacterial strains studied. The results demonstrated the effectiveness of synthesized AgNPs as the antibacterial agent.

Introduction

Over the past decade, considerable research has been conducted on the synthesis, properties and the applications of nanomaterials in various fields such as catalysis, photography, electronics, photonics, information storage, optoelectronics, biological labeling, imaging, and sensing applications (*1*, *2*). The synthesis of noble metal nanoparticles such as gold, silver and platinum has received considerable attention in particular owing to their superior properties in nano sizes as compared to their macrosized counter-parts (*3*). Among the metal nanoparticles, AgNPs have been used extensively as anti-bacterial agents in the medical field, health industry, food storage, textile coatings and a number of environmental applications (*4*, *5*). In the environmental field, the potential use of AgNPs in water and air filtration systems presents a unique solution to a variety of environmental problems (*6*).

AgNPs can be synthesized and stabilized by both physical and chemical methods (*6–8*). The chemical approach involves the use of chemical reduction, electrochemical techniques, and photochemical reduction for the synthesis of nanoparticles (*9–11*). In the chemical reduction, silver ions (Ag^+) are reduced to colloidal AgNPs using common reductants such as borohydride, citrate, ascorbate, polymers and elemental hydrogen (*12*). Biological synthesis of nanoparticles is an attractive approach because it reduces the need of using hazardous chemicals and decreases the downstream processing requirements, which make the whole process cost effective due to less energy consumption and also environmentally benign (*7*, *13*). Several studies have thus employed biomaterial powders such as those from camphor leaf, cinnamon bark and *Sargassum* biomass to synthesize nanoparticles (*14–16*). In recent years, marine algal species have also been attempted to synthesize nanoparticles, based on the earlier studies demonstrating efficient adsorption and removal of metals in water by algae (e.g. Au(III)) (*15*, *17*, *18*).

The present study was stimulated by our previous research on the green recovery of gold through biosorption, bio-crystallization followed by pyro-crystallization using seaweed *Sargassum* sp. (*15*). Although seaweeds have been used to synthesize several types of metal nanoparticles (*15*, *17*, *19*, *20*), their effectiveness in the synthesis of AgNPs has not been successfully demonstrated in the literature yet, to the best of our knowledge. In this work, marine algal extracts were used to synthesize AgNPs. Specifically, algal extracts were obtained by using only water, a green solvent. The aims of the study are: (i) to optimize parameters such as boiling time of water for maximum extraction of phenolic contents from seaweeds, and pH to be maintained during the course of the biological synthesis, (ii) to characterize AgNPs using UV−Vis, FE-SEM, and XRD techniques, and (iii) to assess the antibacterial activity of AgNPs against four opportunistic bacterial pathogens that commonly occur in tropical reservoirs, namely, *Klebsiella pneumonia*, *Aeromonas hydrophila*, *Pseudomonas aeruginosa*, and *Escherichia coli*.

Materials and Methods

Preparation of Sargassum and Turbinaria Powder and Extract

Sargassum sp. and *Turbinaria* sp. were collected from the coast of Bay of Bengal (Mandapam), India and washed repeatedly with copius amounts of de-ionized water to remove debris. The clean biomass was then sundried for a week. The dried algae were ground fine and sieved using a sieve with 150 µm spacing. The particles under the size of 150 µm were used for the study. The algal extracts were prepared by boiling 1 g of the corresponding algal powders in 100 mL of de-ionized (DI) water for various time intervals. The time at which maximum bioactive components could be leached was optimized.

Biosynthesis of Nano-Scale Silver Particles and UV–Vis Spectral Analysis

Silver nitrate ($AgNO_3$, Sigma-Aldrich (Singapore)) was used as precursor material for the biosynthesis of nano-scale silver particles. Both the *Sargassum and Turbinaria*powder and their corresponding extracts were examined for the synthesis of AgNPs. For the direct synthesis of AgNPs with the algal powders, two Vulcan tubes with 10 mL of 1 mM $AgNO_3$ solution were used, to which 0.1 g of powdered *Sargassum* sp. or *Turbinaria* sp. (SP and TP) was added. For the synthesis using extracts, 2.5 mL of *Sargassum* or *Turbinaria* extracts (SE and TE) was added to the two Vulcan tubes containing 7.5 ml of $AgNO_3$ solution. The tubes were then placed on a rotary shaker and agitated at 150 rpm and 25 °C. All the experiments were done in dark. The syntheses of the AgNPs were monitored periodically using a UV–Vis Spectrophotometer (Hitachi U2800). For the spectral analysis, 0.2 mL of the reaction mixture diluted to 1 mL with DI water was used. The UV–visible spectra of the resulting diluents were monitored as a function of reaction time, boiling time, and extract dosage at a resolution of 1 nm.

Characterization of Synthesized Nanoparticles

Images of AgNPs were captured using Field Emission Scanning Electron Microscope (FESEM Model: JEOL JSM-6700F). The samples for the FESEM measurements were prepared by placing a drop of the reaction mixture containing AgNPs from the extract onto a copper strip stuck on a stub, dried and then introduced into the FESEM instrument for imaging. Prior to taking the FESEM images, the stubs were coated with platinum at 20 eV for 30 seconds.

Determination of Total Phenolics

The total phenolic contents of SP and TP and the corresponding powders were determined as described by Pavagadhi et al. (*21*), and results are expressed as gallic acid equivalent. A known quantity of the algal extract was dissolved in a mixture of methanol and water (6:4 v/v). The resulting samples (0.1 mL) were taken in test tubes and made up to the volume of 10 mL with de-ionized water. Then 0.5 mL of Folin-Ciocalteu reagent (1:1 with water) and 0.1 mL of Na_2CO_3 solution (7.5%) were subsequently added to each tube and mixed well by a vortex mixer. After allowing the homogeneous solution to stand for 30 min at room temperature, the absorbance was measured at 760 nm against the reagent blank using visible spectrophotometer (Hitachi U2800). The amount of total phenolic content was calculated as gallic acid equivalents (GAE) from the calibration curve. The estimation of the phenolic content in the fractions was carried out in triplicate and the average of the absorbance values was calculated.

Radical Scavenging Activity Using DPPH Method

Radical scavenging activity of the SE and TE was determined as per the procedure described by Blois (*22*). The algal extracts were prepared by controlled heating. To evaluate the effect of heating time, the algal extracts were withdrawn at regular intervals between 5 seconds to 5 min. Triplicates were maintained for all the samples studied. Briefly, 100 μL of the extract was removed at all intervals for both *Sargassum* sp. and *Turbinaria* sp. and collected in Vulcan tubes. 5.0 mL of 0.1 mM methanolic solution of DPPH (2,2-diphenyl-1-picrylhydrazyl) was added to these tubes and well shaken. The tubes were allowed to stand at 27 °C for 20 min. The control was prepared as mentioned above without any extract and methanol was used for the baseline correction. The changes in the absorbance of the algal extract samples were measured at 517 nm. The radical scavenging activity is expressed as the inhibition percentage and was calculated according to the following formula where OD refers to the optical density:

$$\% \text{ radical scavenging activity} = (1 - \text{sample OD}/\text{Control OD})/100$$

Bactericidal Studies

Bactericidal studies were performed on four different pathogenic bacteria, namely, *Pseudomonas aeruginosa*(ATCC 15692), *Klebsiella pneumonia* (ATCC 13883), *Aeromonas hydrophila* (ATCC 35654), and *Escherichia coli*(ATCC 700790) purchased from the American Type Culture Collection (ATCC), Rockville, USA. The growth inhibition studies were done on the above-mentioned Gram-negative bacteria using Nutrient Broth (Becton-Dickinson and Co., USA). The study was conducted in 100 mL Erlenmeyer flasks containing 100 mL of Nutrient Broth and the concentration of AgNPs ranging from 0.1 to 50 mg/L. Each of the Erlenmeyer flasks was inoculated with the desired bacteria on which the inhibition was required to be observed. The flasks were then incubated in a temperature controlled rotary shaker at 37 °C. The flasks were monitored for bacterial growth every hour for 16 h by measuring the optical density using a UV–Vis Spectrophotometer (Hitachi U2800). Experiments were also conducted with a control flask containing only media and the bacteria in the absence of AgNPs.

Results and Discussions

Synthesis of AgNPs

Initially, synthesis experiments were performed using two types of marine algae, namely, *Sargassum* sp. and *Turbinaria* sp. in powder forms. Algal powders did not show any characteristic color of the AgNPs even after 72 hours, indicating that there was no chemical reaction between Ag(I) ion and algal powders. This observation was different from the earlier studies on Au nanoparticles, which could be synthesized by reducing Au(III) by marine alga *Sargassum* sp. and *Turbinaria conoides* (*15*, *17*, *19*, *23*). The lack of reaction between Ag(I) and algal powders is not surprising considering that the reduction potential of Au(III) ($E°$ = 1.5 V) is higher than that of Ag(I) ($E°$ = 0.8 V) (*24*). This suggests that a higher amount of reducing agents in algal species would be required to synthesize AgNPs compared to Au nanoparticles. Hence, reduction of Ag(I) was carried out using algal extracts in which bioactive components may be present in sufficient concentrations to drive the reduction process. In these experiments performed in dark, after 10 hours following the addition of algal extracts to the $AgNO_3$ solution, a visible color change from pale brown to darkbrown was noticed. The intensity of the absorbance increased with increasing dosage of the algal extract. Based on these observations, one can conclude that, in the presence of extracts of both the algal species i.e. *Sargassum* sp. and in *Turbinaria* sp. bioreduction of Ag(I) took place, leading to the formation of AgNPs, but not with the algal powders. Hence, it was decided to conduct further experiments using the algal extracts only for the synthesis of AgNPs. In order to gain a better understanding of the process parameters, an optimization study was conducted involving extraction time, dosage of algal extracts, reaction time and pH.

Bioactive compounds such as phenolic acids and other reducing sugars can reduce the ionic form of a metal to its non-ionic form and in turn get themselves oxidized during the process. Phenolic components in biomaterials are believed to be mainly responsible for inducing the bioreduction of metallic compounds to the corresponding zero-valent metals (*14*, *16*, *25*). Therefore, in the present work, total phenolic contents in extracts of algal species were analyzed. The values determined were 144.1 ± 1.3 and 165.2 ± 1.3 μg GAE/ mg of SE and TE, respectively.

The temperature of the aqueous medium and the length of extraction are invariably recognized as other important process parameters. Therefore, the extracts in the study were prepared from the two marine algae by varying the duration of extraction from as low as 5 seconds to 5 min while the temperature and the reaction time were kept constant at 110 °C and 72 hours, respectively. The upper limit of temperature was fixed at 110 °C because phenolic compounds are prone to undergo degradation at temperatures higher than 120 °C (*16*). Algal extract samples were withdrawn at different time intervals and analyzed for their radical scavenging activities as well as for AgNPs synthesis in order to optimize the length of extraction. The yield of AgNPs followed the following trend: 30 > 60 > 300 > 10 > 5 seconds. Representative UV−Vis spectra of AgNPs are shown in Figure 1. The sharp peak observed at ~ 435 nm is the absorbance spectrumthat is characteristic of AgNPs (*14*).

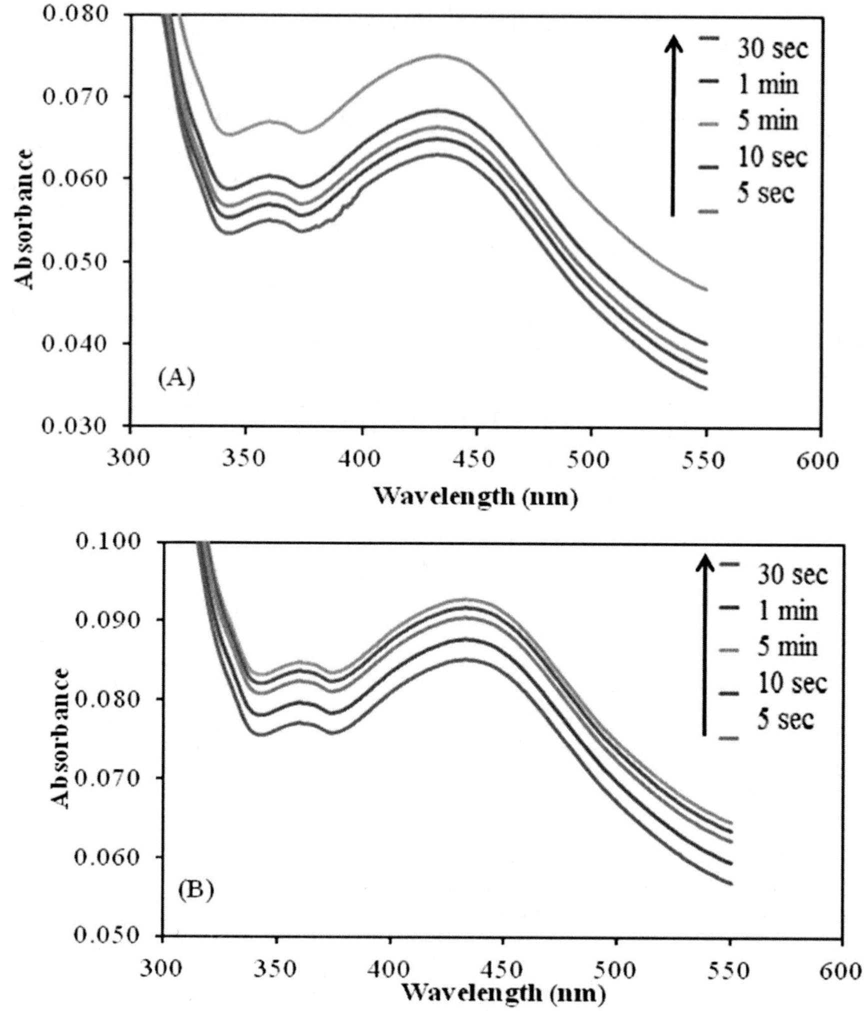

Figure 1. Effect of extraction time of SP (A) and TP (B) on formation of AgNPs; the synthesis of AgNPs was carried out for 72 hours.

At low boiling times (5 and 10 seconds) maintained for extraction of bioactive components, it is possible that the phenolic substances were not leached from the biomaterials sufficiently, resulting in a small yield of AgNPs. On the other hand, at higher extraction times, it is possible that phenolic compounds were partially degraded and bioreduction of Ag(I) to Ag(0) could not proceed efficiently. It is also possible that there could be oxidation of phenolics at higher residence times (*26*). Thus the synthesis of AgNPs was found to be optimal when the powders were boiled for 30 s. To examine if phenolic contents were causing the reduction of Ag(I), we estimated the radical scavenging activity using a free radical-based method for the algal extracts at all extraction times (Figure 2). The same trend as observed for the synthesis of AgNPs was noticed here as well with 30 s giving the highest radical scavenging activity.

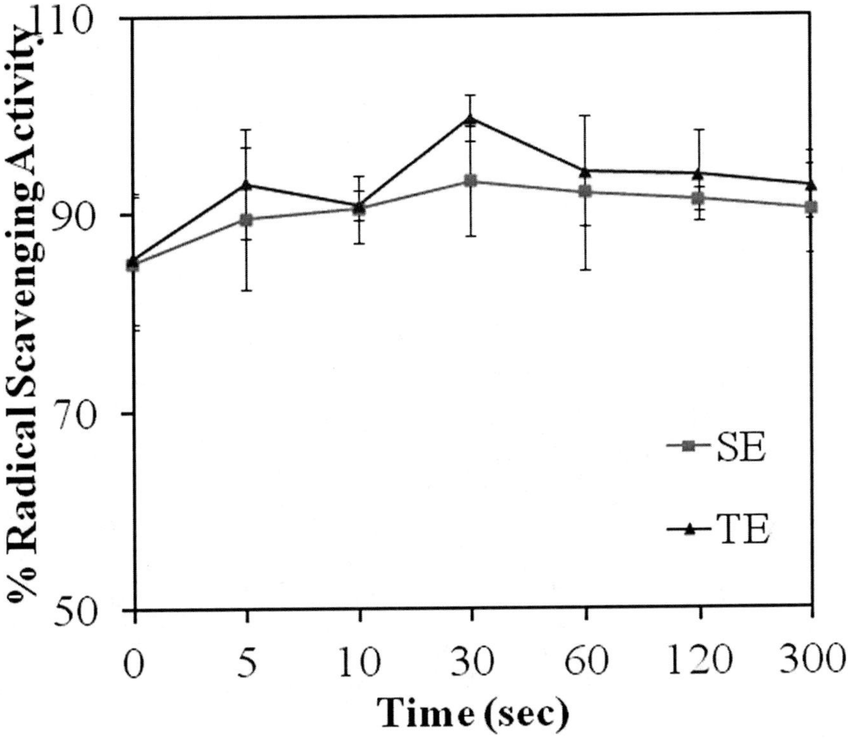

Figure 2. Radical scavenging activity of extracts as a function of extraction time.

An increase in absorbance at 435 nm with respect to time was noticed in the UV−Vis spectra in the case of the algal extracts which were similar to those observed during the course of the extraction time optimization as shown in Figure 1. The absorbance peak was centred ~ 435 nm as shown in Figure 1. While comparing the peak area between the extracts of the two algal species studied, *Turbinaria* sp. showed a larger peak than that of *Sargassum* sp. In the time-dependent absorbance values, a significant increase in the formation rate of AgNPs was observed after 48 h and 72 h in extracts of *Sargassum* sp. and *Turbinaria* sp., respectively (Figure 3).

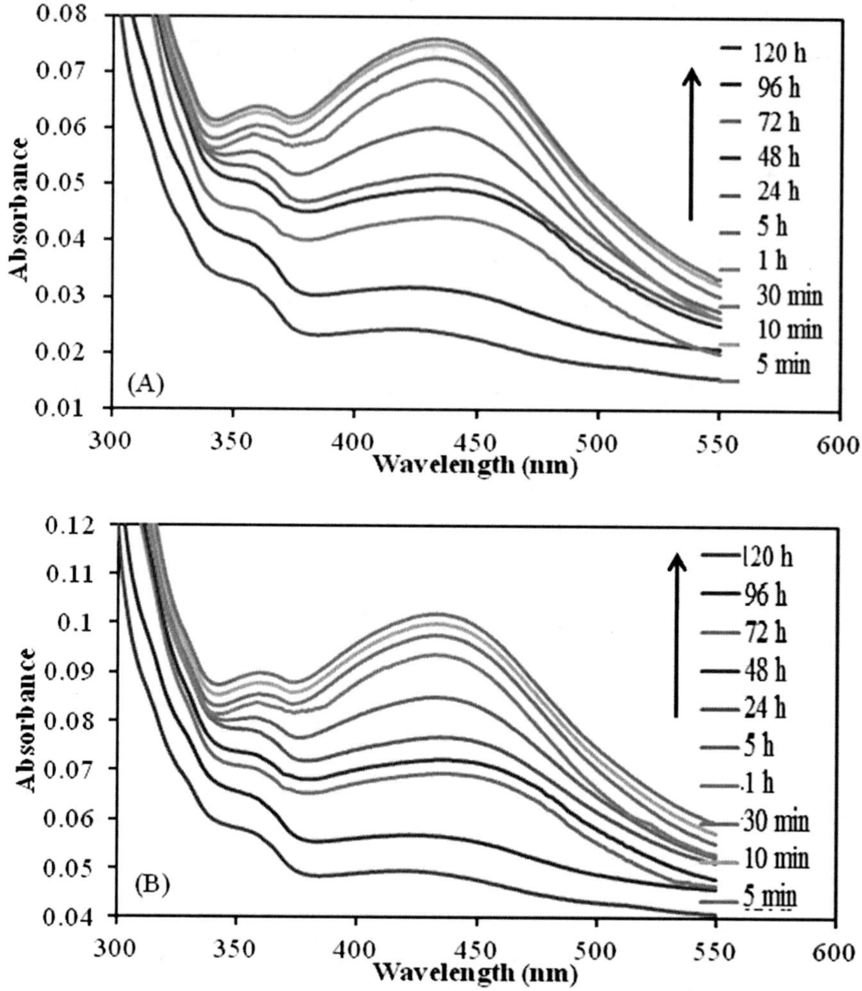

Figure 3. Absorbance values of AgNPs synthesized using SE (A) and TE (B) extracts at various reaction times.

The effects of the amount of algal extracts on the production of AgNPs at different volumes were studied. Various proportions of $AgNO_3$ solution and algal extracts were prepared with a final volume of 10 mL ranging from 9:1 to 5:5. It was observed that the greater the proportion of the algal extract, the greater the spectral absorbance at around 435 nm, as expected. This observation demonstrates the higher yield of AgNPs at higher dosage of SE and TE, which was presumably due to the availability of more bioactive compounds i.e. a phenolic substance for the reduction of Ag^+ ions at higher dosage.

The formation of AgNPs was further studied at various pH values. Biosynthesis experiments were conducted at pH values ranging from 2.0 to 9.0. Beyond pH 9.0, it was not possible to proceed with the biosynthesis experiments due to precipitation of the reaction mixture. It was observed that at pH 9.0, there was a maximum formation of AgNPs. At low pH values, the aggregation of AgNPs to form larger NPs is believed to be favored over the nucleation to form new NPs. However, at higher pH values, a large number of functional groups available such as anionic species of polyphenol facilitated chemical interactions with higher number of Ag(I), followed by their subsequent reduction to form a large number of NPs with smaller diameters (*16*).

Characterization of AgNPs

Figure 4 shows the FESEM images of the AgNPs produced by SE and TE at various magnifications, respectively. The AgNPs formed by both the algal extracts were irregular in shape. There was no marked difference in the shape of the AgNPs synthesized by different extracts studied. However, the nanoparticles synthesized by *Sargassum* sp. were slightly larger in size compared to *Turbinaria* sp. The different amounts of reducing organics present in algal extracts may be the cause of difference in sizes and thus nucleation potencies of synthesized AgNPs using these extracts. Results from this study indirectly show that higher content of phenolics in SE induces more pronounced nucleation of AgNPs than in the case of TE.

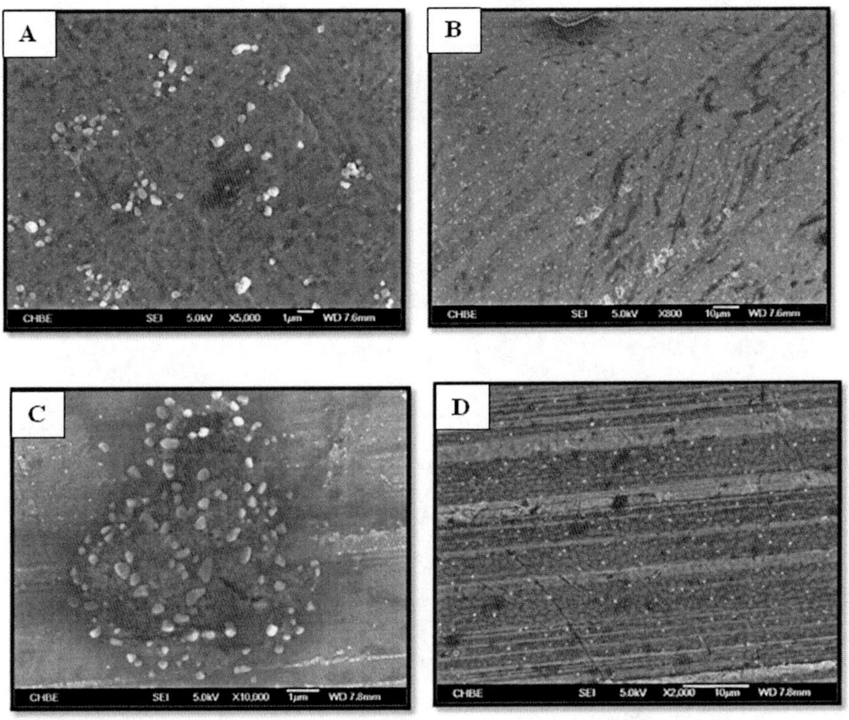

Figure 4. FE-SEM image of silver nanoparticles synthesized using SE (A and B) and TE (C and D) at various magnifications.

The phase of the prepared AgNPs was investigated by X-ray diffraction, and the corresponding XRD-patterns are shown in Figure 5. AgNPs have shown clear peaks of cubic phases (JCPDS No. 03-0921) at 38.3 (111), 44.45 (200), 64.4 (220) and 77.2 (311). The slight shift in the peak positions indicated the presence of strain in the crystal structure which is a characteristic of nanocrystallites (*16*). Thus, the XRD pattern proves to be complementary to the UV−Vis spectra and FESEM images for the presence of Ag nanocrystals.

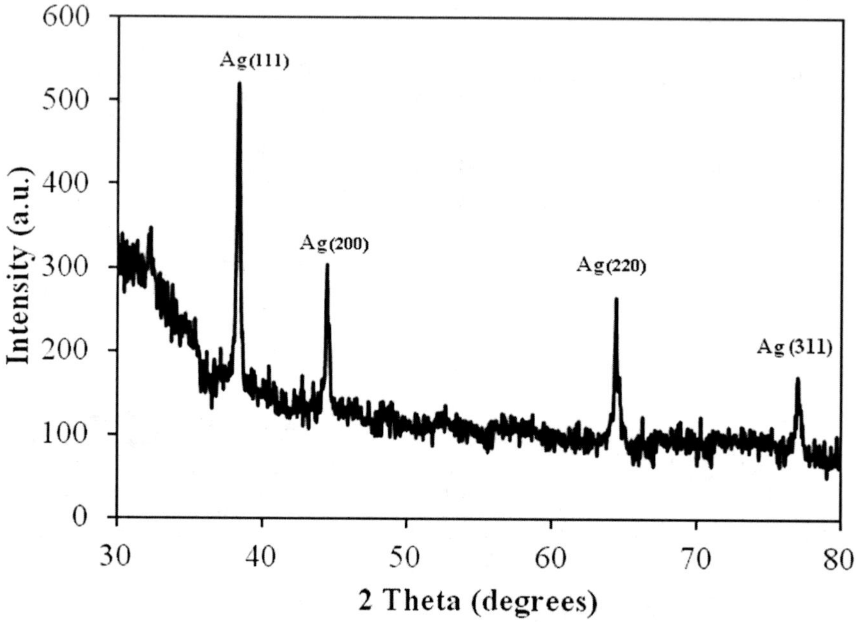

Figure 5. XRD pattern of silver nanoparticles synthesized using TE.

Bactericidal Activity

The various concentrations of AgNPs i.e. 0.1, 0.5, 1.0, 2.0, 5.0, 20.0 and 50.0 mg/L produced inhibition of 4.1, 9.6, 21.2, 29.8, 60.3, 84.5 and 97.5%, respectively against *E. coli* (Figure 6). Minimum inhibitory concentration (MIC) refers to the lowest concentration of material that inhibits 99.9% growth of an organism. In the present study,a MIC of as low as 50 mg/L was found to inhibit all the four bacterial strains studied. A MIC of 40 mg/L for *Staphylococcus aureus* (KCCM 12256) strain has recently been reported (*12*). Various strains of the same species were reported to possess different inhibitory effects towards AgNPs (*27*). In batch studies with *E. coli* and colloidal AgNPs (size range 2–25 nm), MIC was reported to be in the range of 3–25 mg/L for initial bacterial concentrations of 10^5–10^8 CFU/mL (*27–29*); CFU refers to colony forming units. The same *E. coli* (BL 21) strain was found to poses a MIC of 50 mg/L against AgNPs synthesized from cinnamon bark and turmeric tuber extracts (*16*, *30*). A MIC ranging from 40 to 180mg/L and minimum bactericidal concentration (MBC) ranging from 60 to 220mg/L for various strains of *E. coli* were reported (*27*). A growth inhibition of ~ 50% in 18 h was observed for the *E. coli* MTCC 1302 strain when grown with 100 mg/L of AgNPs (*27*). This observation is in agreement with our result of about 99.9% inhibition at 50 mg/L of AgNPs. The difference among MIC values from different studies can be attributed to the fact that the bactericidal effect of NPs is dependent on their concentration. This would also include the starting residual Ag+ concentration in the sampe as presence of molecular oxygen can oxidize the sample to release different amounts of Ag^+ which is known to attribute toxicity (*31*, *32*).

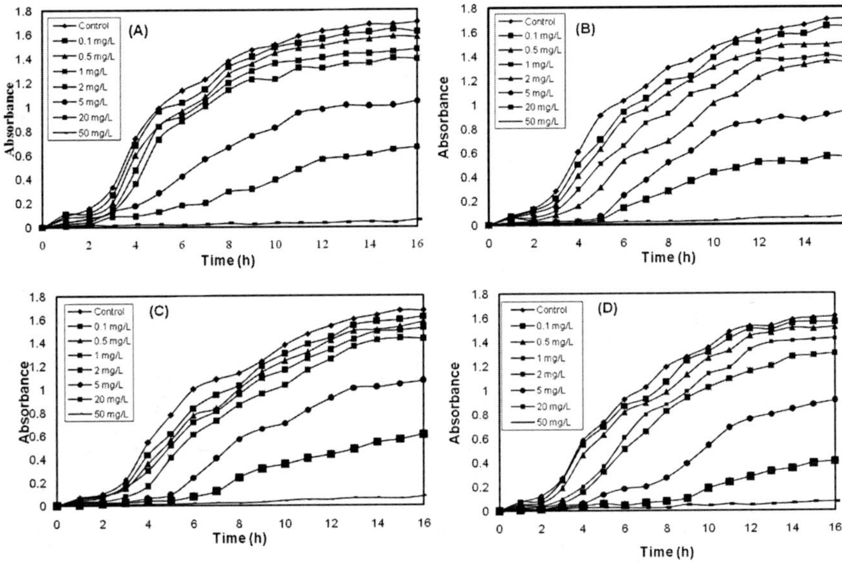

Figure 6. Batch growth profile of Klebsiella pneumonia (A), Aeromonas hydrophila (B), Pseudomonas aeruginosa (C) and Escherichia coli (D) at various silver nanoparticle concentrations synthesized using TE.

Among all the four Gram-negative bacterial strains, *E. coli* and *A. hydrophila* were found to have more sensitivity towards the AgNPs as compared to *P. aeruginuosa* and *K. pneumoniae*. The percentage inhibition for *E. coli*, *A. hydrophila*, *P. aeruginuosa* and *K. pneumoniae* were found to be 97.6, 97.2, 96.8 and 96.4%, respectively(Figure 7). This observation is very similar to the findings of Veerapandian et al. (*33*), who found *K. pneumoniae* and *P. aeruginuosa* to be less sensitive than *E. coli* with the MIC values 16 µg/mL and 512 µg/mL, respectively as compared to that of *E. coli* with 8 µg/mL.

Although the exact mechanism for the growth inhibition by AgNPs has not yet been elucidated, many possible mechanisms have been proposed. The pharmacological action of Ag against bacterial strains is generally reported to involve the interaction of Ag^+ ions with disulfide or sulfhydryl groups of enzymes, causing physiological changes that lead to disruption of metabolic processes followed by cell death (*34*, *35*). The inhibitory action of AgNPs is also based on the release of Ag^+ ion (*36*). These Ag^+ ions from nanoparticles are believed to become attached to the negatively charged bacterial cell wall and rupture it, which leads to denaturation of protein and finally cell death (*37*). The attachment of either Ag^+ ions or nanoparticles to the cell wall causes accumulation of envelope protein precursors, which results in dissipation of the proton motive force. AgNPs also exhibited destabilization of the outer membrane and the rupture of the plasma membrane, thereby causing depletion of intracellular ATP (Adinosine Triphosphate) (*38*). The generation of free radicals from the surface of the nanoparticles may also cause damage to the cellular membrane.

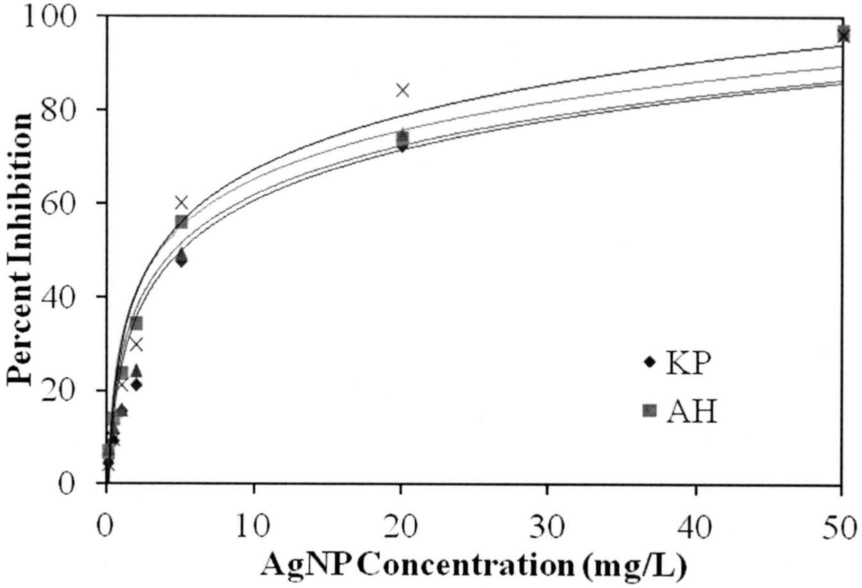

Figure 7. Percent growth inhibition of Klebsiella pneumonia (KP), Aeromonas hydrophila (AH), Pseudomonas aeruginosa (PA) and Escherichia coli (EC) at various silver nanoparticle concentrations synthesized using TE.

Conclusions

The present study demonstrated a simple synthesis of nano-sized silver particles in water using algal extracts. The fresh algal powder with no pre-treatment was not able to convert Ag(I) ion to AgNPs even after 72 hours of contact time. The duration of extraction at 110°C and the phenolic contents were shown to be important parameters for efficient synthesis of AgNPs. This conclusion is supported by the analysis of the radical scavenging activity. The spectroscopic and surface analyses of AgNPs revealed the formation of AgNPs. The synthesized AgNPs were found to possess biocidal activity against four different opportunistic bacterial pathogens. TE showed higher potential for the synthesis of AgNPs with higher antibacterial activity compared to SE. The natural biomaterials used in the study are environmentally-friendly, and therefore have potential for the green synthesis of AgNPs.

Acknowledgments

The authors gratefully acknowledge the support and contributions of this project to the Singapore-Delft Water Alliance (SDWA). The research presented in this work was carried out as part of the Singapore-Delft Water Alliance (SDWA)'s research programme (R-264-001-013-272).

References

1. Hutchison, J. E. *ACS Nano* **2008**, *2*, 395–402.
2. Dallas, P.; Sharma, V. K.; Zboril, R. *Adv. Colloid Interface Sci.* **2011**, *166*, 119–135.
3. Wiley, B.; Sun, Y.; Mayers, B.; Xia, Y. *J. Eur. Chem. A* **2005**, *11*, 454–63.
4. Nino-Martinez, N.; Martinez-Castanon, G. A.; Aragon-Pina, A.; Martinez-Gutierrez, F.; J Martinez-Mendoza, J. R.; Ruiz, F. *Nanotechnology* **2008**, *19*, Article ID 065711.
5. Tang, B.; Wang, J.; Xu, S.; Afrin, T.; Xu, W.; Sun, L.; Wang, X. *J. Coll. Interface Sci.* **2011**, *356*, 513–518.
6. Sharma, V. K.; Yngard, R. K.; Lin, Y. *Adv. Colloid. Interface Sci.* **2009**, *145*, 83–96.
7. Durán, N.; Marcato, P. D.; Durán, M.; Yadav, A.; Gade, A.; Rai, M. *Appl. Microbiol. Biotechnol.* **2011**, *90*, 1609–1624.
8. Philip, D. *Spectrochim. Acta A* **2011**, *78*, 327–331.
9. Chen, W.; Cai, W.; Zhang, L.; Wang, G.; Zhang, L. *J. Colloid Interface. Sci.* **2001**, *2*, 291–295.
10. Sato-Berru, R.; Redon, A.; Vazquez-Olmos, J. M.; Saniger, J. *Raman Spectrosc.* **2009**, *40*, 376–380.
11. Rastogi, A. *Mater. Chem. Phys.* **2011**, *129*, 558–563.
12. Binupriya, A. R.; Sathishkumar, M.; Yun, S. I. *Ind. Eng. Chem. Res.* **2010**, *49*, 852–858.
13. Philip, D. *Spectrochim. Acta A* **2010**, *75*, 1078–1081.
14. Huang, J.; Li, Q.; Sun, D.; Lu, Y.; Su, Y.; Yang, X.; Wang, H.; Wang, Y.; Shao, W.; He, N.; Hong, J.; Chen, C. *Nanotechnology* **2007**, *18*, 105104.
15. Sathishkumar, M.; Mahadevan, A.; Vijayaraghavan, K.; Pavagadhi, S.; Balasubramanian, R. *Ind. Eng. Chem. Res.* **2010**, *49*, 7129–7135.
16. Sathishkumar, M.; Sneha, K.; Won, S. W.; Cho, C.-W.; Kim, S.; Yun, Y.-S. *Colloids Surf., B* **2009**, *73*, 332–338.
17. Vijayaraghavan, K.; Mahadevan, A.; Sathishkumar, M.; Pavagadhi, S.; Balasubramanian, R. *Chem. Eng. J.* **2011**, *167*, 223–227.
18. Kaviya, S.; Santhanalakshmi, J.; Viswanathan, B.; Muthumary, J.; Srinivasan, K. *Spectrochim. Acta A* **2011**, *79*, 594–598.
19. Singaravelu, G.; Arockiamary, J. S.; Ganesh Kumar, V.; Govindaraju, K. *Colloids Surf., B* **2007**, *57*, 97–101.
20. Mata, Y. N.; Torres, E.; Blázquez, M. L.; Ballester, A.; González, F.; Mũnoz, J. A. *J. Hazard. Mater.* **2009**, *166*, 612–618.
21. Pavagadhi, S.; Joseph, G. S.; Jena, B. S. *J. Food Prop.* **2011**, *99*, 1–5.
22. Blois, M. S. *Nature* **1958**, *4617*, 1199–1200.

23. Liu, B.; Xie, J.; Lee, J. Y.; Ting, Y. P.; Chen, J. P. *J. Phys. Chem. B.* **2005**, *109*, 15256–15263.
24. Sharma, H. D. *Geoenvironmental Engineering: Site Remediation, Waste Containment and Emerging Waste Management Technologies*; Reddy, K. R., Ed.; John Wiley and Sons: New York, 2004; Chapter 2.
25. Akaighe, N.; MacCuspie, R. I.; Navarro, D. A.; Aga, D. S.; Banerjee, S.; Sohn, M.; Sharma, V. K. *Environ. Sci. Technol.* **2011**, *45*, 3895–3901.
26. Khanna, S. K.; Vishawanatham, P. N.; Krishnan, P. S.; Sanwai, G. G. *Phytochemistry* **1968**, *7*, 1513–1517.
27. Ruparelia, J. P.; Chatterjee, A. k.; Duttagupta, S. P.; Mukherji, S. *Acta Biomater.* **2008**, *4*, 707–716.
28. Panáček, A.; Kvítek, L.; Prucek, R.; Kolář, M.; Veceřová, R.; Pizúrová, N; Sharma, V. K.; Nevěcná, T; Zbořil, R. *J. Phys. Chem. B* **2006**, *110*, 16248–16253.
29. Gogoi, S. K.; Gopinath, P.; Paul, A.; Ramesh, A.; Ghosh, S. S.; Chattopadhyay, A. *Langmuir* **2006**, *22*, 9322–9328.
30. Sathishkumar, M.; Sneha, K.; Yun, Y. S. *Bioresource Technol.* **2010**, *101*, 7958–7965.
31. Pal, S.; Tak, Y. K.; Song, J. M. *Appl. Environ. Microbiol.* **2007**, *73*, 1712–1720.
32. Xiu, Z. M.; Ma, J.; Alvarez, P. J. *Environ. Sci. Technol.* **2011**, *45*, 9003–9008.
33. Veerapandian, M.; Lim, S. K.; Nam, H. N.; Kuppannan, G.; Yun, K. S. *Anal. Bioanal. Chem.* **2010**, *398*, 867–876.
34. Butkus, M. A.; Edling, L.; Labare, M. P. *J. Water Supply Res. Technol.−AQUA* **2003**, *52*, 407–416.
35. Feng, Q. L.; Wu, J.; Chen, G. Q.; Cui, Z.; Kim, T. N.; Kim, J. O. *J. Biomed. Mater. Res.* **2000**, *52*, 662–668.
36. Morones, J. R.; Elechiguerra, J. L.; Camacho, A.; Holt, K.; Kouri, J. B.; Ramirez, J. T.; Yacaman, M. J. *Nanotechnology* **2005**, *16*, 2346–2353.
37. Lin, Y. E.; Vidic, R. D.; Stout, J. E.; McCartney, C. A.; Yu, Y. L. *Water Res.* **1998**, *32*, 1997–2000.
38. Lok, C. *J. Proteome Res.* **2006**, *5*, 916–924.

Chapter 8

Reactive Peptide Nanofiltration

P. Marchetti,*[,1,2] A. Butté,[1] and A. G. Livingston[2]

[1]Lonza AG, CH-3930 Visp, Switzerland
[2]Department of Chemical Engineering, Imperial College, London, United Kingdom
[*]E-mail: p.marchetti09@imperial.ac.uk

Manufacturing of pharmaceuticals and fine chemicals is usually carried out in batch reactors. Peptide therapeutics by chemical synthesis represent an important class of Active Pharmaceutical Ingredients (APIs), manufactured in large scale. Long-chain peptide therapeutics are often synthesized by fragment condensation of shorter chains, separately produced by solution-phase or solid-phase techniques. The chemical syntheses, implemented in organic solvents or aqueous-organic mixtures, often require large reaction times and subsequent precipitation of the reaction mixture to isolate the target compound from the side products and the reagents in excess, leading to discontinuous processes. Conventional organic solvent-based fragment condensations may furthermore suffer from: (i) presence of poorly soluble compounds that can precipitate or unmix; (ii) high organic solvent consumption; (iii) deleterious side reactions; and (iv) poor yield of the desired product. The integration of membrane separation techniques into the reaction step can help in addressing these problems. The purification strategy presented in this work is proposed for improved peptide fragment condensation, and named "Peptide Reactive Nanofiltration". This strategy is based on the incorporation of nanofiltration units into the reaction step, the separation of small side products from the reaction mixture, the recycle of the solvent after the nanofiltration, and the elimination of time-consuming steps, typical of the conventional strategy. The performances of the new strategy will be presented for one industrial case study and compared to the conventional batch processes by means of techno-economical analysis. The Reactive Peptide Nanofiltration will be demonstrated advantageous in terms of both costs and productivity.

1. Introduction

Peptides are attractive targets for drug discovery because they have been shown to be diagnostically and therapeutically important in many areas of biomedical research (*1*). In the recent years, advances in the areas of formulation and delivery systems have vitalized the research on peptides and their chemical synthesis, leading to several highly successful peptide drugs (*2*). The increasing interest in peptides as pharmaceuticals has been challenging the peptide industry to develop economically competitive methods for manufacturing peptides in large quantities. To this purpose, a variety of methods have been developed for the commercial-scale manufacture of peptides. Peptides can be synthesized by linear (or stepwise) synthesis, when each amino acid is added to the peptide chain, up to the assembly of the entire sequence, or by fragment condensation (or convergent synthesis), when the manufacture of the target structure is done by assembling separately synthesized intermediate segments (*3*). The synthesis can be done in solution-phase, solid-phase or by hybrid methods. Hybrid methods have been proposed to overcome some of the limitations of the single techniques, like the long production times required by the solution-phase, and the necessity of frequent isolation, purification, and re-solubilization steps, required by solid-phase methods. Hybrid approaches have opened up the interesting possibility of manufacturing complex sequences by synthesizing small intermediate fragments by stepwise elongation and assembling them afterwards in solution- or solid-phase by fragment condensation.

Chemical syntheses for peptides are usually carried out in organic solvents or aqueous-organic mixtures. Typical solvents for peptide synthesis are acetonitrile (ACN), dimethylformammide (DMF), N-methylpyrrolidone (NMP), isopropanol, and tetrahydrofuran (THF). Syntheses are carried out in batch reactors and can be composed of different chemical steps, normally separated by intermediate isolations and purifications. Precipitation and re-solubilization significantly increase the solvent consumption and are often cost and time intensive steps. Type and amount of solvent for these steps depend on the kind of solute and its preferential solubilization. Solvent recovery after these steps is not easy, especially if solvent mixtures undergo formation of azeotropes. The final purification of the crude (i.e. impure) peptide mixtures is carried out by preparative chromatography, to reach the high level of purity normally required from the market. ACN / water is the typical solvent mixture used for the final chromatography step.

In this context, membrane technology for organic solvent nanofiltration is an emerging technique to perform separation, solvent exchange and concentration of peptides. Some authors have proposed to use nanofiltration to assist organic synthesis and enhance the production of target compounds (*4*) and to help the production of peptides (*5*).

In this paper, it will be shown how peptide synthesis can be improved by integrating membrane technology into a reaction step, in the so-called Reactive Peptide Nanofiltration scheme. The case study presented here is addressing the fragment condensation between peptide chains by disulfide bond. The peptides involved in this reaction were pre-synthesized by stepwise elongation. The reaction between the protected sulfur atom of one fragment and the unprotected sulfur atom of the other occurs via nucleophile attach to sulfur atom, with the release of the protecting molecule, as side-product of the condensation. This side product is deleterious as its permanence in the reaction mixture causes side reactions or phase separation. Due to its size, in the following this side-product will be simply referred to as small molecule, or SM. Ceramic membranes were identified as the most suitable for peptide solutions. Shorter production time and lower solvent consumption characterize this scheme, with evident advantages from technological, economical and environmental point of view.

2. Case Study: Fragment Condensation by Disulfide Bond

The case study presented in this work concerns the chemical synthesis of a therapeutic peptide (P3), produced by fragment condensation of two pre-synthesized peptides (P1-SH and P2-S-SM, respectively). The reaction of fragment condensation occurs via disulfide bond, with the release of a small molecule, SM, which is originally the protecting group of the S atom of P2-S-SM. The reaction is shown in Eq. 1:

$$P1 - SH + P2 - S - SM \rightarrow P3 + SM \qquad (1)$$

P1-SH and P2-S-SM are composed of almost the same number of amino acids but differ significantly for their hydrophilic / hydrophobic properties. The physico-chemical properties of the species involved in the reaction are reported in Table 1.

Table 1. Physico-Chemical Properties for the Species Involved in the Reaction

Molecule	*MW [Da]*	*pI*	*Hydrophilic / Hydrophobic*	*Water solubility*
P1-SH	2700	12	Hydrophilic	$> 30\ g\ l^{-1}$
P2-S-SM	1500	4	Hydrophobic	$< 0.1\ g\ l^{-1}$
P3	4050	11	Hydrophilic	$> 30\ g\ l^{-1}$
SM	150	3.5	–	n/a

2.1. Conventional Production

The conventional production of P3 is composed of three main steps:

- batch reaction between P1-SH and P2-S-SM in organic solvent (ACN or THF) / water mixture;

- precipitation in 90%v ACN / water, followed by drying and re-solubilization in 10%v ACN / water;
- final purification by preparative chromatography.

The scheme for the conventional production is shown in Figure 1.

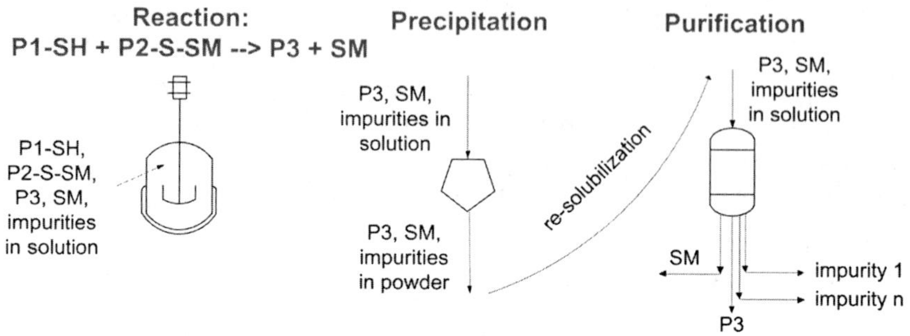

Figure 1. Conventional process for the production of P3.

The main issues concerning this scheme are:

(1) The small molecule, SM, is continuously produced during the reaction and its permanence in the mixture causes secondary reactions; (2) high content of organic solvent is needed in the reactor to solubilize P2-S-SM (hydrophobic) and to avoid its phase separation; (3) the process is globally discontinuous, with time-consuming intermediate steps, like precipitation, drying and re-solubilization.

2.2. Reactive Peptide Nanofiltration

An alternative approach to the production of P3 is proposed in this work. The scheme was named Reactive Peptide Nanofiltration, as it is based on the coupling between reaction and nanofiltration in one unique scheme, as shown in Figure 2.

The process works in semi-batch mode: P1-SH (the hydrophilic reagent) is loaded into the reactor in 10%v ACN / water and P2-S-SM (the hydrophobic reagent) is slowly added by diluted stream B. The reaction mixture is pumped afterwards on the nanofiltration membrane, which separates the small molecules from the three peptides. SM permeates through the membrane in stream D, due to its small size, while P1-SH, P2-S-SM and P3 are completely retained in stream E. SM is finally captured from stream D by an ion exchange resin, in order to allow for solvent recycling. Alternatively to the semi-batch mode, it is possible to operate the system in continuous mode, by continuously adding P1 to the reactor by stream A.

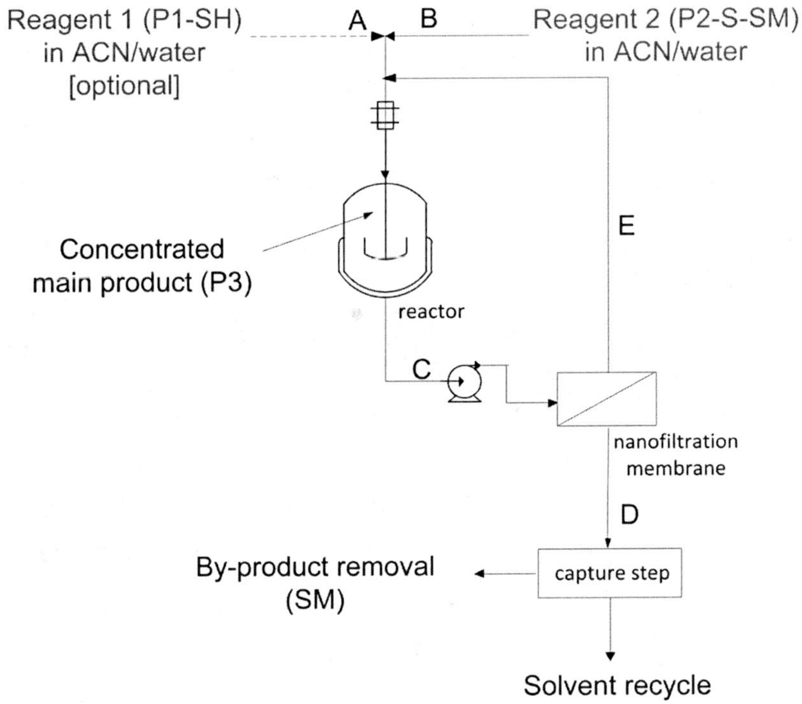

Figure 2. Reactive Peptide Nanofiltration for the synthesis of P3.

The advantages of this scheme are:

(1) SM is continuously removed by the membrane, thus minimizing secondary reactions;
(2) the organic content in the reactor can be low, since the hydrophobic reagent, P2-S-SM, is added slowly and diluted;
(3) the outgoing reaction mixture can be directly injected into the preparative chromatography for the final purification, without necessity of solvent exchange (10 - 20%v ACN / water);
(4) time consuming steps, like precipitation, drying and resolubilization, are not required.
(5) the amount of organic solvent required for the whole process is lower than for the conventional process, with consistent economical and environmental benefits.

3. Experimental

3.1. Materials

P1-SH and P2-S-SM were synthesized by solid-phase chain elongation and purified for kinetic and permeation studies by preparative chromatography. P3 used in the feasibility experiments was produced by fragment condensation of P1-SH and P2-S-SM in liquid-phase, according to the conventional strategy (cf. Paragraph 2.1). Pure SM was purchased. The physico-chemical properties of the species are reported in Table 1. The solvents used in this study were distilled water and ACN. THF was additionally used for testing the reaction kinetics.

The cross-flow filtration system used in this study contained a tubular module, suitable for multichannel tubular membranes with a length of 250 mm and an outer diameter of 25 mm. The membranes used in this study were produced by HITK (Hermsdorf, Germany) under the commercial names of Inopor® Nano 450 and Inopor® Nano 750. The active nano-filtration layer in TiO_2 is supported by a ultra-filtration layer of Al_2O_3. The membranes have 19 channels with an internal diameter of 3.5 mm. The nominal Molecular Weight Cut-Off (MWCO) and the nominal pore dimension provided by the supplier are reported in Table 2.

Table 2. Physico-Chemical Properties for the Membranes Used in This Study

Membrane	*MWCO [Da]*	*Pore dimension [nm]*	*Material*
Inopor® Nano 450	450	0.9	TiO_2
Inopor® Nano 750	750	1	TiO_2

3.2. Methods

3.2.1. Reaction Tests

The reaction kinetics of the fragment condensation between P1-SH and P2-S-SM was studied in batch conditions. Solutions of P1-SH and P2-S-SM of different concentrations were prepared and mixed together. The consumption rate of the reagents and the production rate of the products were monitored at regular intervals by means of HPLC analysis. Three different concentration of ACN / water (10%v, 30%v and 50%v), two different temperatures (10 and 20 °C) and the effect of the solvent (ACN vs. THF) were investigated.

3.2.2. Rejection Tests for Membrane Selection

Solutions of pure P1-SH, P2-S-SM, P3 and SM in ACN / water mixtures were prepared and their rejection and permeation tested in a Natan cross-flow system (Natan GmbH, CH). Rejection quantified the amount of solute that is retained by the membrane and is calculated as $1\text{-}C_p/C_r$, where C_p and C_r represent the solute concentrations at permeate and retentate side, respectively.

The %v ACN / water between 20 and 40 and pH between 2 and 4 were investigated. TFA was used to set the pH. Samples of retentate and permeate were taken at regular intervals, the permeate collected in a graduated cylinder and the flux monitored. Rejection and permeation tests of pure substances were used to choose the best membrane to perform the separations of interest, and namely, retain P1-SH, P2-S-SM and P3 and permeate SM.

3.2.3. Rejection and Permeation Tests for P3

Rejection and permeation of P3 through Inopor® Nano 450 were studied as function of five operating conditions, reported in Table 3. Solutions of P3 were prepared in ACN / water mixtures. 100 mM ammonium sulfate was added as buffer and the pH adjusted with TFA . The nanofiltration performances were studied by means of Design of Experiments (DoE), and specifically by using a Fractional Factorial Experimental design.

Table 3. Operating Parameters Investigated by DoE for Rejection and Permeation of P3

Factor	*Operating parameter*	*Range of variation*
A	P3 concentration [g l^{-1}]	2 – 10
B	pH	2 – 5
C	Pressure [bar]	5 – 15
D	Pump frequency [Hz]	25 – 45
E	%v ACN / water	10 – 50

Samples of retentate and permeate were taken at regular intervals, the permeate collected in a graduated cylinder and the flux monitored. Experimental data for permeation flux and rejection of P3 were analyzed by Analysis of Variance (ANOVA) and a statistical model was found for the two responses as function of the five influencing parameters. Both design planning and data analysis were done by the software Design Expert 7.0.3.

3.2.4. Reactive Peptide Nanofiltration Feasibility

The feasibility of the Reactive Peptide Nanofiltration scheme was tested in the Natan cross-flow system in semi-batch configuration. To check the effect of pH, simultaneous loading of both reagents was adopted. Solution of P1-SH and P2-S-SM were prepared in ACN / water mixtures of different compositions and mixed together. The reaction mixture was immediately loaded into the reaction loop (tank and nanofiltration chamber) and the NF started. The operating pressure was set at 8 bars.

To check the effects of reagents concentration and organic content, separate loading of reagents was adopted. Solution of P1-SH and P2-S-SM were prepared in ACN / water mixtures of different composition. The solution of P1-SH was loaded in the reaction loop (tank and nanofiltration chamber) and the solution of

P2-S-SM put in an external tank. P2-S-SM was added slowly to the reaction loop, at the same velocity at which the permeate was collected in the cylinder. The operating pressure was set at 5 bars.

For both cases of simultaneous and separated loading of reagents, the temperature was fixed at 19 ± 2 °C. Samples of retentate and permeate were collected at different time points, to monitor the evolution of the reaction, and the permeate collected in a graduated cylinder, to monitor the flux decline.

3.2.5. Model Equations for the System

Both conventional process and Reactive Peptide Nanofiltration were described with model equations. The conventional process was described by the mass balance for batch reactor. The reaction rate for the species *i* was given by:

$$\frac{dC_i}{dt} = R_i \tag{2}$$

C_i is the concentration of the species ($C_i = m_i / V_R$, where V_R is the reactor volume) and R_i the reaction rate ($R_i = \nu_i \cdot r$, where ν_i is the stoichiometric coefficient of species *i* and r is the reaction rate).

The Reactive Peptide Nanofiltration was described by the mass balance for semi-batch reactor. The derivative of the concentration in time for the species *i* was given by Equation 3:

$$\frac{dC_i}{dt} = Q^{in} \cdot C_i^{in} - Q^{out} \cdot C_i^{out} + R_i \tag{3}$$

Q^{in} and Q^{out} are the inlet and outlet volumetric flows for the semi-batch reactor, respectively. When P1-SH and P2-S-SM were loaded together into the reaction loop, no external stream was successively provided and Q^{in} was set to zero. In the case of Reactive Peptide nanofiltration, outlet flow Q^{out} is representative of the permeation rate thorugh the membrane.

Kinetic constants, volumetric flow and rejection were generally function of the operating parameters:

$$k_r = k_r\left(C_i, pH, \%vACN / water\right) \tag{4}$$

$$Q^{in} = Q^{in}\left(C_i, pH, P, Hz, \%vACN / water\right) \tag{5}$$

$$rej = rej\left(C_i, pH, P, Hz, \%vACN / water\right) \tag{6}$$

Detail about the reaction kinetics will be provided in Paragraph 4.1, while considerations about the effects of operating parameters on flux and rejection will be discussed in Paragraph 4.2.

4. Results and Discussion

4.1. Reaction Kinetics

4.1.1. Kinetic Scheme

The main reaction, represented by Equation 1, is composed of the three following steps:

$$P1-SH \underset{k_{1,i}}{\overset{k_{1,d}}{\rightleftarrows}} P1-S^{-}+H^{+} \quad (7)$$

$$P1-S^{-}+P2-S-SM \xrightarrow{k_{2,d}} P3+SM^{-} \quad (8)$$

$$SM^{-}+H^{+} \underset{k_{3,i}}{\overset{k_{3,d}}{\rightleftarrows}} SM \quad (9)$$

Equations 7 and 9 are reactions at equilibrium. Their equilibrium constants are:

$$K_{1,a}=\frac{k_{1,d}}{k_{1,i}} \quad (10)$$

$$K_{3,a}=\frac{k_{3,d}}{k_{3,i}} \quad (11)$$

Equation 8 is an irreversible reaction, with p3 = P1-S-S-P2.

The reaction rate for reaction 1 is:

$$r_{main}=k_{2,d}\cdot\frac{K_{1,a}}{C_{H^{+}}}\cdot C_{P1-SH}\cdot C_{P2-S-SM}=k_{main}\cdot C_{P1-SH}\cdot C_{P2-S-SM} \quad (12)$$

k_{main}, the kinetic constant of the main reaction, is function of the acid-base equilibrium for P1-SH ($K_{1,a}$), the velocity of the nucleophile attack to S atom ($k_{2,d}$) and the pH (C_{H+}).

The two main side reactions identified in this process are the permutation of SM from P2-S-SM to P1-SH (Equation 13) and the dimerization of P2-S-SM (Equation 14):

$$P1 - SH + SM \xrightarrow{k_{perm}} P1 - S - SM \quad (13)$$

$$P2 - S - SM + P3 \xrightarrow{k_{\dim}} P1 - S - SM + P2 - S - S - P2 \quad (14)$$

For both side reactions 13 and 14, the same kinetic mechanism was considered as for the main reaction (i.e. nucleophile attack to S atom). The reaction rates proposed for the side reactions are therefore:

$$r_{perm} = k_{perm} \cdot C_{P1-SH} \cdot C_{SM} \quad (15)$$

$$r_{\dim} = k_{\dim} \cdot C_{P2-S-SM} \cdot C_{P3} \quad (16)$$

4.1.2. Calculation of Kinetic Constants

Solutions of P1-SH + SM and P2-S-SM + P3 were prepared and the variation of the concentration monitored during time, to test secondary reactions 13 and 14, respectively. The kinetic constants of secondary reactions could not be obtained by fitting of experimental data, due to lack of pure material for the calibration of P1-S-SM and P2-S-S-P2. It was possible to observe, however, that secondary reaction 13 was negligible and the consumption of P3 and P2-S-SM to give P2-S-S-P2 and P1-S-SM in secondary reaction 14 was significantly slower in time, compared to the consumption rate of the same reagents during the main reaction. This means that k_{dim} was expected to be much smaller than k_{main}.

The variation of the concentration was monitored during time for solutions of P1-SH and P2-S-SM. The kinetic constant of reaction 1 and 14 were calculated from fitting of experimental profiles of P3 concentration vs. time.

The kinetic constant of the main reaction was found to be a function of pH, as expected from Equation 12. Figure 3 shows the profiles of P3 for four different values of pH at constant concentration of starting reagents ($C_{P1\text{-}SH}$ = 1 mM and $C_{P2\text{-}S\text{-}SM}$ = 1 mM) in 50%v ACN / water. Kinetics that considers the occurrence of side reaction 14 (continuous line) described the experimental data better than the kinetics that considers the main reaction only (dashed line). This observation allowed the fitting of k_{dim}, which was found to be 5 to 6 times slower than k_{main}. Therefore the hypothesis of less significant side reactions with respect to the main reaction was validated. The existence of a relationship between the kinetic constants was explained by the same kinetics mechanism (i.e. nucleophile attack) that influences both reactions.

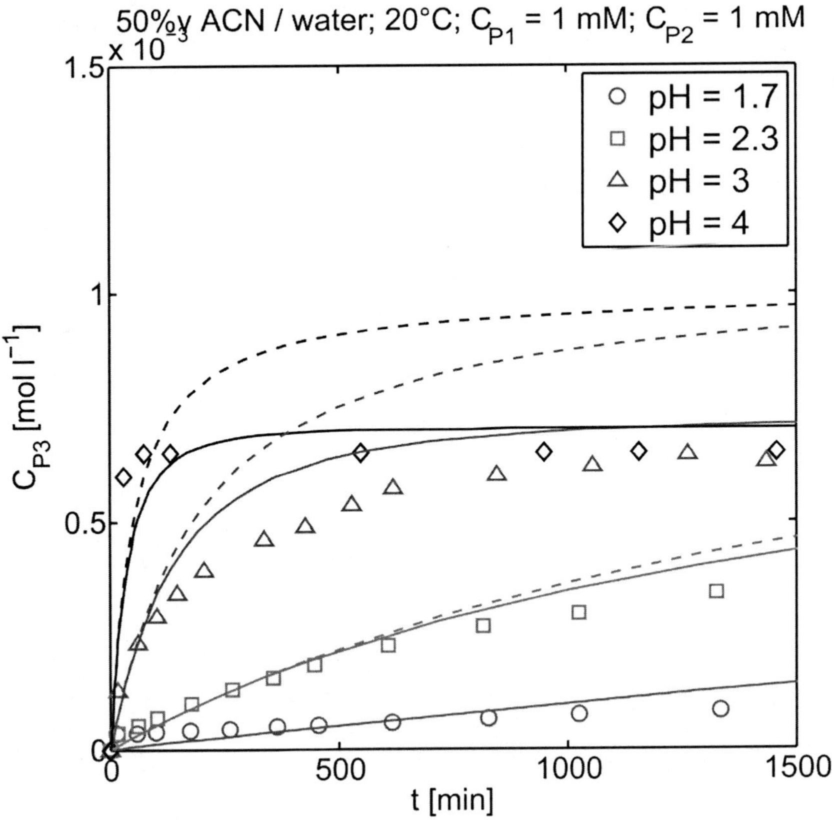

Figure 3. Batch production of P3 as function of pH. (○) experimental data; (−) complete kinetics; (− −) simplified kinetics.

For each profile, the corresponding k_{main} value was found.

Figure 4 and fitting function 17 describe the variation of the kinetic constant with the pH.

$$k_{main}\left[\frac{l}{mol \cdot \min}\right] = 0.0067 \cdot pH^{2.003} \tag{17}$$

The same dependency was extended to the range 10%v - 50%v ACN / water, since no significant effect of the organic content in this range was found on the reaction rate, as shown in Table 4.

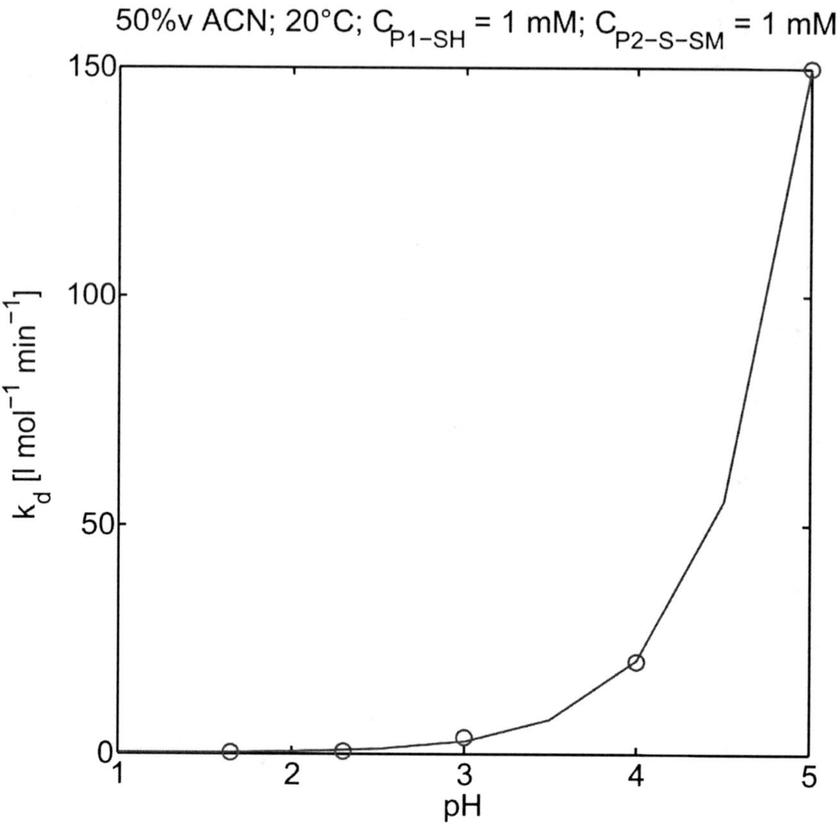

Figure 4. k_{main} as function of pH. (○) experimental data; (−) fitting function 17.

Table 4. Kinetic Constant as Function of Organic Content and Temperature (pH = 4)

$C_{P1\text{-}SH}$ / $C_{P2\text{-}S\text{-}SM}$	%v organic / water	T [°C]	k [l mol^{-1} min^{-1}]
1 mM / 1 mM	50%v ACN	10	16
1 mM / 1 mM	50%v ACN	20	20
1 mM / 1 mM	50%v THF	20	19

The kinetic constant of the main reaction was not significantly affected by temperature in the range 10 - 20 °C (cf. Table 4). It was concluded that only the pH had a significant influence on the kinetics of the main reaction.

The production / consumption rates of all the species are described by Equations 18 - 21:

$$\frac{dC_{P1-SH}}{dt} = -k_{main} \cdot C_{P1-SH} \cdot C_{P2-S-SM} \tag{18}$$

$$\frac{dC_{P2-S-SM}}{dt} = -k_{main} \cdot C_{P1-SH} \cdot C_{P2-S-SM} - k_{\dim} \cdot C_{P2-S-SM} \cdot C_{P3} \tag{19}$$

$$\frac{dC_{P3}}{dt} = k_{main} \cdot C_{P1-SH} \cdot C_{P2-S-SM} - k_{\dim} \cdot C_{P2-S-SM} \cdot C_{P3} \tag{20}$$

$$\frac{dC_{SM}}{dt} = k_{main} \cdot C_{P1-SH} \cdot C_{P2-S-SM} \tag{21}$$

4.2. Rejection Tests for Membrane Selection

Ceramic membranes were chosen, for the advantages they offer, compared to polymeric membranes. They are inert, robust, suitable to handle mixtures of water and organic solvents in any compositions and with large fluxes. The majority of commercially available membranes developed for water solutions, in fact, can stand only small amount of organic solvents, and vice versa, polymers developed for organic solvents are destroyed by water or show no flux. For all these reasons, ceramic membranes have been largely used with peptides (*5*, *6*).

Two ceramic NF membranes were tested: Inopor® Nano 450 and Inopor® Nano 750, respectively. Rejection and flux for P1-SH, P2-S-SM, P3 and SM are reported in Tables 5, 6, 7 and 8, respectively.

Rejection and permeability of P1-SH in different operating conditions are reported in Table 5.

Table 5. Rejection of P1-SH

C_{P1-SH} *[g l⁻¹]*	*%v ACN / water*	*pH*	*P [bar]*	*Membrane MWCO [Da]*	*Rejection [%]*	*Permeability [l m⁻² h⁻¹ bar⁻¹]*
1	20	2	3	750	99.0	30.5
1	30	2	3	750	99.2	31.0
1	30	3	3	750	96.1	30.9
1	30	4	3	750	95.9	29.0
1	40	2	3	750	100.0	27.5
1	30	2	3	450	100.0	9.5

Rejections were always large for both membranes. A slight effect of pH was observed: at pH 3 and 4 the rejection was 96%, while at pH 2 the rejection was complete. This can be attributed to the larger steric dimension and hydrophobicity that the peptide has at lower pH (due to the adsorption of more TFA anions). With the Inopor® Nano 450, the rejection was complete , but the permeability was lower, as expected from the smaller pore dimension of this membrane (last experiment in Table 5).

Rejection and permeability of P2-S-SM in different operating conditions are reported in Table 6.

Table 6. Rejection of P2-S-SM

$C_{P2\text{-}S\text{-}SM}$ $[g\ l^{-1}]$	*%v ACN / water*	*pH*	*P [bar]*	*Membrane MWCO [Da]*	*Rejection [%]*	*Permeability* $[l\ m^{-2}\ h^{-1}\ bar^{-1}]$
0.1	20	2	3	750	63.2	11.2
0.1	20	3	3	750	60.3	10.2
0.1	30	2	3	750	75.5	16.9
0.1	30	2	5	750	73.2	14.6
0.1	30	4	3	750	8.7	2.0
0.1	40	2	3	750	81.9	13.2
0.1	40	3	3	750	75.8	7.1
0.1	20	1.5	5	450	85.6	3.0
0.1	20	2	5	450	94.4	2.66
0.1	40	2	5	450	96.1	5.8
0.1	40	3	5	450	95.9	5.26

An effect of the organic content on the rejection of P2-S-SM was found for the Inopor® Nano 750. The larger the %v ACN / water, the larger the rejection, while the permeability was not significantly affected. pH between 2 and 3 gave almost the same rejection and permeability for P2-S-SM. A strong difference was found at pH 4, which is close to the isolelectric point of this peptide. At this point, the rejection dropped down to 8.7%. This can be explained by the complete neutralization of the peptide charge and therefore the absence of TFA ions in the molecule shell. The lower the amount of TFA linked to the peptide, the smaller its size. Rejection was larger and permeability lower for the Inopor® Nano 450, as observed for P1-SH.

It is worth to note that permeation fluxes are lower for P2-S-SM than for P1-SH, although P1-SH is bigger. This was explained by the larger friction experienced by the hydrophobic peptide in the hydrophilic membrane pore, compared to that experienced by the hydrophilic peptide, due to difference in chemical affinity.

Rejection and permeability of P3 with the two NF membranes are reported in Table 7.

Rejection of P3 was large for both membranes, but in the case of Inopor® Nano 750 insufficient for the Reactive Peptide Nanofiltration: rejection of 95-97%, or in turn, permeation of 3-5%, means too high product loss from the retentate loop. No significant effect of pressure on both flux and rejection was detected at constant concentration, while a slight effect of concentration was found at constant pressure: at higher concentration, both flux and rejection were slightly lower than at lower concentration. Rejection was almost complete with the Inopor® Nano 450 and not significantly influenced by concentration, pressure and organic solvent. This membrane was, therefore, found suitable to perform the Reactive Peptide Nanofiltration.

Table 7. Rejection of P3

C_{P3} *[g l⁻¹]*	*%v ACN / water*	*pH*	*P [bar]*	*Membrane MWCO [Da]*	*Rejection [%]*	*Permeability [l m⁻² h⁻¹ bar⁻¹]*
1	30	3.5	3	750	97.1	6.6
1	30	3.5	6	750	97.3	7.9
2	30	3.5	3	750	95.2	3.2
6	30	3.5	10	450	99.8	1.5
2	10	2	5	450	99.9	2.8

It is interesting to note that the rejection of P3 was comparable with that of P1-SH in almost the same operating conditions, although the MW of P3 was much larger (cf. Tables 5 and 7). Retention mechanism, in fact, was influenced not only by steric properties of the peptide chain but also by the additional steric and hydrophobic properties caused by TFA anions. P3 and P1-SH have very similar behavior towards TFA anions (they possess almost the same number of free sites available for forming the salt). Since the contribution of P2-S-SM in terms of available adsorption sites is negligible, the presence of P2-S-SM did not influence the effect of the operating parameters on the retention mechanism of P3, and the behavior of P3 was similar to that of its hydrophilic fragment, P1-SH.

Rejection and permeability of SM with the two NF membranes are reported in Table 8. Concentration of 0.2 g l^{-1} was chosen as representative of the concentration of SM in the process.

Rejection of pure SM was always low for both membranes. No significant effect of pH was detected between pH 2 and 4. For the Inopor® Nano 450, it was observed that the rejection of SM increased and the permeability decreased with the increase of the organic content.

By comparing the data from Tables 5 - 8, it was concluded that Inopor® Nano 450 was the best membrane to perform the Reactive Peptide Nanofiltration. It showed complete rejection for P1-SH and P3, almost complete (96%) for P2-S-SM and null for SM. Inopor® Nano 750 showed larger fluxes, however rejections of P2-S-SM and P3 were not sufficiently high.

Table 8. Rejection of SM

C_{SM} *[g l⁻¹]*	*%v ACN / water*	*pH*	*P [bar]*	*Membrane MWCO [Da]*	*Rejection [%]*	*Permeability [l m⁻² h⁻¹ bar⁻¹]*
0.2	30	2	3	750	0.0	18.6
0.2	30	4	3	750	0.0	30.0
0.2	20	2	4	450	0.0	12.0
0.2	30	2	4	450	0.0	9.4
0.2	40	2	4	450	10.0	7.5
0.2	30	4	4	450	3.0	11.0

4.3. Rejection and Permeation Tests for P3

Nanofiltration tests for membrane selection (cf. Paragraph 4.2) gave useful information about the feasibility of selected membranes to perform the Reactive Peptide Nanofiltration. However, no systematic insight was given to the effect of operating parameters on nanofiltration performances. An efficient investigation of the main effects is normally desired at this stage, in order to understand how flux and rejection can be manipulated by changing the operating parameters.

Rejection and flux for solutions of P3 in ACN / water mixtures were studied as function of different operating parameters by DoE (cf. Table 3). Operating parameters were discussed in terms of positive or negative effects on the responses. Positive effects occur when the increase in the operating factor causes an increase in the response, differently negative effects occur when the increase in the operating factor causes a decrease in the response. Effects of interactions between operating parameters could be also detected by this analysis. Interactions represent the combined change in two factors that produces an effect greater (or less) than the additive effect expected from the factors alone. In other words, interactions occur when the effect of one factor changes with the level of the other factor.

In the following, the experimental results will be reported, the effects of the operating parameters on the responses and the statistical models discussed, and finally the optimal working conditions determined.

4.4. Experimental Results

Experimental design for P3 is reported in Table 9.

The flux was found to vary between 3.67 l m^{-2} h^{-1} (run 4) to 61.17 l m^{-2} h^{-1} (run 17), which in turn means that the permeability was between 1 and 4 l m^{-2} h^{-1} bar^{-1}. The rejection was always higher than 98.0 %.

Table 9. Experimental Design for P3

Run	C_{P3} *[g l⁻¹]*	*pH*	*P [bar]*	*Pump freq. [Hz]*	*%v ACN*	Rej_{P3}	*Flux [l m⁻² h⁻¹]*
1	6	3.5	10	35	30	99.59	19.35
2	2	2	5	45	10	99.89	13.66
3	2	2	15	25	10	99.76	30.59
4	10	5	5	25	50	99.06	3.67
5	10	5	15	45	50	99.09	10.80
6	6	3.5	10	35	30	99.74	17.58
7	10	2	15	45	10	99.88	40.97
8	10	2	5	25	10	99.93	12.53
9	10	5	5	45	10	99.83	6.41
10	10	5	15	25	10	99.45	14.57
11	2	5	5	45	50	99.30	5.06
12	2	5	15	25	50	98.83	11.67
13	6	3.5	10	35	30	99.82	15.27
14	10	2	5	45	50	99.88	27.09
15	10	2	15	25	50	99.84	64.17
16	2	2	5	25	50	99.78	17.50
17	2	2	15	45	50	99.75	61.17
18	2	5	5	25	10	99.73	8.80
19	2	5	15	45	10	99.79	26.79
20	6	3.5	10	35	30	99.73	21.66

4.4.1. Statistical Analysis

The statistical models for rejection of P3 and permeation flux in terms of coded factors (lower level: -1; upper level: +1; cf. Table 3) are the following:

$$Rej = 99.68 - 0.05 \cdot A - 0.18 \cdot B - 0.11 \cdot E - 0.075 \cdot A \cdot B - 0.088 \cdot B \cdot E \quad (22)$$

$$Flux^{0.5} = 4.34 - 1.17 \cdot B + 1.06 \cdot C + 0.20 \cdot D + 0.13 \cdot E - 0.24 \cdot A \cdot B + \\ - 0.302 \cdot B \cdot C - 0.59 \cdot B \cdot E \quad (23)$$

The statistical models for rejection of P3 and permeation flux in terms of actual factors are the following:

$$Rej = 99.76 + 0.03 \cdot C_{P3} + 0.05 \cdot pH + 4.58 \cdot \%vACN - 0.01 \cdot C_{P3} \cdot pH + \\ - 2.92 \cdot pH \cdot \%vACN \tag{24}$$

$$Flux^{0.5} = -0.22 + 0.45 \cdot pH + 0.35 \cdot pressure + 0.02 \cdot Pump\ freq. + \\ + 0.06 \cdot \%vACN + -0.04 \cdot C_{P3} \cdot pH - 0.04 \cdot pH \cdot pressure + \\ - 0.02 \cdot pH \cdot \%vACN \tag{25}$$

The performances of the model were analyzed by Analysis of Variance (ANOVA). The coefficients of correlation for this model are reported in Table 10.

Table 10. Statistical Analysis of the Models for Flux and Peptide Rejection

	Rejection	*Flux*
Response transformation	None	$Flux^{0.5}$
Lack of Fit p-vaue	0.2170	0.5283
R^2	0.8217	0.9790
Predicted R^2	0.7532	0.9111
Adjusted R^2	0.5573	0.9622

Response transformation for flux was suggested by ANOVA. The "Lack of Fit p-value" higher than 0.05 implies the Lack of Fit is not significant relative to the error. Non-significant lack of fit is desired for the model to fit the experimental data.

The importance of each parameter in affecting the responses is quantified by its normalized effect, i.e. the model parameter representative for the coded factor in the statistical model (cf. Equations 22 and 23). These are shown in Figure 5, respectively.

4.4.2. Physical Interpretation of Statistical Models

Analysis of operating parameters was done simultaneously on flux and rejection, as the responses are coupled in the transport mechanism.

As shown in Figure 5, the main influencing parameters for the rejection are pH and %v ACN, both with negative effect. pH shows a negative effect on the flux too, while %v ACN is non significantly affecting it.

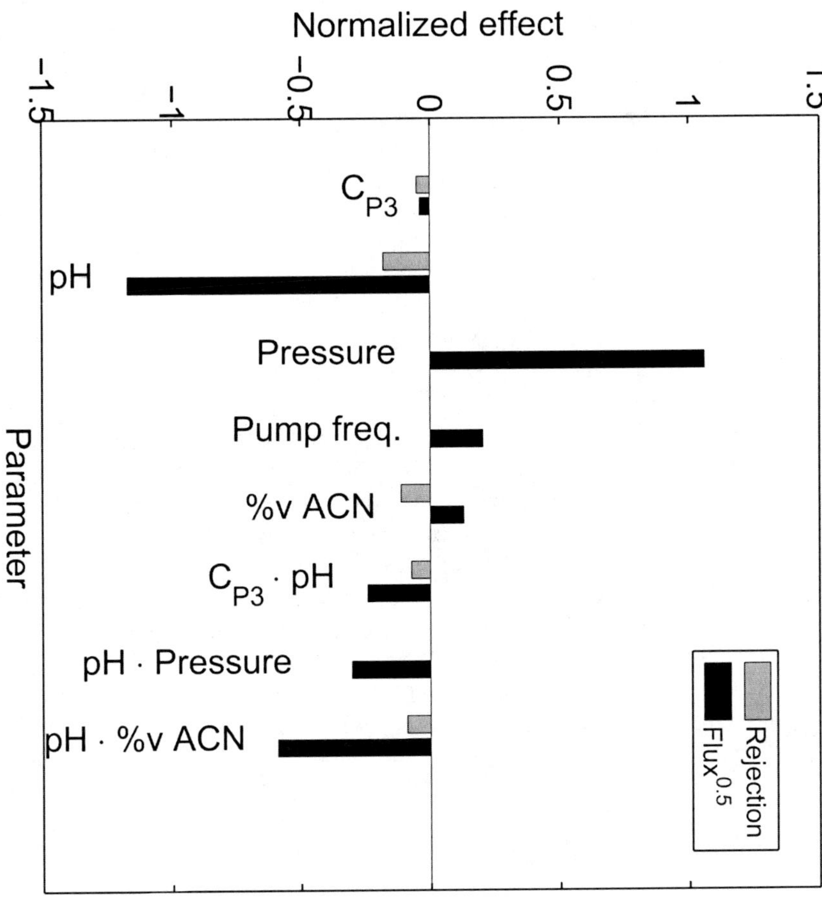

Figure 5. Normalized effects of operating parameters for rejection and flux.

The strong contribution of pH (i.e. TFA) is explained by its influence on steric and hydrophobic properties of the peptide: the increase in TFA content determines the increase in peptide dimension and hydrophobicity, that in turn contributes to the ousting the molecule from the pore. Solvent flux is larger through the pore in correspondence of larger rejection, since friction due to solute flux is less important.

The negative effect of ACN content on rejection is explained by its poor solubilization of P3. By increasing the ACN content, the hydrophobicity of the solvent increases, thus contributing to the decrease in relative solvent - membrane affinity, as compared to the solute - membrane one. When the solvent - membrane affinity is lower than the solute - membrane affinity, preferential permeation of solute compared to solvent occurs through the membrane, thus reducing rejection (*7*).

Peptide concentration has no significant effect on both flux and rejection. This is limited to the investigated experimental range only. It is plausible that the increase of concentration to larger values will decrease the flux, as concentration polarization and adsorption may occur. Concentration polarization occurs in a thin film layer at the membrane surface, when the back diffusion of the solute towards the bulk of the retentate, caused by the difference in concentration between film layer and bulk, cannot balance the convection flow of the solute towards the surface, caused by the applied pressure. Consequence is the accumulation of solute at the membrane surface and a larger effective retentate concentration experienced by the membrane, as compared to the bulk one.

Pressure shows a remarkable positive effect on flux, as expected from the driving force of the process, and no significant effect on the rejection. Effect of pressure on concentration polarization is therefore negligible.

Pump frequency, that in turn represents the linear velocity on the membrane surface, has a positive effect on the flux and a non significant effect on the rejection. This is explained by the effect of linear velocity on concentration polarization (*8*): the larger the linear velocity, the lower the solute accumulation at the membrane surface and the less significant the concentration polarization layer. Similarly to what observed for the effect of pressure, changing in fluid dynamics conditions is affecting the flux but not the retention performances.

Effects of interactions between pH and %v ACN and between pH and peptide concentration affect both rejection and flux in the same way (negative effects, cf. Table 11), while interaction between pH and pressure affect the flux only.

It is worth to note that pressure and pump frequency affect the flux and not the peptide rejection, both as single and interactions effects. These parameters, in fact, affect the fluid dynamics of the system but do not influence solute-membrane interface interactions, as pH and %v ACN / water do. In other words, it is plausible to assume that rejection of P3 is governed by solute-membrane interactions at the surface level only, while flux is influenced by both solute-membrane interactions and fluid-dynamics in the module.

4.4.3. Numerical Optimization

The best operating conditions were found, based on the results from the experimental investigation shown in Table 9, using Eqs. 24 and 25. The concentration was kept fixed at the maximum value (C_{P3} = 10 g l^{-1}), as the goal was to explore the high concentration range. The optimum for two different numerical optimization criteria is reported in Table 11. In Case I, flux and rejection were maximized, while all other parameters were left free to assume any value in the operating range. In case II, the amount of ACN was fixed at 10%v, as this is the condition to have a direct injection in the HPLC column for purification.

The desirability function was used to characterize the quality of the optimization when dealing with multiple response. It has to be 1 when all products characteristics are on target, and 0 when one or more characteristics are unacceptable.

Table 11. Numerical Optimization

Parameter	*Case I*	*Case II*
C_{P3} [g l^{-1}]	10	10
pH	2	2
Pressure [bar]	15	15
Pump Frequency [Hz]	45	45
%v ACN / water	46	10
Flux [l m^{-2} h^{-1}]	61.17	42.98
Rejection [%]	99.86	99.90
Desirability [%]	97.2	88.6

The numerical optimization for the case of free concentration of ACN (Case I) gave a desirability function of 97.2%. The rejection in correspondence of the optimum is 99.86% and the flux 61.17 l m^{-2} h^{-1}. Optimal pH is low (pH = 2) and 46%v ACN / water is the best amount of organic solvent.

Considering that the objective of the Reactive Peptide Nanofiltration is to work with the lowest amount of organic solvent possible, a second optimization was done setting the %v ACN / water at 10 (Case II). The desirability function is always large, peptide concentration high, pH low and pressure high. Flux and peptide rejection obtained from this simulation are satisfactory with the process requirements.

It has to be noted than the flux obtained from the DoE is representative for, but not the same as, the flux expected during the Reactive Peptide Nanofiltration. During the Reactive Peptide Nanofiltration, in fact, the reaction mixture is much more complex than the mixture used for the DoE, as contains P1-SH, P2-S-SM and side products, too. It is plausible to assume, however, the occurrence of the same observed dependences.

4.4.4. Choice of Working Conditions for the Reactive Peptide Nanofiltration

As discussed, reaction kinetics was found to be influenced by reagents concentration and pH, and not significantly influenced by organic content and temperature. Permeation rate was found to be function of peptide concentration, pH, pressure, pump frequency and organic content. The main observation from kinetic and permeation studies were combined together to find the best working conditions for the Reactive Peptide Nanofiltration.

The principal issue was the choice of the operating pH, which showed opposite effects on reaction and permeation: it had positive effect on the reaction rate and negative effect on the permeation. At the desired working pH, both reaction and flux had to be sufficiently fast, in order to have high productivity. pH 4 was chosen as best compromise value.

Low organic content was advantageous for having high flux. Since the effect of the solvent on the reaction rate was negligible, 10%v ACN / water was chosen as reaction solvent. The great advantage of this choice was the possibility of exiting the Reactive Peptide Nanofiltration with a solution ready for the chromatographic step and that minimizes the solvent consumption.

The pump frequency affected positively the permeation flux and showed no significant effects on the peptide degradation. High value (45 Hz) was therefore preferred.

The concentration of P1-SH in the reactor was an adjustable parameter, differently the concentration of P2-S-SM was limited by the solubility limit in 10%v ACN / water ($\leq 0.1\ g_{crude}\ l^{-1}$).

4.5. Reactive Peptide Nanofiltration Feasibility

The Reactive Peptide Nanofiltration was studied in the Natan cross-flow NF system with the Inopor® Nano 450 in the range 10 - 50 %v ACN / water. Firstly, the effect of pH was tested in semi-batch conditions, constant pressure (8 bar) and constant organic content (30%v ACN / water). In order to compare the occurrence of the reaction at constant pressure, simultaneous loading of P1-SH and P2-2-SM solutions was adopted. The effect of pH is shown in Figure 6(a-c).

It is evident that the reaction was slower at pH 2.5 than at pH 5, from both the profiles of P3 production and P2-S-SM consumption. In the case of pH = 2.5 (cf. Figure 6(a)), the concentration of SM increased slower than the concentration of P3, as expected from its lower rejection by the NF membrane. Differently, for the case of pH = 5 (cf. Figure 6(b)), SM reached the maximum value after some minutes, and then started to decrease, when the reaction reached the completion. Similar effect of pH was found for 10%v ACN / water (data not shown). The flux was larger at pH 2.5 than at pH 5 (cf. Figure 6(b)), as expected from the results of DoE (cf. Paragraph 4.4).

The effect of concentration was tested in semibatch conditions, with separate loading of reagents (cf. Paragraph 3.2.4 for experimental detail), at pH = 4. This test was meant to to compare the occurrence of the reaction in the nanofiltration loop with that found in batch conditions. Two different starting concentrations for P1-SH and P2-S-SM in 50%v ACN / water and two different solutions, 30 and 50%v ACN / water were tested. The profiles of SM, P2-S-SM and P3 at pH = 4 are reported in Figures 7(a-c).

The concentration of SM increased quickly at the beginning of the reaction and then slower during time (cf. Figure 7(a)), since it was continuously removed by the membrane. The concentration values were significantly higher for the case of higher concentration of starting reagents and only slightly affected by the organic content. The rejection of SM was constant at 50%v ACN / water for both the concentration of starting reagents, while it was slightly lower at 30 %v (cf. Figure 8).

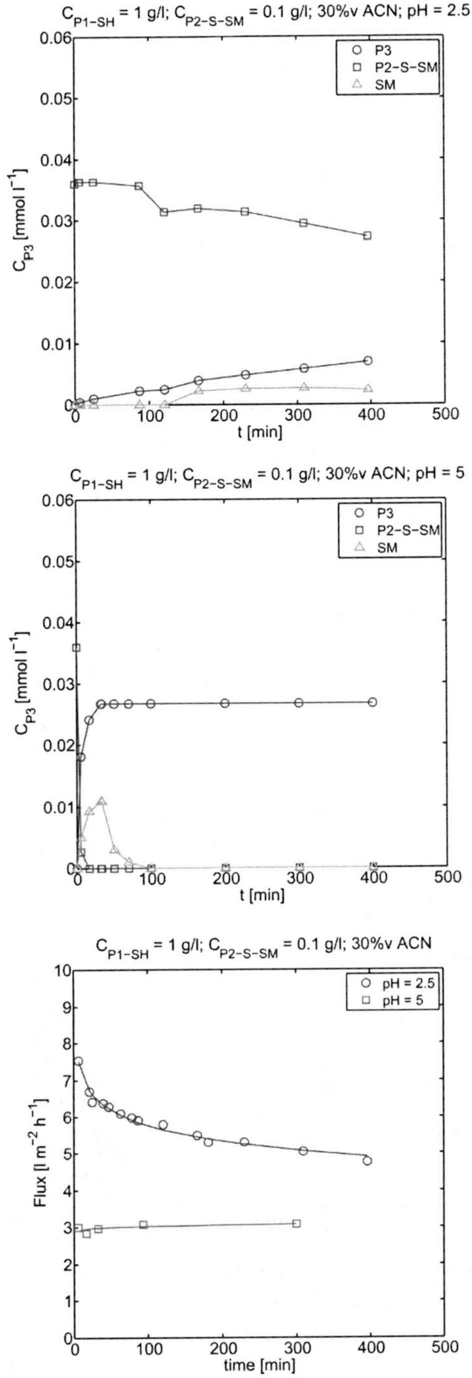

Figure 6. Effect of pH during Reactive Peptide Nanofiltration.

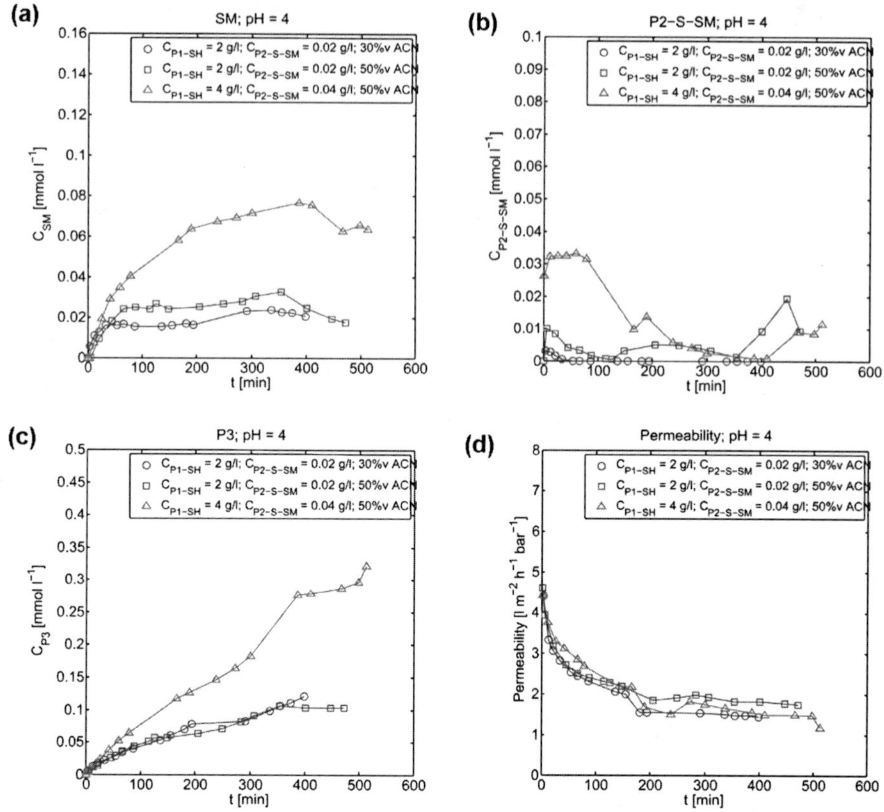

Figure 7. Reaction and flux profiles for the Reactive Peptide Nanofiltration at pH = 4.

The concentration of P1-S-SM was always low, since it was continuously added to the system and immediately consumed by the reaction (cf. Figure 7(b)). The concentration of P3 increased during time, together with the consumption of the reagents (cf. 7(c)). The rejection of P1-SH, P2-S-SM and P3 by the NF membrane was complete. The difference in concentration and organic solvent was not significant to determine a noticeable difference in the permeability decline during time (cf. Figure 7(d)).

4.6. Process Modelling and Cost Evaluation

The mass balances for the Reactive Peptide Nanofiltration were introduced in Paragraph 3.2.5. The process model was validated on the experimental data and the Reactive Peptide Nanofiltration compared with the conventional process in terms of performances and costs.

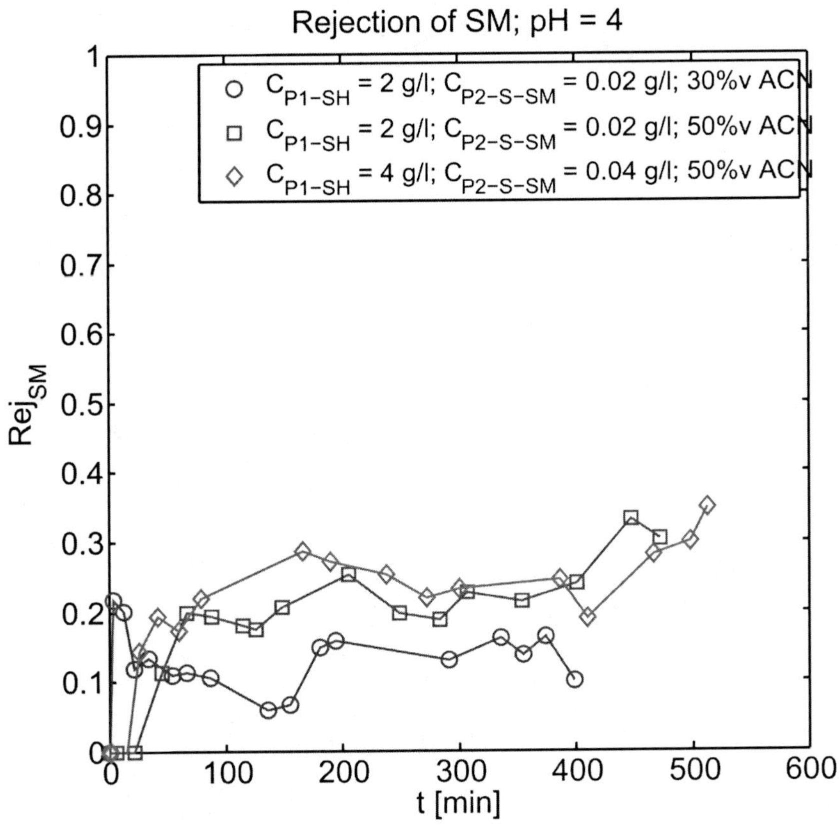

Figure 8. Rejection of SM during Reactive Peptide Nanofiltration.

4.6.1. Model Validation

Firstly, the effect of pH in semibatch conditions with simultaneous loading of reagents was simulated. The kinetics described well the profiles for the cases of pH = 2.5 (cf. Figure 9(a)) and pH = 5 (cf. Figure 9(b)). Equation 17 was used to describe the kinetic constant as function of pH.

Afterwards, semibatch conditions with separated loading of reagents were simulated for the cases of two different concentrations of organic solvents (Figure 10(a) vs. 10(b)) and two different starting concentrations of reagents (Figure 10(b) vs. 10(c)).

Experimental fluxes were used for the description of the permeation rate ($Q_{in} = Q_{out}$ in Equation 3). They were shown in Figures 6(b) and 7(d), for the simulation of Figures 9 and 10, respectively.

The matching between simulated and experimental profiles was satisfactory. Experimental kinetics obtained in batch conditions was representative for the kinetics in the NF loop and its combination with the mass balances for the semi-batch reactor successfully described the Reactive Peptide Nanofiltration.

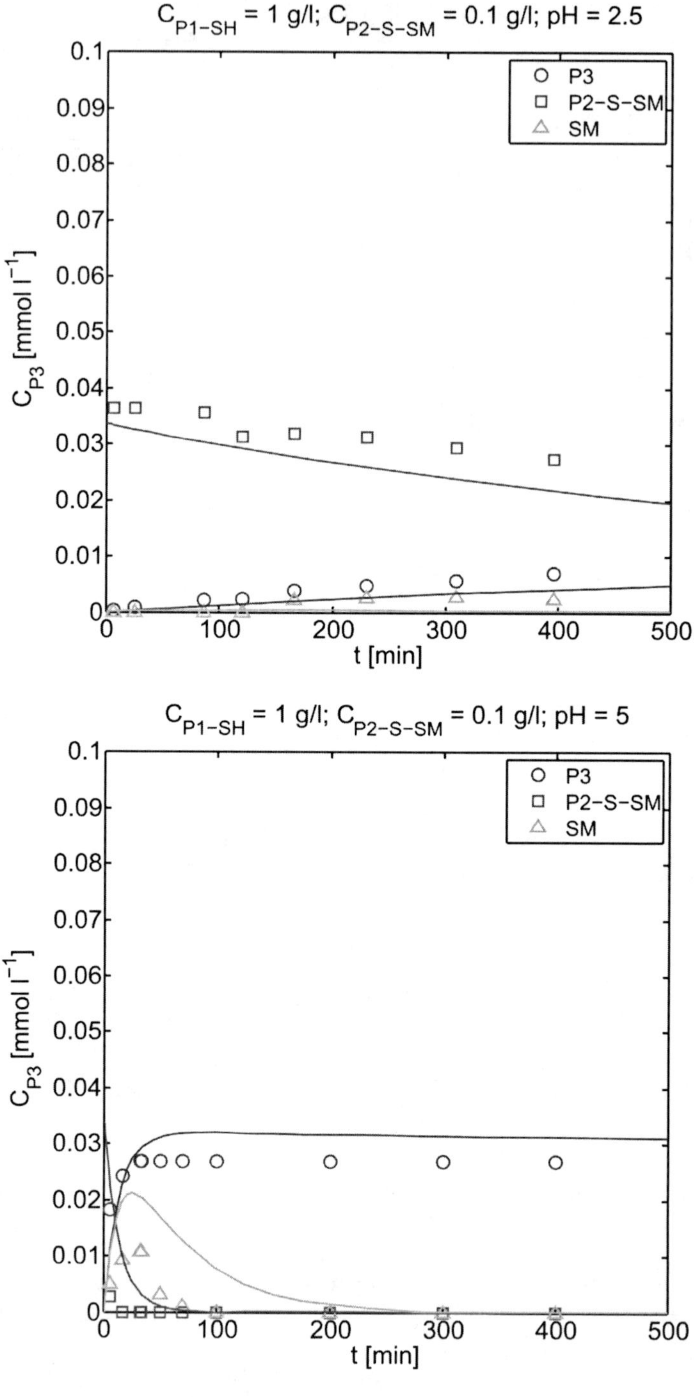

Figure 9. Effect of pH during Reactive Peptide Nanofiltration. (○) experimental data; (−) simulation.

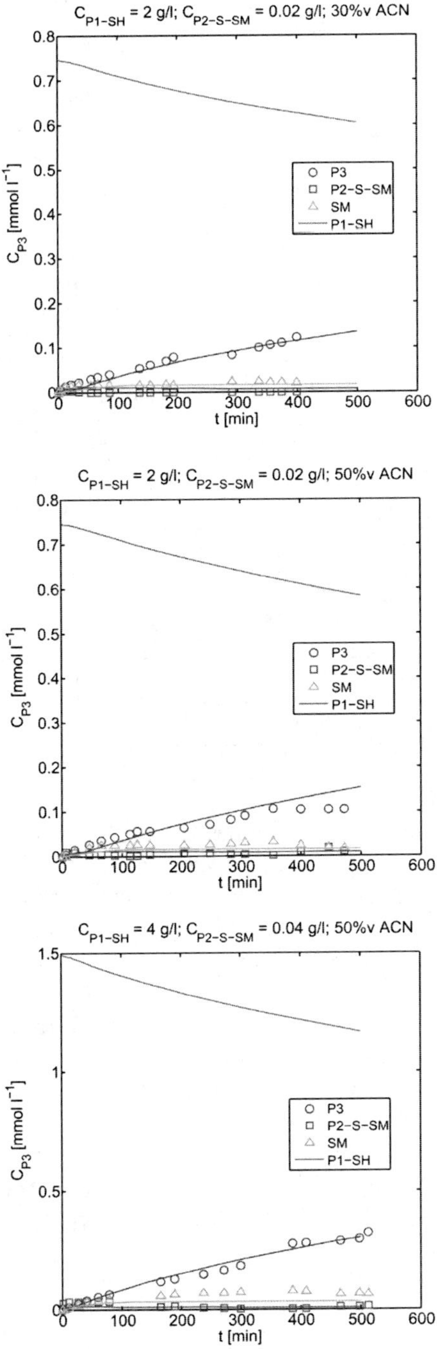

Figure 10. Effect of peptide concentration and organic content during Reactive Peptide Nanofiltration. (○) experimental data; (−) simulation.

4.6.2. Reactive Peptide Nanofiltration vs. Conventional Production: Process Performances and Cost Comparison

Reactive Peptide Nanofiltration and conventional process were compared each other in terms of performances and costs. The simulation was done at large scale: 26 kg pf P3 was set as goal for the production. Production steps and relative operating conditions for the conventional process are reported in Table 12.

Table 12. Conventional Process and Relative Working Conditions

Step	*Working value*	*Time*
Preparation of THF / water = 75 / 25 solution	170 l	1 h
Addition of p2-S-SM	10.6 kg	2 h
Temperature adjustment	0° C	1 h
Addition of P1-SH	17.4 kg	4 h
Reaction		12 h
Addition of ACN for precipitation	1370 l	2 h
Wash x 3	400 l	3 h
Drying		36 h
Dissolution time for chromatography		4 h

The conventional reaction was done with P1-SH in slight defect, compared to P2-S-SM. High amount of organic solvent was necessary to solubilize P2-S-SM. The time required for the reaction was long, almost 12 h. After the reaction, ACN was added to precipitate the main product. Three steps of washing and drying followed the precipitation. The time required for the drying was significantly long. Finally, the product was re-solubilized for the final chromatography step. The total reaction volume was 1540 l and the total time required for the production of P3 was 67 h.

Production steps and relative operating conditions for the Reactive Peptide Nanofiltration are reported in Table 13.

The number of steps was significantly reduced for the Reactive Peptide Nanofiltration. Permeability of 2 l m^{-2} h^{-1} bar^{-1} and pressure of 20 bar were assumed as representative values for real working conditions. Filtration area of 25 m^2 was assumed as representative for commercially available ceramic modules. The time required for addition of P2-S-SM / reaction / dialysis (i.e. 15 h) was due to the necessity of having 10 dialysis volumes with permeation flow of 1000 l h^{-1}. The total time required for the production of P3 was 18 h.

Costs for the raw materials were similar, since the same amount of starting reagents was used. Costs for the solvents were ten times higher for the conventional process than for the Reactive Peptide Nanofiltration, since the organic solvent could not be recovered at the end of the former process, while could be recycled at the end of the latter. Production costs were estimated to be three times higher for the conventional process, due to longer production time and larger manpower required. Costs for membranes are present for the Reactive Peptide Nanofiltration only, but they were not significant compared to the others cost items. According to the cost evaluation for this business case, total costs foreseen for the Reactive Peptide Nanofiltration were almost the 50% of those for the conventional process. The advantages from technical, economical and environmental point of view were significant for this kind of technology, which was therefore demonstrated promising for application in the large-scale production of peptides.

Table 13. Reactive Peptide Nanofiltration and Relative Working Conditions

Step	*Working value*	*Time*
Preparation of ACN / water = 10 / 90 solution	1450 l	1 h
Addition of P1-SH	17.4 kg	2 h
Addition of p2-S-SM / Reaction / Dialysis	10.6 kg	15 h

5. Conclusion

Peptide therapeutics by chemical synthesis represent an important class of Active Pharmaceutical Ingredients (APIs), manufactured in large scale. Integration of membrane separation techniques into the reaction step was presented in this work to address some of the drawbacks of the conventional peptide production by fragment condensation. The strategy was named Peptide Reactive Nanofiltration and the main features were: (i) the incorporation of nanofiltration unit into the reaction step; (ii) the separation of the small side product from the reaction mixture (iii) the recycle of the solvent after the nanofiltration (which means in turn lower solvent consumption for the global process); and (v) the elimination of time-consuming steps, typical of the conventional strategy.

The performances of the new strategy were presented for one industrial case study. Reaction kinetics was studied as function of operating conditions and rejection and permeability investigated for the membrane selection. Permeation through the membrane was studied as function of different operating parameters, to select the best operating conditions. Experimental kinetics obtained in batch conditions was representative for the kinetics in the NF loop and its combination with the mass balances for the semi-batch reactor successfully described the Reactive Peptide Nanofiltration.

The Reactive Peptide Nanofiltration was compared to the conventional batch processes by means of techno-economical analysis. The Reactive Peptide Nanofiltration demonstrated advantageous in terms of both costs and productivity.

Acknowledgments

We would like to thank the NEMOPUR project (as part of the Marie Curie Initial Training Network) for funding.

References

1. Ghosh, A.; Fan, E. *Tetrahedron Lett.* **2000**, *41*, 165–168.
2. Andersson, L.; Blomberg, L.; Flegel, M.; Lepsa, L.; Nilsson, B.; Verlander, M. *Biopolymers* **2000**, *55*, 227–250.
3. Sewald, N.; Jakubke, H. *Peptides: Chemistry and Biology*; Wiley-VCH Verlag GmbH & Co. KGaA: Weinheim, Germany, 2008
4. Whu, J. A.; Balzis, B. C.; Sirkar, K. K. *J. Membr. Sci* **1999**, *163*, 319–331.
5. So, J.; Peeva, L.; Tate, E. W.; Leatherbarrow, R. J.; Livingston, A. G. *Org. Process Res. Dev.* **2010**, *14*, 1313–1325.
6. Martin-Orie, C.; Bouhalla, S.; Garem, A. *J. Membr. Sci.* **1998**, *142*, 225–233.
7. Darvishmanesh, S.; Degrève, J.; Van der Bruggen, B. *Phys. Chem. Chem. Phys.* **2010**, *12*, 13333–13342.
8. Peeva, L. G.; Gibbins, E.; Luthra, S. S.; White, L. S.; Stateva, R. P.; Livingston, A. G. *J. Membr. Sci.* **2004**, *236*, 121–136.

Chapter 9

HCl Effect on Two Types of Ag Nanoparticles Utilizable in Detection of Low Concentrations of Organic Species

Karolina Siskova,* Ondrej Becicka, Klara Safarova, and Radek Zboril

RCPTM, Department of Physical Chemistry, Faculty of Science, Palacky University, Slechtitelu 11, 783 71 Olomouc, Czech Republic
*E-mail: karolina.siskova@upol.cz

In this chapter, we deal with the impact of HCl in the concentration range of 0.1 mM – 0.1 M on characteristics of two different kinds of Ag colloids. These Ag colloids were prepared by chemical reduction of Ag^+ using either borohydride, or glucose in alkaline medium (i.e. the latter known as a modified Tollens method). Both types of Ag colloids are treated with HCl and characterized by UV-visible absorption spectroscopy, dynamic light scattering, zeta potential measurements, transmission and scanning electron microscopies, energy dispersive X-ray spectrometry, and surface-enhanced resonance Raman scattering spectroscopy (the last mentioned by employing a model pollutant in the concentration of 1×10^{-7} M). Distinct differences in HCl effect on these two types of silver colloids are encountered and related to the specificity of stabilizing species surrounding Ag nanoparticles in each of the colloids under study.

Introduction

Silver nanoparticles are well-known for their unique optical properties which can be fully exploited in surface-enhanced Raman scattering (SERS) spectroscopy (*1*, *2*). Recently, several papers applying SERS for qualitative and quantitative analysis of aqueous and airborne organic and inorganic contaminants as well as pathogens have been published (*3*, *4*). This direct way of monitoring of the environmental pollution (even at very low concentrations of pollutants) is very important for the concept of sustainability. Like any analytical technique, SERS may seem challenging namely due to (i)·changes in enhancement mechanism contributions when switching from one to another excitation wavelength of laser beam and (ii) a complex character of observed vibrational spectra.

Basically, there are two main mechanisms of Raman scattering enhancement: electromagnetic and chemical (*1*, *2*). While the former originates from a simultaneous interaction of light with surface plasmon localized on Ag nanostructures and a molecular species placed at their surfaces; the latter is strongly dependent on the type of a molecular species (particularly, on its transfers between energy levels being close to or far away from the resonance with the incident laser light) and its interactions with Ag nanoparticle surface. Obviously if both enhancement mechanisms are exploited, the highest Raman signal of a particular molecular species will be gained and the spectroscopy is dubbed as surface-enhanced resonance Raman scattering (SERRS). For instance, having a chromophoric molecule which absorbs in the visible region between 420-600 nm and is placed to the close vicinity of Ag nanoparticle surface and, simultaneously, being in resonance with surface plasmon extinction of these Ag nanoparticles (e.g., employing the 532 nm excitation wavelength of a laser), a huge enhancement of Raman-scattered light from the studied molecule has been observed (*5*). Indeed, the porphyrin concentrations as low as 100 fM in the final system have been repeatedly detected due to SERRS spectroscopy (*5*). In general, maximizing the contributions of both enhancement mechanisms is a necessary condition for the detection of molecular species at very low concentrations (*3*).

Concerning the complexity of observed vibrational spectra, SERS/SERRS techniques are not only selective for a molecular species, but also very sensitive to slight changes in molecular structure (*1–3*). Therefore, the orientation of molecule toward Ag nanoparticle surface, local environment changes, and presence of impurities among others result in different SERS/SERRS spectral features (*6*, *7*).

Around the second millennium, several seminal studies achieving SERS/SERRS spectral detection of species at the single-molecule level appeared (*8–13*). It turned out that in single-molecule SERS/SERRS detection, the "activation" of Ag nanoparticles by addition of namely chlorides play a very important role (*8–13*). Since that, many other research teams have dealt with the effect of chloride ions on Ag nanoparticles and SERS/SERRS signal of a particular molecular species in neutral medium (*14–23*). It should be noted at this time that only a few studies focused on the influence of HCl on Ag nanoparticles, i.e., chloride anions added simultaneously with H^+ ions into a Ag colloid (*19*, *24*).

As it has been stated quite recently (*23*), there are two possible ways of chlorides effect (introduced in the form of a water-soluble salt) on citrate-stabilized Ag nanoparticles depending on the final chloride concentration: (a) slow oxidative decomposition of Ag nanoparticles at NaCl concentrations lower than 27 mM resulting in silver chloride precipitate, and (b) aggregation of Ag nanoparticles at and above the 27 mM NaCl concentration. Similarly, the aggregation of citrate-stabilized Ag nanoparticles has been observed either in ref. (*14*) when the 10 mM chloride anion concentration together with a dicationic adsorbate are present, or in ref. (*17*) where the 10 mM KCl concentration lead to an enormous increase of brilliant green SERS signal. On the other hand, considering low chlorides concentrations, Bell and Sirimuthu (*15*) spectroscopically evidenced that already the 0.27 mM $MgCl_2$ concentration present in Ag colloids (again citrate-stabilized ones) revealed chlorides adsorption on Ag nanoparticle surface resulting in AgCl formation (*15*). In fact, chlorides competitively adsorbed on Ag nanoparticle surface and only AgCl dominated the SERS spectrum at the 8 mM $MgCl_2$ final concentration although a rather high concentration of the studied molecule (dipicolinic acid) was used (1.5 mM) (*15*).

On the contrary, Glaspell and co-workers (*16*) observed the maximum enhancement of the studied molecule SERS signal when the concentration of chlorides was approximately equal to that of the studied molecule (rhodamine 6G) in Ag colloids prepared via laser vaporization controlled condensation. They concluded (*16*) that the co-adsorption of chlorides and studied molecules took place. This conclusion was made by Hildebrandt and Stockburger for the same molecule under study (rhodamine 6G), however, on a different kind of Ag nanoparticles (citrate-stabilized) in 1984 (*25*). Moreover, they (*25*) attributed the SERS activating effect of chlorides to the formation of new, specific adsorption sites on Ag nanoparticle surface.

Sloufova and co-workers (*19*) contributed to the subject by a series of experiments in which not only the formation of new adsorption sites, but also distinct morphology changes have been demonstrated for borohydride-reduced Ag nanoparticles treated by NaCl and/or HCl ranging from 50 μM to 110 mM. Indeed, the fusion of Ag nanoparticles into small or large compact aggregates caused by adsorbed chlorides (above a certain threshold of chlorides surface coverage) was followed as a function of chloride concentration in both acidic and neutral media (*19*). Several years later, Prucek and co-workers (*22*) reported on a re-crystallization of silver nanoparticles from 30-nm spheres to one-order larger crystallites (~400 nm). The process in ref. (*22*) was induced by a rather high NaCl concentration (400 mM) introduced into Ag colloids generated by a modified Tollens method using maltose as a reduction agent. On the contrary, Yang and co-workers (*26*) achieved the transformation of polycrystalline Ag nanospheres (which were sodium dodecyl sulphate-stabilized) into single crystalline triangular Ag nanoplates by using only the 10 mM NaCl concentration. The same NaCl concentration (10 mM) induced the conversion of spherical citrate-stabilized Ag nanoparticles (ca. 32 nm in diameter) to Ag particles of various shapes such as ellipsoids, rods, spheres or their aggregates with wide range of sizes (30-150 nm) (*18*). Interestignly, no aggregation or dissolution of citrate-stabilized Ag nanoparticles purged by nitrogen was observed (*18*).

From all the above mentioned, it is thus obvious that there are several different parameters which influence the chlorides impact on Ag nanoparticles. Except from the (i) chlorides concentration effect, (ii) absence and/or presence of another adsorbing molecule (which is in focus of SERS/SERRS studies), (iii) neutral vs. acidic ambient, (iv) nitrogen purging or full air access, there is most probably also a distinct effect of Ag nanoparticles origin. In other words, the preparation procedure of Ag colloids and, consequently, the occurrence of different kinds of stabilizing species on Ag nanoparticle surface play a key role in the chlorides impact. To the best of our knowledge, only one comparative study evaluating the influence of the 20 mM NaCl concentration on three kinds of Ag colloids have been published so far (*21*). However, the authors in ref. (*21*) concentrated namely on SERS-activities of their three colloids.

Therefore, in the present work, we focus our attention on the influence of HCl in the concentration range of 0.1 mM – 0.1 M on two types of Ag nanoparticles with significantly different stabilizing surface species: borates vs. gluconic and/or glucuronic acids. Using SERRS signal of a model pollutant, 5, 10, 15, 20-tetrakis(1-methyl-4-pyridyl)21H, 23H-porphine (TMPyP) which is potentionally dangerous because it can intercalate into DNA (*27*), we estimate the HCl impact on both Ag colloids under study. Possible exploitation of HCl-treated Ag nanoparticles for the detection of low concentrations of the organic species is discussed. Moreover, we compare morphological changes of both types of Ag nanoparticles induced as a consequence of the HCl impact.

Experimental

Materials

Analytical grade chemicals and distilled deonized water were used for all samples preparations. $AgNO_3$, $NaBH_4$, aqueous solution of NH_3, NaOH, glucose, 37% aqueous solution of HCl, and 5,10,15,20-tetrakis(1-methyl-4-pyridyl)21H,23H-porphine (TMPyP) were used as received without any further purification. The glassware was cleaned by diluted nitric acid (1:1), extensively rinsed with distilled water and subsequently with deionized water. Alternatively, the mixture of sulphuric acid and hydrogen peroxide (1:1) was used, particularly when the glassware was polluted with porphyrin molecules.

Ag Nanoparticles Were Prepared by Two Distinctly Different Methods

(i) direct reduction of $AgNO_3$ by sodium borohydride – a detailed preparation procedure can be found in ref. (*28*), and (ii) modified Tollens method using glucose as a reducing agent – the procedure is described in details in ref. (*5*). Concentrated HCl (11.6 M) and diluted HCl (by deionized water) to the 1 M and/or 1 x 10^{-2} M aqueous solutions were employed in order to generate the final HCl concentrations of a particular value in the investigated systems. The stock solution of the cationic porphyrin TMPyP of the 1 x 10^{-5} M concentration was appropriately diluted to the final concentration of 1 x 10^{-7} M in the investigated systems.

Instrumentation

UV-Vis spectra were measured in a quartz 1-cm cuvette using Libra S6 (Biochrom, Cambridge, UK) spectrometer operating in the range from 230 to 800 nm. Dynamic light scattering (DLS) and zeta potential were recorded on Zetasizer Nano Series (Malvern Instruments). Transmission electron microscopic (TEM) images were measured on a JEOL JEM-2010 microscope equipped with a LaB_6 cathode (accelerating voltage of 160 kV; a point-to-point resolution of 0.194 nm). A drop of a particular sample was place onto a holey-carbon film supported by a copper TEM grid and allowed to dry at room temperature. Energy dispersive X-ray spectrometry (EDX) together with scanning electron microscopic (SEM) measurements were obtained on a Hitachi SU 660 microscope working at 5 kV. SERRS spectra were recorded at room temperature with a Raman spectrometer using the 532 nm excitation line of a Verdi V5 laser (DPSS technology, Coherent Inc., Santa Clara, CA, USA, power ~100 mW at the sample). Scattered radiation was collected at 90° relative to the excitation beam and detected with a nitrogen cooled CCD detector Symphony (Horiba Jobin Yvon S.a.S., Longjumeau, France) with 2048x512 pixels, coupled to a iHR320 imaging spectrometer (Horiba Jobin Yvon S.A.S., Longjumeau, France) with a 1200 grooves/mm grating. System was controlled by LabSpec 5 software (Horiba Jobin Yvon S.A.S., Longjumeau, France). Accumulation time was set to 90x 1s and slit to 110 μm. Precise calibration of Raman shift axis was done due to toluene signal acquisition.

Results and Discussion

To investigate the effect of HCl on Ag nanoparticle solutions and to demonstrate that it strongly depends on stabilizing molecular species present on Ag nanoparticle surface, we have chosen two completely different chemical preparation procedures of Ag nanoparticles. First, we used the well-known borohydride reduction (*28*) and, second, we employed the modified Tollens method for Ag nanoparticles generation (*5*). The as-prepared stock solutions of Ag nanoparticles are labeled as B and G colloids for borohydride-reduced and modified Tollens method prepared Ag nanoparticles, respectively.

We show UV-visible extinction spectra of B and G colloids in Figure 1. While a narrow surface plasmon resonance (SPR) band located at 405 nm is clearly distinguished in B colloid (curve a in Figure 1A); a rather broad SPR band with a maximum at around 440 nm is characteristic for G colloid (curve a in Figure 1B). According to the previously published literature data (*1*, *2*, *5*, *28*) and on the basis of our DLS measurements (Table 1), the mean size of nanoparticles present in B colloid is around 12 nm, while G colloid contains nanoparticles of ~86 nm in diameter. Therefore, the differences in SPR band positions observable in Figure 1 can be explained namely by different mean particle size and size distribution of Ag nanoparticles.

The addition of HCl in the final concentration of 1 x 10^{-4} M induces almost no changes in SPR band of both colloids (cuves b in Figures 1A and 1B). On the other hand, the effect of 1 x 10^{-2} M HCl concentration on both colloids is more pronounced: it results in lowering of SPR band and generation of a shoulder in the case of B colloid (curve c in 1A); while only a slight decrease of SPR band is observed in the case of G colloid (curve c in Figure 1B). The highest HCl concentration, 1 x 10^{-1} M, has lead to a complete disappearance of SPR band of B colloid (thus, no spectrum corresponding to such HCl concentration is shown in Figure 1A) and an abrupt decrease of SPR band intensity of G colloid together with a red tail in the region between 600-800 nm (curve d in Figure 1B). The decrease of SPR band intensity and occurrence of the red tail can be caused by two processes, either nanoparticles aggregation, or their fusion and mutual intergrowth (*19*). The distinction between these two processes can be made on the basis of DLS measurements (evaluating mean particle size in solution) and/or TEM imaging (determining particles morphology and size from a droplet of a dried sample) .

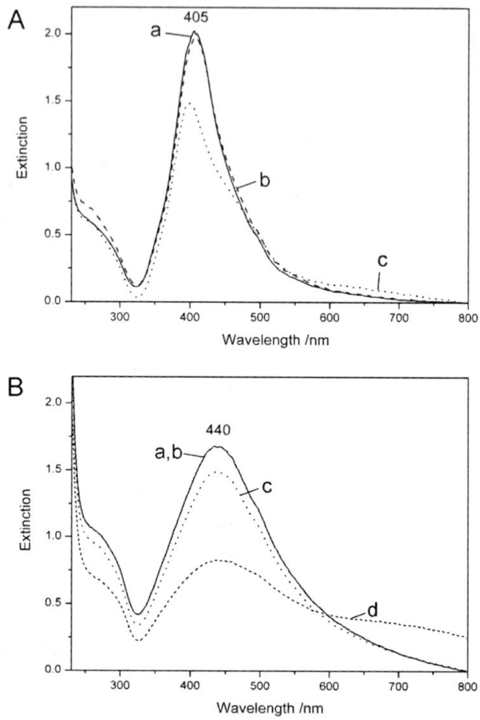

Figure 1. UV-Vis spectra of (A) borohydride-reduced and (B) modified Tollens method prepared Ag nanoparticles HCl-untreated (curves a) and treated by HCl of 1 x 10^{-4} M (curves b), 1 x 10^{-2} M (curves c), and 1 x 10^{-1} M (curves d) concentrations.

While the morphology changes observable by TEM have been previously visualized for B colloid after the treatment with HCl in ref. (*19*); TEM observation of HCl-induced changes of nanoparticles morphology in G colloid is presented in Figure 2 for the very first time. Obviously, nanoparticles in B colloid have started to fuse at HCl concentrations as low as 5 x 10^{-5} M (*19*) and created more and more compact aggregates with HCl concentration increase. This behavior has not been observed in the present study for nanoparticles of G colloid when HCl is added in the final concentration of 1 x 10^{-4} M. Although it seems that the 1 x 10^{-4} M HCl concentration has induced an aggregation of Ag nanoparticles (Figure 2B), DLS measurement has not confirmed any aggregation at all (Table 1), i.e., ~86 nm average diameter has been still encountered. Apparently, the observed agglomerates of Ag nanoparticles in Figure 2B stem from the evaporation of a droplet of G colloid placed on a TEM grid. The nanoparticle size and shape visualized in Figure 2B revealed the same features as those in the initial G colloid (Figure 2A). This observation well corroborates the UV-Vis spectra where no changes in the SPR band have been observed for Ag nanoparticles of G colloid treated by the 1 x 10^{-4} M HCl concentration (curve b in Figure 1B). The 1 x 10^{-2} M HCl concentration induced size changes of Ag nanoparticles in G colloid as evidenced indirectly by DLS (Table 1) and UV-Vis spectral (curve c in Figure 1B) measurements, directly then by TEM imaging (Figure 2C).

Table 1. Average Size (determined by DLS) and Zeta Potential Values of Ag Nanoparticles Untreated and Treated by HCl of a Particular Concentration

Sample	*Size average [nm]*	*Polydispersity*	*Zeta potential [mV]*	*Zeta deviation [mV]*
B colloid	11.6	0.11	-31.6	6.1
G colloid	85.7	0.18	-65.5	11.3
G colloid + 1 x 10^{-4} M HCl	85.1	0.17	-64.3	13.1
G colloid + 1 x 10^{-2} M HCl	95.0	0.19	-47.2	13.2
G colloid + 1 x 10^{-1} M HCl	706.1	0.26	-1.9	10.9

The addition of HCl to the final concentration of 1 x 10^{-1} M caused Ag nanoparticles dissolution as it is clearly seen in Figure 2D. The partial dissolution of the initial Ag nanoparticles in G colloid treated by the 1 x 10^{-1} M HCl concentration and possible aggregation of the just formed small nanoparticles can explain a sudden decrease of the SPR band and occurrence of the red tail in the UV-Vis spectrum (curve d in Figure 1B). The presence of nanoparticles aggregates in this particular sample is further confirmed by DLS measurements (Table 1). Elemental composition of the nanoparticle aggregates in G colloid treated by the 1 x 10^{-1} M HCl concentration has been also measured and the results are presented in Figure 3. Apparently, elements such as silver, chloride, carbon (from glucose and gluconic/glucuronic acids, partially from the holey carbon TEM grid), oxygen (exactly from the same compounds as in the case of carbon, plus from borates), sodium (from borates), and copper (from a TEM grid) dominate the EDX spectrum (Figure 3).

Figure 4 shows SERRS spectral signal of a watersoluble cationic porphyrin (TMPyP) in the same final concentration value introduced to B and/or G colloids untreated as well as treated by HCl. Intentionally, we present the SERRS spectra without any smoothing and baseline correction. Obviously, the SERRS spectral signals of the same porphyrin qualitatively differ on Ag nanoparticles in B (Figure 4A) and G (Figure 4B) colloids. While so-called metalated forms of the porphyrin (Ag^+-TMPyP and Ag^0-TMPyP) (*29*, *30*) are detected in HCl-untreated and HCl-treated B colloid (Figure 4A); free-base (*5*, *30*) and diacidic (*31*) forms of the same porphyrin are distinguished in HCl-untreated and HCl-treated G colloid, respectively (Figure 4B). For the sake of clarity, free-base porphyrin is recognized due to the presence of the bands located at around 335, 965, 1005, 1336, 1353, 1555 cm^{-1} (*5*, *30*). On the contrary, the metalation markers appear at around 398, 1017, 1343 and 1544 cm^{-1} for Ag^+-TMPyP (*5*, *29*, *30*). The other metalated form, Ag^0-TMPyP, can be distinguished by the bands placed at ~386, 1366, 1568 cm^{-1} (*30*). The formation of diacidic form of TMPyP is characterized by a pronounced relative intensity increase of the bands located in the region of higher wavenumbers (*31*); simultaneously, by generating low pH values, the positions of several porphine-core-sensitive bands are shifted to 1544, 1371, 332 cm^{-1} – as it is summarized for pH values between 0.2 – 1.2 in Table III in ref. (*31*).

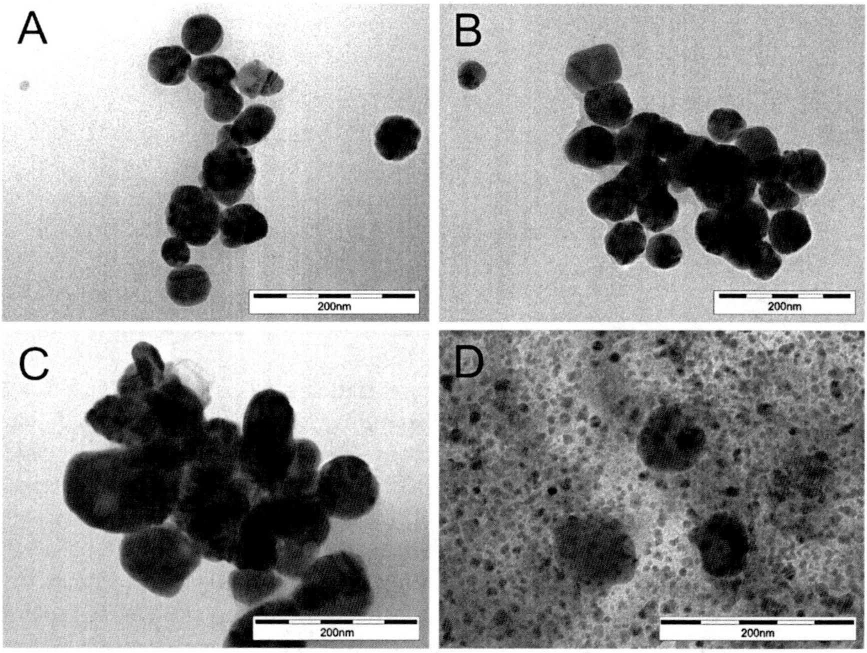

Figure 2. TEM images of (A) G colloid (i.e., modified Tollens method prepared Ag nanoparticles) and the same colloid influenced by (B) 1 x 10^{-4} M HCl, (C) 1 x 10^{-2} M HCl, and (D) 1 x 10^{-1} M HCl.

Furthermore, from the mutual comparison of the SERRS spectra of the porphyrin in HCl-untreated and HCl-treated B colloid (Figure 4A), it can be concluded that the SERRS signal diminishes in the overall intensity with an increasing HCl concentration. Simultaneously, the peak positioned at 243 cm^{-1} appears and starts to dominate to the spectrum when the 1 x 10^{-2} M HCl concentration is used. This peak can be attributed to Ag-Cl bond (*15*, *32*). Therefore, we hypothesize that AgCl is present on the surface of Ag nanoparticles of B colloid treated by the 1 x 10^{-2} M HCl concentration. AgCl is also detected in B colloid treated by the 1 x 10^{-4} M HCl concentration, however, to a much lower extent (Figure 4A). The presence of AgCl can hinder the porphyrin to approach closer to the Ag nanoparticle surface resulting thus in the porphyrin SERRS signal decrease.

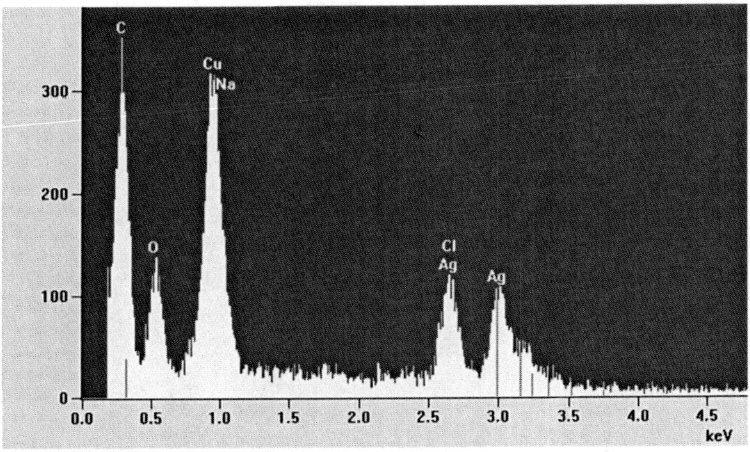

Figure 3. Elemental composition of G colloid treated by the 1 x 10^{-1} M HCl determined on the basis of EDX analysis.

The situation is totally different in the case of G colloid (Figure 4B): (i) no 243 cm^{-1} peak is observed (thus, no AgCl present) in the HCl-treated G colloid; (ii) the overall SERRS spectral intensity of the porphyrin significantly increases going from HCl-untreated G colloid to that treated by the 1 x 10^{-4} M HCl concentration, but then it drops down when the 1 x 10^{-2} M HCl concentration is present. Considering the carbonaceous layer around Ag nanoparticles of G colloid as evidenced and described in details in ref. (*5*) and taking into account the zeta potential values measured for HCl-treated G colloid (Table 1), it can be concluded that electrostatic interaction between the cationic porphyrin and negative charges surrounding Ag nanoparticles in G colloid is most probably responsible for the observed intensity fluctuations.

For the sake of clarity, we have drawn a schematic depiction of HCl effect on the two types of Ag nanoparticles under study in Figure 5. According to our results, Ag nanoparticles generated by borohydride reduction possess an electric bilayer created by borate and silver ions. When HCl is added and disproportionates in aqueous dispersions of Ag nanoparticles, borate anions can be quite easily replaced by chlorides until an equilibrium is reached. Since chlorides interact with silver cations forming AgCl, which is an insoluble species in aqueous solutions, this compound is detected in the SERRS spectra of the porphyrin. We imagine that there are islands of AgCl and Ag^0 places simultaneously present on Ag nanoparticle surface in B colloid treated with HCl. It should be mentioned that AgCl presence was not observed in ref. (*19*) because the SERS/SERRS signal was not measured in the full range between 50-1800 cm^{-1} as it is the case in the present work. With an increasing concentration of HCl, AgCl starts to dominate on the surface suppressing thus Ag-TMPyP SERRS signal similarly as it has been documented in ref. (*15*). However, on the contrary to ref. (*15*), in our case we do not observe a complete disappearance of the porphyrin SERRS signal.

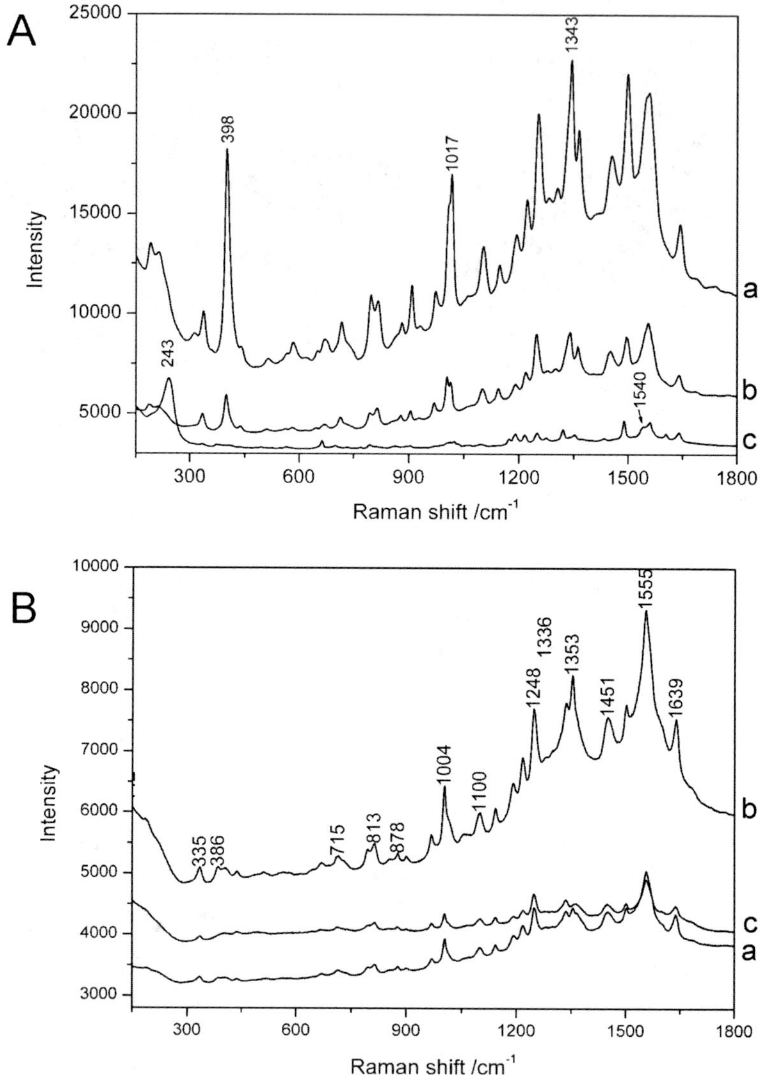

Figure 4. SERRS spectra of the 1 x 10^{-7} M TMPyP concentration in (A) borohydride-reduced and (B) modified Tollens method prepared Ag nanoparticles untreated by HCl (curves a), treated by HCl of 1 x 10^{-4} M (curves b) and/or 1 x 10^{-2} M (curves c).

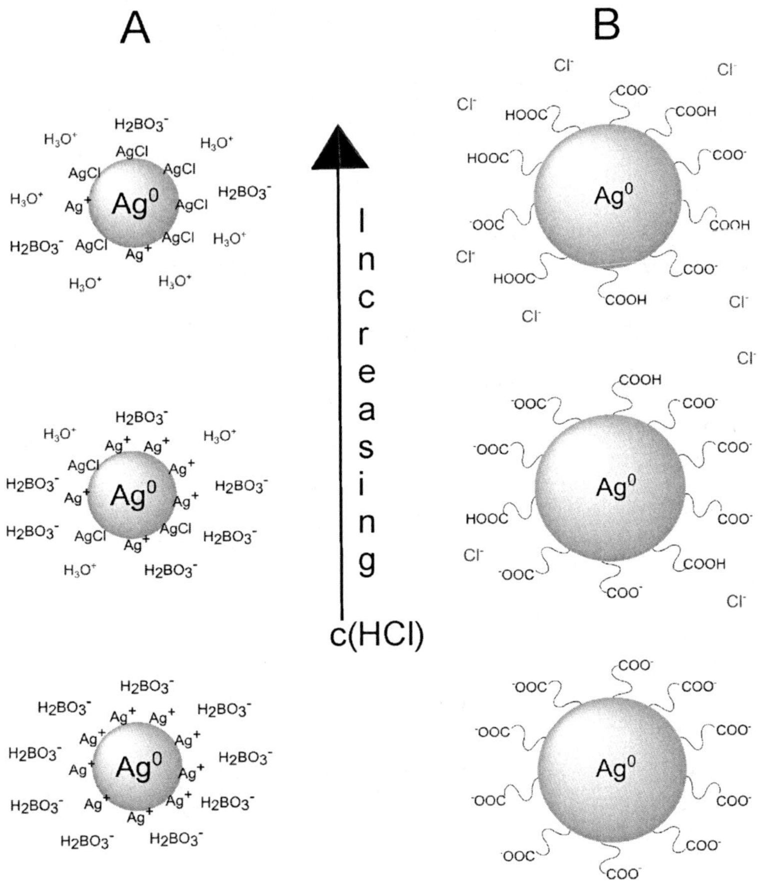

Figure 5. Schematic depiction of HCl concentration increase (in the range investigated in the present study) on (A) borohydride-reduced and (B) modified Tollens method prepared Ag nanoparticles. Schemes of two different types of Ag nanoparticles are not mutually in scale, however, for the sake of comparison, a smaller size of borohydride-reduced ones is indicated.

The addition of HCl to G colloid does not induce AgCl formation due to the fact that Ag nanoparticles in G colloid are fully covered by a carbonaceous layer consisting of gluconic and glucuronic acids (*5*) which are prone to protonization in acidic medium as it is schematically shown in Figure 5. The HCl concentration increase results in the decrease of zeta potential absolute value of Ag nanoparticles in HCl-treated G colloid which is initially of a very negative value (Table 1) and, consequently, a lower attraction of the cationic porphyrin toward Ag nanoparticle surface is reflected by the porphyrin SERRS signal decrease. The dissolution of Ag nanoparticles in G colloid treated by 1 x 10^{-1} M HCl concentration is not schematically depicted in Figure 5, however, it obviously takes place according to our results (Figures 1B and 2D) and previously published observations (*24*). Interestingly, the influence of such a high HCl concentration can be even exploited for the generation of ultrafine silver nanoparticles which have been otherwise synthesized by sonoelectrochemical methods from bulk silver substrates (*33*).

Conclusions

It has been demonstrated that Ag nanoparticles prepared by two different reductive procedures provide completely different behavior upon HCl addition. Based on our results for a commercially available porphyrin serving as a model pollutant, it can be envisaged that Ag colloids preparations can be appropriately tailored for unequivocal SERS/SERRS detection of organic species at low concentrations even in acidic conditions. These findings are very important for monitoring of the environmental contamination by various organic species.

Acknowledgments

Assoc. Prof. Jiří Bok is kindly thanked for the software enabling the precise calibration of Raman shift axis. We thank to Dr. Vlastimil Masek from the Dept. of Pharmacology, Faculty of Medicine and Dentistry, Palacky University, Olomouc, Czech Republic, for the access to Raman spectrometer purchased from Horiba Jobin Yvon. The authors gratefully acknowledge the support by the Operational Program Research and Development for Innovations – European Regional Development Fund (project CZ.1.05/2.1.00/03.0058 of the Ministry of Education, Youth and Sports of the Czech Republic). Dr Siskova also acknowledges the financial support by GAČR P108/11/P657.

References

1. Le Ru, E. C.; Etchegoin, P. G. *Principles of Surface-Enhanced Raman Spectroscopy and Related Plasmonic Effects*, 1st ed.; Elsevier: Oxford, UK, 2009.
2. Etchegoin, P. G.; Le Ru, E. C. In *Surface Enhanced Raman Spectroscopy: Analytical, Biophysical and Life Science Applications*; Schlucker, S., Ed.; Wiley-VCH Verlag & Co. KGaA: Weinheim, DE, 2011; pp 1−37.
3. Halvorson, R. A.; Vikesland, P. J. *Environ. Sci. Technol.* **2010**, *44*, 7749–7755.
4. Guerrini, L.; Leyton, P.; Campos-Vallette, M.; Domingo, C.; Garcia-Ramos, J. V.; Sanchez-Cortes, S. In *Surface Enhanced Raman Spectroscopy: Analytical, Biophysical and Life Science Applications*; Schlucker, S., Ed.; Wiley-VCH Verlag & Co. KGaA: Weinheim, DE, 2011; pp 103−128.
5. Siskova, K.; Becicka, O.; Masek, V.; Safarova, K.; Zboril, R. *J. Raman Spectrosc.* **2012**, *43*, 689–691.
6. Moskovits, M. In *Surface-Enhanced Raman Scattering*; Kneipp, K., Moskovits, M., Kneipp, H., Eds.; Springer: Berlin, DE, 2006; pp 1−17.

7. Vlckova, B.; Pavel, I.; Sladkova, M.; Siskova, K.; Slouf, M. *J. Mol. Struc.* **2007**, *834-836*, 42–47.
8. Kneipp, K.; Wang, Y.; Kneipp, H.; Perelman, L. T.; Itzkan, I.; Dasari, R. R.; Feld, M. S. *Phys. Rev. Lett.* **1997**, *78*, 1667–1670.
9. Michaels, A. M.; Jiang, J.; Brus, L. *J. Am. Chem. Soc.* **1999**, *121*, 9932.
10. Michaels, A. M.; Jiang, J.; Brus, L. *J. Phys. Chem. B* **2000**, *104*, 11965–11971.
11. Weiss, A.; Haran, G. *J. Phys. Chem. B* **2001**, *105*, 12348.
12. Doering, W. E.; Nie, S. *J.Phys.Chem. B* **2002**, *106*, 311–317.
13. Ishikawa, M.; Maruyama, Y.; Ye, J. Y.; Futamata, M. *J. Lumin.* **2002**, *98*, 81.
14. Millan, J. I.; Garcia-Ramos, J. V.; Sanchez-Cortes, S.; Rodriguez-Amaro, R. *J. Raman Spectrosc.* **2003**, *34*, 227–233.
15. Bell, S. E. J.; Sirimuthu, N. M. S. *J. Phys. Chem. A* **2005**, *109*, 7405–7410.
16. Glaspell, G. P.; Zuo, C.; Jagodzinski, P. W. *J. Cluster Sci* **2005**, *16*, 39–51.
17. Wang, Y.; Li, D.; Li, P.; Wang, W.; Ren, W.; Dong, S.; Wang, E. *J. Phys. Chem. C* **2007**, *111*, 16833–16839.
18. Futamata, M.; Maruyama, Y. *Appl. Phys. B* **2008**, *93*, 117–130.
19. Sloufova, I.; Siskova, K.; Vlckova, B.; Stepanek, J. *Phys. Chem.Chem. Phys.* **2008**, *10*, 2233–2242.
20. Impellitteri, C. A.; Tolaymat, T. M.; Scheckel, K. G. *J. Environ. Qual.* **2009**, *38*, 1528–1530.
21. Dong, X.; Gu, H.; Kang, J.; Yuan, X.; Wu, J. *Colloids Surf., A* **2010**, *368*, 142–147.
22. Prucek, R.; Panacek, A.; Fargasova, A.; Ranc, V.; Masek, V.; Kvitek, L.; Zboril, R. *CrystEngComm* **2011**, *13*, 2242–2248.
23. Espinoza, M. G.; Hinks, M. L.; Mendoza, A. M.; Pullman, D. P.; Peterson, K. I. *J. Phys. Chem. C* **2012**, *116*, 8305–8313.
24. Li, L.; Zhu, Y.-J. *J. Colloid Interface Sci.* **2006**, *303*, 415–418.
25. Hildebrandt, P.; Stockburger, M. *J. Phys. Chem.* **1984**, *88*, 5935–5944.
26. Yang, J.; Zhang, Q.; Lee, J. Y.; Too, H.-P. *J. Colloid Interface Sci.* **2007**, *308*, 157–161.
27. Izbicka, E.; Wheelhouse, R. T.; Raymond, E.; Davidson, K. K.; Lawrence, R. A.; Sun, D.; Windle, B. E.; Hurley, L. H.; Von Hoff, D. D. *Cancer Res.* **1999**, *59*, 639–644.
28. Vlckova, B.; Matejka, P.; Simonova, J.; Pancoska, P.; Cermakova, K.; Baumruk, V. *J. Phys. Chem.* **1993**, *97*, 9719.
29. Cotton, T. M.; Schultz, S. G.; Van Duyne, R. P. *J. Am. Chem. Soc.* **1982**, *104*, 6528–6532.
30. Procházka, M.; Turpin, P.-Y.; Štìpánek, J.; Vlèková, B. *J. Raman Spectrosc.* **2002**, *33*, 758–760.
31. Blom, N.; Odo, J.; Nakamoto, K.; Strommen, D. P. *J. Phys. Chem.* **1986**, *90*, 2847.
32. Nakamoto, K. *Infrared and Raman Spectra of Inorganic and Coordination Compounds, Part B, Applications in Coordination, Organometallic, and Bioinorganic Chemistry*, 6th ed.; Wiley & Sons, Inc.: Hoboken, NJ, 2009; pp 193–194, 283–286.
33. Liu, Y.-C.; Lin, L.-H. *Electrochem. Commun.* **2004**, *6*, 1163–1168.

Chapter 10

Stability and Toxicity of Silver Nanoparticles in Aquatic Environment: A Review

Virender K. Sharma*

Department of Chemistry, Florida Institute of Technology, Melbourne, Florida 32901, U.S.A.
***E-mail: vsharma@fit.edu**

The stability of silver nanoparticles (AgNPs) influences their fate, transport, and toxicity in the aquatic environment. This review discusses solutions parameters, which are largely responsible for the stability of AgNPs in biological and natural environments. The coating of chemical material onto AgNPs leads to their stabilization in which the surface charge plays a significant role. High molecular weight materials (e.g. polyvinyl pyrrolidone, PVP and natural organic matter, NOM) tend to stabilize the AgNPs. Citrate-coated AgNPs are usually stable for a few days while AgNPs in solutions of Ag^+ and humic acid (or fulvic acid) are stable for several months. Solution pH, ionic strength, and ionic constituents control the aggregation of AgNPs. Numerous studies on the toxicity of AgNPs to microorganisms, plants, and human cells have been performed, but understanding the mechanisms of toxic effects are still in progress. Release of reactive oxygen species (ROS) during dissolution and interactions of particles with surface cell components may be involved in the toxic effects of AgNPs. Progress on the applications of proteomic and toxiconomic techniques in understanding the mechanism of toxic effects of AgNPs is briefly discussed.

Introduction

In recent years, there has been an exponential growth in the production of nanoparticles (NPs) and the estimated economy of NPs will be about a trillion dollar by 2015 (*1*, *2*). NPs have a wide range of applications in industries such as textiles, household products, foods, aerospace, health care, energy, communication, and chemicals (*3*). The textile industry has recognized the importance of using NPs in enhancing the strain and water resistance and in increasing the ability of molecules to absorb dyes (*4*). Examples of NPs in the textile industry are silver nanoparticles (AgNPs), which possess antimicrobial, and antibacterial activities (*4*, *5*). The use of NPs in the environmental remediation field has also been demonstrated (*6*). The presence of NPs in the wastewater could thus be from human activities. There are also naturally occurring NPs (iron oxides and carbon black) (*7*). Among the NPs consumer products, AgNPs-containing materials represent the largest group (more than 50 %) (*8*).

The widespread use of AgNPs may have unintended effects on ecosystems and human health (*9*–*12*). For example, AgNPs could be taken by zebrafish embryos with potential negative effects at a minimum level of 1.9×10^{-10} mol L^{-1} (*13*). Currently, environmental amounts of AgNPs are not known, but concentrations between $6.7 \times 10^{-12} - 7.4 \times 10^{-10}$ mol L^{-1} have been estimated in Swiss surface water (*14*, *15*). Furthermore, no potential risks have been suggested based on the risk quotient between predicted environmental concentrations and predicted no-effect concentrations for aquatic systems (*16*). From the literature, it is not clear if AgNPs have any potential risk from the perspective of environmental health and safety (EHS). Hence, a number of programs have been initiated by agencies of different countries to assess whether the nano forms of materials including AgNPs possess any potential risks. The major focus is on the fate, transport, and health effects of NPs in the environment. Transport and toxic effects of AgNPs largely depend on their stability (i.e. aggregation state of NPs) in the environment they exist (*17*).

This chapter first summarizes the stability of AgNPs in biological and natural waters media, followed by their toxic effects to organisms and human cells. The toxic mechanism is also briefly described.

Stability

There have been several reports on the stability of AgNPs in different media. Various analytical techniques have been applied to assess the extent of aggregation in order to determine the stability of AgNPs. These techniques include dynamic light scattering (DLS), atomic force microscopy (AFM), transmission electron microscopy (TEM), nanoparticle tracking analysis (NTA), fluorescence correlation spectroscopy (FCS), and flow field fractionation (FFFF). Zeta potential and BET surface area measurements allowed understanding the role of surface charge and area in stabilization of AgNPs. Besides the surface charge, steric stabilization of AgNPs could also occur (*17*). The coating of chemical materials onto AgNPs generally determines the stability of these particles in aqueous solutions. Most studies on the stability have been performed using citrate-coated AgNPs (*18–22*). A comprehensive study of citrate-coated AgNPs in natural freshwater, stimulated estuarine waters, and synthetic seawater indicated a stability of a few days (*18*). Polymer (polyvinylpyrrolidone-PVP 360) and surfactants (sodium dodecyl sulfate-SDS and polyoxyethylenesorbitane monooleate-Tween 80) as coating agents have shown to stabilize the AgNPs (*23*, *24*).

More recently, a comprehensive study on the stability of citrate, PVP, and PEG (polyethylene glycol) coated AgNPs has been performed in ecotoxicological media. This to understand the dynamics of exposure with respect to a change in concentration of media and the possible alteration in properties of AgNPs during the study period of the toxicity tests (*25*). Citrate coated particles are considered charged stabilized NPs while PVP and PEG are sterically stabilized NPs (*25*). The media tested was standard OECD (Organization for Economic Co-operation and Development) for *Daphnia sp* for acute and chronic studies. Results for the stability of AgNPs, monitored for 21 days are presented in Figures 1 and 2 (*25*). The studied time period is equivalent to the time period for the toxicity tests for OECD chronic daphnia. The standard OECD was donated as CM-1 and dilution of the media by 10 times was donated as CM-10. When chloride was replaced with nitrate and sulfate at the same ionic strength, terms of NM-1 and NM-10 or SM-1 and SM-10 were used (Figures 1 and 2). The concentrations of AgNPs were 2.4, 4.8, and 3.0 mg L^{-1} for citrate, PVP, and PEG coated particles, respectively. The surface plasmon resonance (SPR) of citrate-coated AgNPs showed a decrease with time for all media (Figure 1). Total loss of SPR peaks was observed between 0-3 days in all concentrated media (CM-1, NM-1, and SM-1) (Figure 1). The decrease in intensity of SPR peaks of citrate-coated AgNPs in CM-10 suspensions was approximately 40 % in 21 days. NM-10 and SM-10 also showed a continuous decrease and a shift to the longer wavelength was observed. Results of Figure 1 suggest that aggregation of citrate stabilized AgNPs would occur during the toxicity test period and data of the dose-response relationship need careful investigation.

Significant decreases in peaks were found for PEG stabilized AgNPS in all media over 21 days (*25*). The diluted media showed less decrease compared to the concentrated media. Decreases were 30 % and 40 % in CM-10 and CM-1 media, respectively. Complete loss of SPR peaks in NM-1 media was observed. PVP stabilized AgNPs were stable in all media even at high ionic strength (Figure 2). In chloride media, no loss in absorbance due to AgNPs occurred while minor decreases in peak areas (8-13 %) were determined in all chloride free media (Figure 2). The results suggest that PVP-coated AgNPs should be the choice for conducting chronic exposure studies (*25*).

Recently, the stability of carbonate-coated AgNPs in freshwater was studied to gain understanding of the effect of pH, ionic strength, and humic substances in freshwater systems (*26*). The results are shown in Figure 3. In the pH experiments, no change in absorbance was noticed at pH 7.0 and 12.0, however, some change occurred at pH 5 (Figure 3A). Significantly, a large decrease in the intensity of the absorbance at pH 4.0 was seen with a shift of the absorbance peak to longer wavelength. High negative Zeta potentials (ZP) were reported between pH 5.0 and 12.0. Comparatively, the ZP approached zero below pH 5.0 and had even a positive value at pH 2.0. The agglomeration at pH 4.0 may thus be causing the instability of AgNPs. The presence of Ca^{2+} caused destabilization of the AgNPs (Figure 3B). Zeta potentials of the solutions suggested that the surface charge played a role in the destabilization of AgNPs in the presence of cations. Humic acids (HA) in solutions showed no significant influence on the stability of carbonate-coated AgNPs (Figure 3C). The stability of carbonate-coated AgNPs in natural water has also been examined (Figure 3D). The natural water had a pH of 7.8 and contained 0.84 mM Na^{+}, 0.11 mM K^{+}, 2.59 mM Ca^{2+}, 0.62 mM Mg^{2+}, 0.97 mM? Cl^{-}, 0.36 mM NO_3^{-}, 0.25 mM SO_4^{2-} (ionic strength = 8.05 mM), and 3.72 mg C/L dissolved organic carbon). The spectra of Figure 3D indicate that the agglomeration of AgNPs happened within a few days. This is consistent with previously reported studies in simulated estuarine waters, synthetic seawater, and pond water (*18*, *21*, *27*).

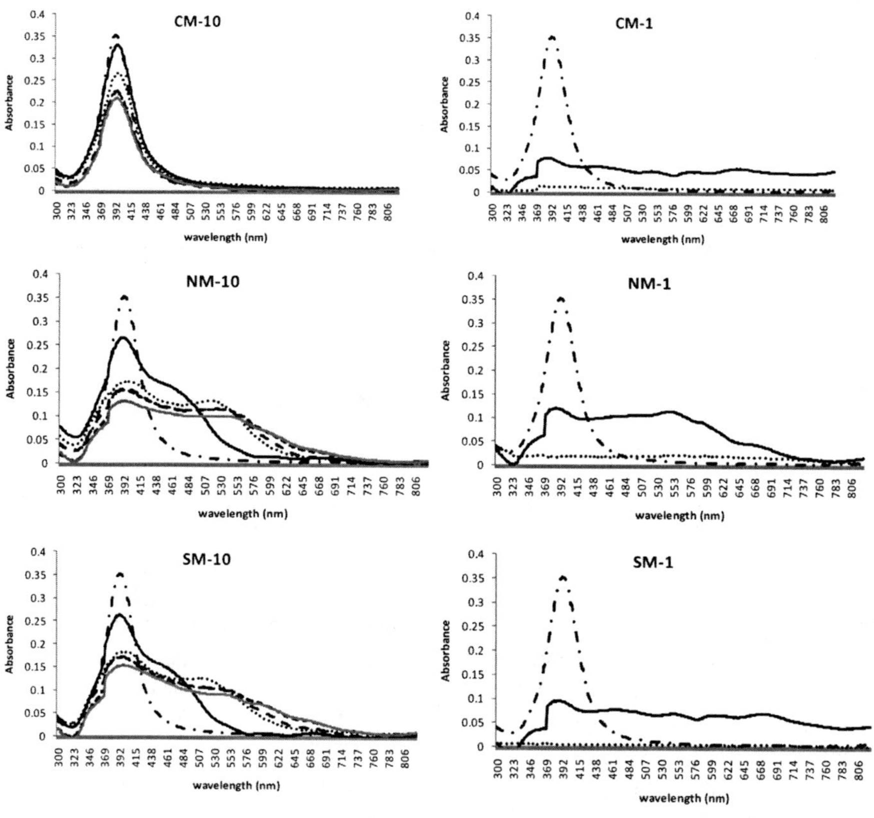

Figure 1. SPR stability of citrate coated AgNPs over 21 days in diluted media and within 3 days in concentrated media (– ■ stock solutions, – 0 day, ····· 3 days, — 7 days, ■■■■■ 14 days, (gray line) 21 days). SPR stability of citrate coated AgNPs over 21 days in diluted media and within 3 days in concentrated media (– ■ stock solutions, – 0 day, ····· 3 days, — 7 days, ■■■■■ 14 days, (gray line) 21 days). (Adapted from (25) with the permission of the American Chemical Society).

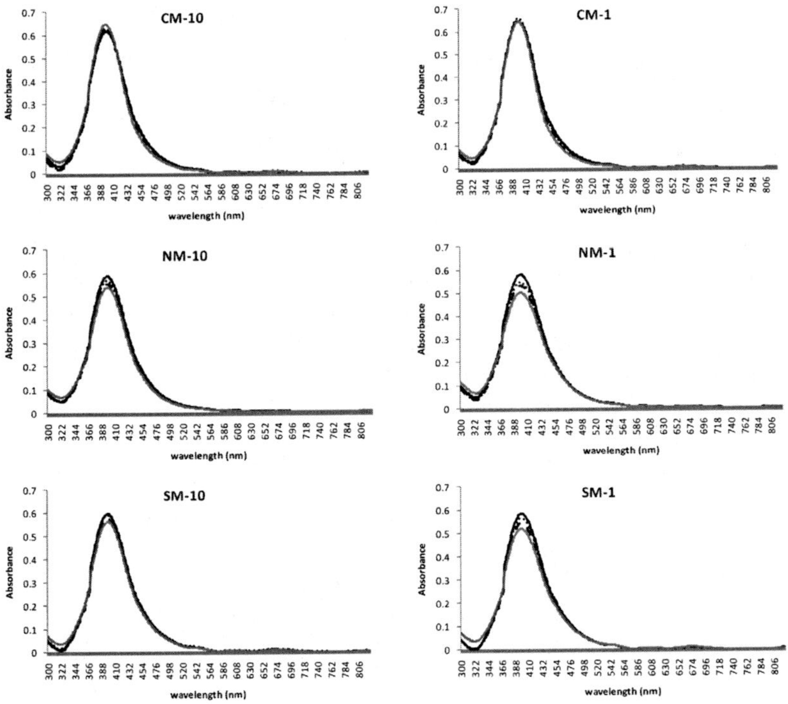

Figure 2. SPR stability of PVP coated AgNPs up to 21 days in media (– 0 day, ···· 7 days, – – 13 days, (gray line) 21 days). (Adapted from (25) with the permission of the American Chemical Society).

A recent study in our laboratory suggested that the AgNPs, formed from reductions of Ag^+ ions by HA under various environmental conditions, were highly stable (Figure 4) (*28*). AgNPs formed using Lake Delancy (LD), Florida freshwater sediments, showed only a 7 % decrease in the SPR peak intensity over two months (Figure 4). However, the peak was blue shifted from 423 to 410 nm during this time period. Similar stability was observed in AgNPs, obtained from the reaction of Ag^+ with HA from the Sebastian River, Florida (*28*). Comparatively, a 25 % decrease in the SPR peak intensity was determined in AgNPs formed from the interaction of Ag^+ with Suwannee River HA (*28*). A more recent study on the stability was performed on AgNPs in a solution mixture of Ag^+ and Fulvic acid (FA) at pH 8.0 (Figure 5). This study was conducted over several months and AgNPs were obtained by reactions at room temperature (AgNPs-RT) and at 90 °C (AgNPs-90) (*29*). AgNPs-RT had no significant changes in 48 days after their formation and a decrease in absorbance of the SPR peak was only ~15%. Furthermore, no broadening of the SPR peak was seen during this period of time (Figure 5A). However, AgNP-90 had continuous broadening of the SPR peak as a function of time (Figure 5B). Thedifference in stability of AgNPs-RT and AgNPs-90 may be related to the surface passivation of the particles and their modes of binding FA. A significant similar percentage decrease in AgNPs-RT and AgNPs-90 was seen after 210 days and 180 days, respectively (Figure 5). The stability of AgNPs, shown in Figures 4 and 5, clearly demonstrates the potential for the direct formation of AgNPs under environmental conditions such as thermal vents or hot springs, silver mine tailings, and photographic wastewater (*30*). Once these naturally formed AgNPs exist in the environment, they can be transported for long distances from their locations of origin.

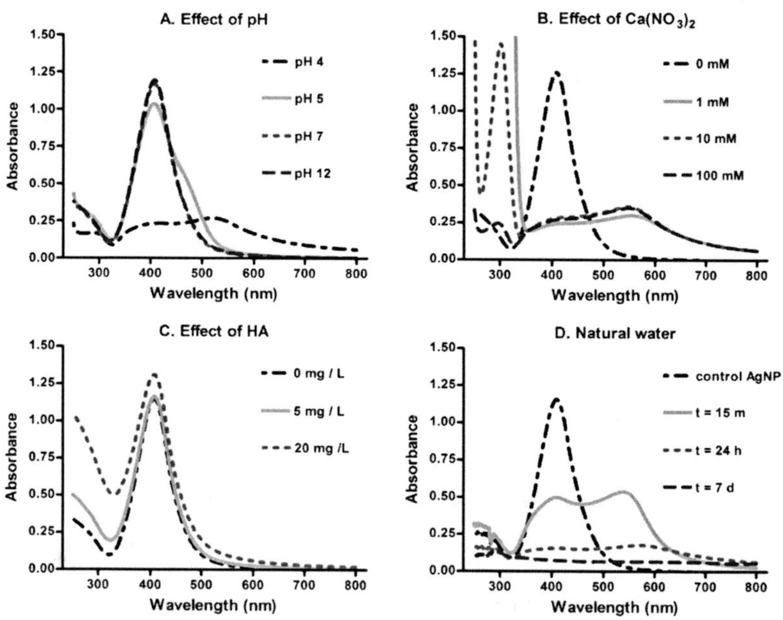

Figure 3. UV–vis absorbance of 100 μM AgNP in 10 mM MOPS buffer as function of pH (A), of $Ca(NO_3)_2$ concentration (B), and of humic acids (C), and over time in Chriesbach water (D). (Adapted from (26) with the permission of the American Chemical Society).

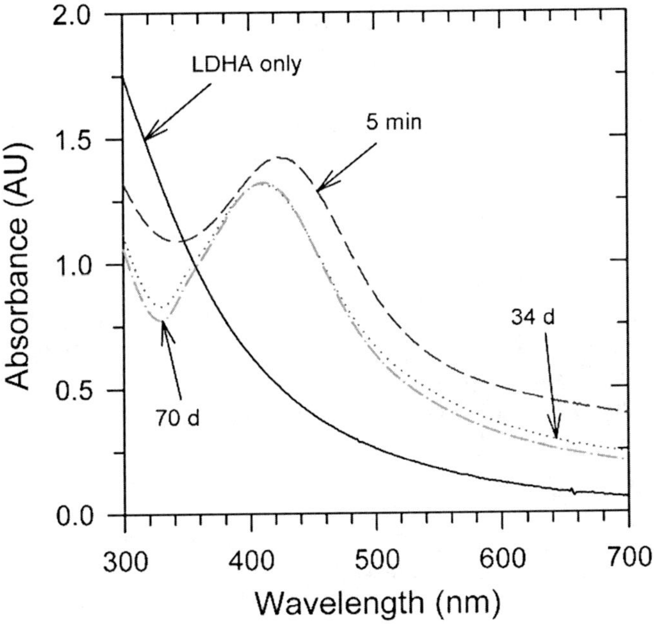

Figure 4. Stability studies of UV-vis absorption spectra of AgNPs following reduction of silver nitrate solution with humic acid. Absorption spectra of the AgNPs were acquired 5 min, 34 min, and 70 days after formation of the AgNPs. (Adapted from (28) with the permission of the American Chemical Society).

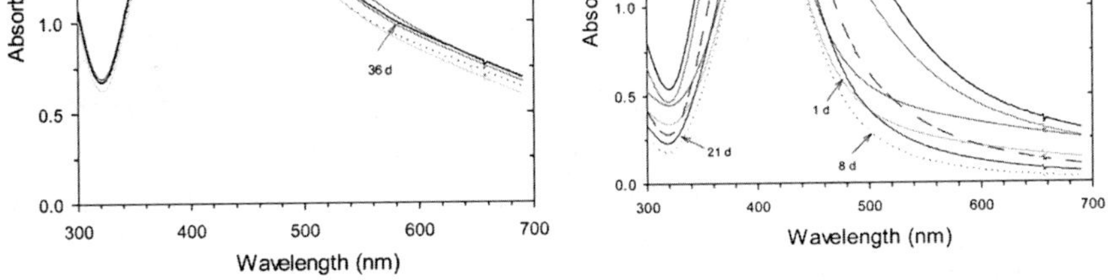

Figure 5. UV-Vis measurements of ageing of AgNPs prepared at RT (A) and 90 °C (B) following reduction of Ag^+ ion by SRFA. (Conditions: $[Ag^+] = 1 \times 10^{-3}$ mol L^{-1}, 100 mg L^{-1} SRFA; pH 8.0. (Days in (A) present time after initial mixing of Ag^+-SRFA while days in (B) represents time after forming of AgNPs at 90 °C in 2 hours). (Adapted from (29) with the permission of the American Chemical Society).

Toxicity

A number of studies using different organisms have been carried out to determine the toxicity of AgNPs (*17*). Organisms studied include human cells, terrestrial and aquatic algae, fungi, and plants as well as vertebrates and invertebrates microorganisms (*17*, *26*, *31–36*). The mechanism of toxicity involves the oxidative damage to the cells in which reactive oxygen species (ROS) are largely responsible for the toxic effects (*12*). The electrostatic interactions and interaction with the thiol groups localized in the cell wall proteins may also be responsible for toxicity of AgNPs. Numerous gene-expression techniques and biomarkers have been applied to better understand the mechanism of toxicity. Below are a few examples to demonstrate advances made in the last few years.

The toxic effects of AgNPs on reference and isolated strains of bacteria as well as on the variety of human and biological organisms have been investigated (*17*, *23*, *37*, *38*). The *in vitro* experiments with AgNPs demonstrated a decrease in the antibacterial activity with an increase in the particle diameter (*23*). The minimum inhibitory concentrations (MIC) of different AgNPs against the tested bacteria and yeasts are presented in Table 1 (*23*). Generally, most of the modified AgNPs showed lower MICs than the unmodified AgNPs. The ranges of MICs for bacterial strains were 1.69 – 3.38 mg L^{-1} and 3.38 – 6.75 mg L^{-1} for modified and unmodified AgNPs, respectively. Similar results of MICs were observed for yeasts in which ranges were 0.052 – 0.84 mg L^{-1} and 0.42 – 1.69 mg L^{-1} for modified and unmodified AgNPs, respectively. A comparison between Ag^+ ion and AgNPs with respect to antibacterial activity and toxicity was made (*23*). In this study, AgNPs were prepared by a modified Tollens process in which reduction of $[Ag(NH_3)_2]^+$ was done by D-maltose (*34*). Ag^+ ions inhibited the bacterial and the yeast growth at 1 mg L^{-1}, which was comparable with the inhibition concentration of AgNPs (1 – 3 mg L^{-1}). In the cases of human fibroblasts, *Monorapaphidium sp., Drosophila melanogaster*, and *Paramecium caudatum*, the LC_{100} were at least 20 times lower for Ag^+ than those for AgNPs. The range for Ag^+ was 0.4 – 3 mg L^{-1} while 30 – 60 mg L^{-1} was the range for AgNPs.

Table 1. Minimum Inhibitory Concentrations (MICs) of the Non-Stabilized and Stabilized Silver NPs against the Tested Bacteria and Yeasts. (Adapted from (*23*) with the permission of IOP Sciences).

Tested strain	*Minimum inhibitory concentrations (mg/L Ag)*				
	Ag^+	*Non-stabilized AgNPs*	*Stabilized AgNPs SDS*	*Tween 80*	*PVP 360*
Staphylococcus aureus CCM 3953	1.69	3.38	1.69	3.38	3.38
Pseudomonas aeruginosa	0.84	3.38	3.38	1.69	1.69
Staphylococcus aureus MRSA	0.84	3.38	1.69	3.38	1.69
Enterococcus faecium VRE	1.69	6.75	3.38	3.38	3.38
Candida albicans I	0.42	0.42	0.052	0.1	0.1
Candida parapsilosis	1.69	1.69	0.84	0.84	0.84
Candida tropicalis	0.84	0.84	0.42	0.42	0.42

An investigation on the charge-dependent toxicity of AgNPs was performed using highly negative to highly positive charged AgNPs (*8*). The particles studied were uncoated (H_2-AgNPs, hydrodynamic diameter, HDD =18 nm, and zeta (ζ) potential= -22 mV), citrate-coated (Cit-AgNPs, HDD = 10 nm, and ζ = -38 mV), PVP-coated (PVP-AgNPs, HDD = 12 nm, and ζ = -10 mV), and branched polyethyleimine coated (BPEI-AgNPs, HDD = 18 nm, and ζ = -22 mV). The tested microbe was a *bacillus* species and the oxygen consumption as well as the percent live bacteria were monitored (Figure 6). H_2-AgNPs with the largest HDD showed the lowest toxicity compared to other AgNPs. BPEI-AgNPs with HDD of 10 nm had the highest toxicity. Cit-AgNPs and BPEI-AgNPs possessing a similar HDD showed two very different toxicity profiles (Figure 6). The results of Figure 6 clearly demonstrate that the particle size was not the only parameter, but the surface charge was also influencing the toxicity of AgNPs. A correlation between the toxicity and the surface charges was observed (Figures 6A and 6B). The increasing order of toxicity was Cit-AgNPs < H_2-AgNPs < PVP-AgNPs < BPEI-AgNPs, which follows the order of surface charge from more negatively charged Cit-AgNPs to more positively charged BPEI-AgNPs (Figure 6A and 6B). The Cit-AgNPs had a negative ζ of -38 mV, which was similar to the -37 mV ζ potential for the *bacillus* species; thus reducing the toxicity. The positively charged BPEI-AgNPs was thus expected to show the highest toxicity (Figure 6). It should be pointed out that relationships shown in Figure 6 are strictly for the *bacillus* species and several additional physical and chemical interactions may also be involved in the toxicity of AgNPs against other species. For example, levels of molecular oxygen and the nature of binding ligands to Ag^+ have shown to affect the antibacterial activity of AgNPs (*39*).

In the last few years, studies using proteomic and genomic approaches are forthcoming to understand the ecotoxicological impact of AgNPs (*40–44*). A proteomic study on the interaction of *Escherichia coli* with bare AgNPs (bAgNPs) and carbonate-coated AgNPs (cAgNPs) identified a number of proteins that could bind to nanoparticles (*40*). Identified proteins were mostly enzymes (65 %) and membrane porins, chaperones, or periplasmic peptide-binding proteins which were nonenzyme proteins. Significantly, tryptophanase (TNase), which is present in low abundance in *E. coli* had a high affinity to AgNPs, which was confirmed by matrix-assisted laser desorption/ionization time-of-light (MALDI-TOF). This interaction of protein-AgNPs influenced greatly the enzymatic activity of TNase (Figure 7). Both kinds of AgNPs showed inhibitory effects on the enzyme activity and decreases in activities were 90 % and 50 % for bAgNPs and cAgNPs, respectively. Binding of proteins to AgNPs may be a potential mechanism for the biocidal effect of AgNPs (*44*).

The evaluation of *in vitro* toxicity of AgNPs, polystyrene nanoparticles, and $AgCO_3$ against the human hepatoma cell line HepG2 has been performed (*41*). The results of cell viability after 24 h exposure of tested materials are presented in Figure 8. All tested materials at a dose up to 0.5 mg L^{-1} had no significant toxicity (Figure 8). A significant cytotoxicity was seen with AgNPs at higher doses (> 1.0 mg L^{-1}) while exposure to other materials did not have any toxic effects. This study also showed that exposure of AgNPs at a dose of 1 mg L^{-1} induced abnormal cellular morphology. Other changes included cellular shrinkage and acquisition of an irregular shape (*41*).

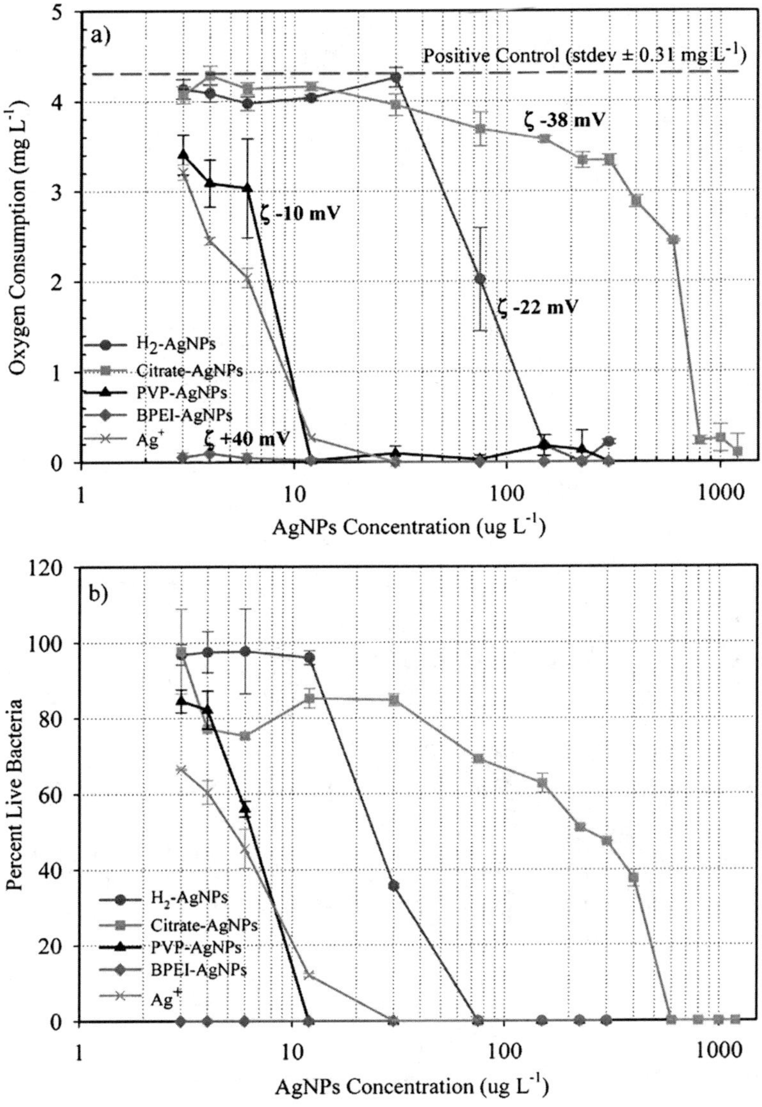

Figure 6. AgNPs toxicity shows correlation with surface charge. (a) Oxygen consumption after 5 days. (b) Percent live bacteria after 5 days. (Adapted from (8) with the permission of the American Chemical Society).

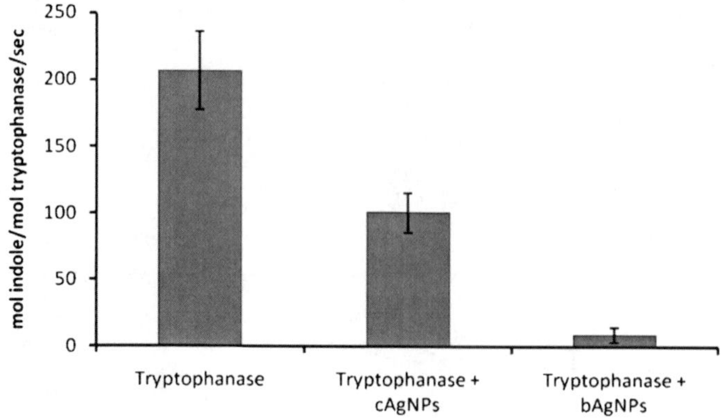

Figure 7. Enzymatic activity of TNase in the absence of AgNPs and after incubation with cAgNPs and bAgNPs. (Adapted from (40) with the permission of the American Chemical Society).

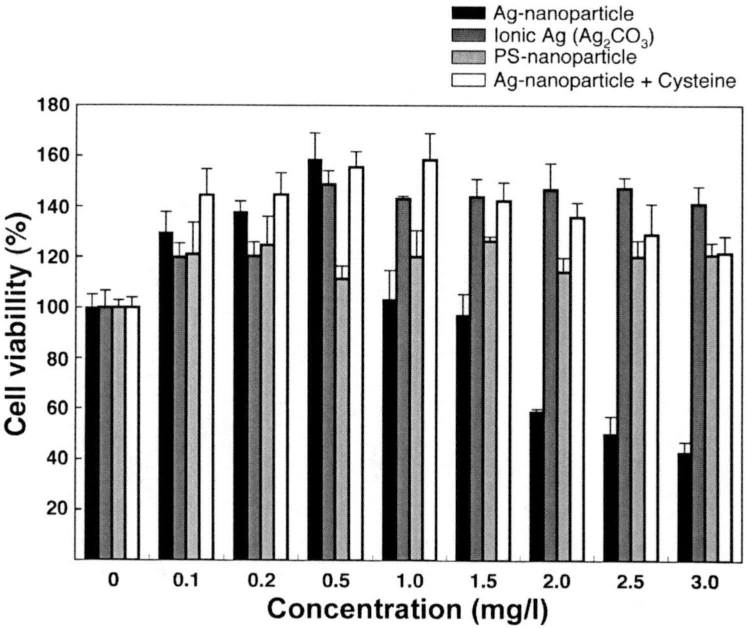

Figure 8. Effects of three test material exposures on cell viability of HepG2 cells. Cells were exposed to different concentrations of Ag-nanoparticles, PS-nanoparticles and Ag_2CO_3 for 24 h. Cell viabilities were determined by the neutral red uptake assay. The data are expressed as mean ±SD of eight wells of a cell culture plate. The concentrations of Ag_2CO_3 are expressed as the amounts of ionic silver (Ag^+). (Adapted from (41) with the permission of the American Chemical Society).

Conclusions

Solution parameters that need to be considered in understanding the stability of AgNPs and addressing their toxicity are reviewed. The surface charge appears to play a crucial role in the stabilization of AgNPs against aggregation and thus in influencing the toxic effects of AgNPs. Citrate-coated engineered AgNPs are likely to aggregate in a few days. Comparatively, high molecular weight compounds including polymers and natural organic matter coating stabilize the AgNPs to a large extend. The humic and fulvic acids coated AgNPs are stable for several months and are likely to be transported over a great distance from their points of origin. Besides coatings, solution parameters such as pH, ionic strength, and concentrations of organic matter also determine the stability of AgNPs. Studies on the influence of solar light on the stability of engineered AgNPs as well as naturally formed AgNPs are forthcoming and future studies include characterizing AgNP transformations under the conditions of solar light. Toxic studies of AgNPs against organisms suggest important roles of surface charges (i.e. zeta potentials), interactions of the components of organisms, and the surfaces of the particles. Proteomic and genomic studies would enhance understanding the mechanism of the toxicity of AgNPs. A very few studies on the toxic effect of engineered AgNPs in the presence of natural organic matter have been performed and future work may include learning any adverse effects of natural organic matter-coated AgNPs on organisms.

Acknowledgments

Author wishes to thank Dr. Ria Yngard for her useful comments on the original version of the manuscript.

References

1. Project on Emerging Nanotechnologies. Woodrow Wilson International Center for Scholars, 2010. www.nanotechproject.org/inventories/consumer/.
2. NSF and EPA Establish Two Centers for Environmental Implications of Nanotechnology. National Science Foundation Press Release 08-155. http://www.nsf.gov/news/news_summ.jsp?cntn_id=112234 (accessed November 16, 2010).
3. Dallas, P.; Sharma, V. K.; Zboril, R. *Adv. Colloid Interface Sci.* **2011**, *166*, 119–135.
4. Rezić, I. *TrAC, Trends Anal. Chem.* **2011**, *30*, 1159–1167.
5. Sharma, V. K.; Yngard, R. A.; Lin, Y. *Adv. Colloid Interface Sci.* **2009**, *145*, 83–96.
6. Sánchez, A.; Recillas, S.; Font, X.; Casals, E.; González, E.; Puntes, V. *TrAC, Trends Anal. Chem.* **2011**, *30*, 507–516.
7. Silva, B. F. D.; Pérez, S.; Gardinalli, P.; Singhal, R. K.; Mozeto, A. A.; Barceló, D. *TrAC, Trends Anal. Chem.* **2011**, *30*, 528–540.
8. El Badawy, A. M.; Luxton, T. P.; Silva, R. G.; Scheckel, K. G.; Suidan, M. T.; Tolaymat, T. M. *Environ. Sci. Technol.* **2010**, *44*, 1260–1266.

9. Fabrega, J.; Luoma, S. N.; Tyler, C. R.; Galloway, T. S.; Lead, J. R. *Environ. Int.* **2011**, *37*, 517–531.
10. Handy, R. D.; Von Der Kammer, F.; Lead, J. R.; Hassellöv, M.; Owen, R.; Crane, M. *Ecotoxicology* **2008**, *17*, 287–314.
11. Klaine, S. J.; Alvarez, P. J. J.; Batley, G. E.; Fernandes, T. F.; Handy, R. D.; Lyon, D. Y.; Mahendra, S.; McLaughlin, M. J.; Lead, J. R. *Environ. Toxicol. Chem.* **2008**, *27*, 1825–1851.
12. Lapresta-Fernández, A.; Fernández, A.; Blasco, J. *TrAC, Trends Anal. Chem.* **2012**, *32*, 40–59.
13. Lee, K. J.; Browning, L. M.; Nallathamby, P. D.; Desai, T.; Cherukuri, P. K.; Xu, X. -. N. *Chem. Res. Toxicol.* **2012**, *25*, 1029–1046.
14. Mueller, N. C.; Nowack, B. *Environ. Sci. Technol.* **2008**, *42*, 4447–4453.
15. Gottschalk, F.; Sonderer, T.; Scholz, R. W.; Nowack, B. *Environ. Toxicol. Chem.* **2010**, *29*, 1036–1048.
16. Gottschalk, F.; Nowack, B. *J. Environ. Monit.* **2011**, *13*, 1145–1155.
17. Levard, C.; Hotze, E. M.; Lowry, G. V.; Brown, J. G. E. *Environ. Sci. Technol.* **2012**, *46*, 6900–6914.
18. Chinnapongse, S. L.; MacCuspie, R. I.; Hackley, V. A. *Sci. Total Environ.* **2011**, *409*, 2443–2450.
19. Zook, J. M.; Long, S. E.; Cleveland, D.; Geronimo, C. L. A.; MacCuspie, R. I. *Anal. Bioanal. Chem.* **2011**, *401*, 1993–2002.
20. Zook, J. M.; MacCuspie, R. I.; Locascio, L. E.; Halter, M. D.; Elliott, J. T. *Nanotoxicology* **2011**, *5*, 517–530.
21. MacCuspie, R. I. *J. Nanopart. Res.* **2011**, *13*, 2893–2908.
22. MacCuspie, R. I.; Allen, A. J.; Hackley, V. A. *Nanotoxicology* **2011**, *5*, 140–156.
23. Kvitek, L.; Panacek, A.; Prucek, R.; Soukupova, J.; Vanickova, M.; Kolar, M.; Zboril, R. *J. Phys.: Conf. Ser.* **2011**, *304*, 012029.
24. Kvítek, L.; Panáček, A.; Soukupová, J.; Kolář, M.; Večeřová, R.; Prucek, R.; Holecová, M.; Zbořil, R. *J. Phys. Chem. C* **2008**, *112*, 5825–5834.
25. Tejamaya, M.; Romer, I.; Merrifield, R. C.; Lead, J. R. *Environ. Sci. Technol.* **2012**, *46*, 7011–7017.
26. Piccapietra, F.; Sigg, L.; Behra, R. *Environ. Sci. Technol.* **2012**, *46*, 818–825.
27. Stebounova, L. V.; Guio, E.; Grassian, V. H. *J. Nanopart. Res.* **2011**, *13*, 233–244.
28. Akaighe, N.; MacCuspie, R. I.; Navarro, D. A.; Aga, D. S.; Banerjee, S.; Sohn, M.; Sharma, V. K. *Environ. Sci. Technol.* **2011**, *45*, 3895–3901.
29. Adegboyega, N. F.; Sharma, V. K.; Siskova, K.; Zbořil, R.; Sohn, M.; Banerjee, S. *Environ. Sci. Technol.* **2012**Submitted.
30. Liu, J.; Sonshine, D. A.; Shervani, S.; Hurt, R. H. *ACS Nano* **2010**, *4*, 6903–6913.
31. Pokhrel, L. R.; Dubey, B. *Environ. Sci. Technol.* **2012**, *46*, 7755–7762.
32. Hayashi, Y.; Engelmann, P.; Foldbjerg, R.; Szabó, M.; Somogyi, I.; Pollák, E.; Molnár, L.; Autrup, H.; Sutherland, D. S.; Scott-Fordsmand, J.; Heckmann, L. *Environ. Sci. Technol.* **2012**, *46*, 4166–4173.
33. Panáček, A.; Kolář, M.; Večeřová, R.; Prucek, R.; Soukupová, J.; Kryštof, V.; Hamal, P.; Zbořil, R.; Kvítek, L. *Biomaterials* **2009**, *30*, 6333–6340.

34. Panáček, A.; Kvítek, L.; Prucek, R.; Kolář, M.; Večeřová, R.; Pizúrová, N.; Sharma, V. K.; Nevěčná, T.; Zbořil, R. *J Phys Chem. B* **2006**, *110*, 16248–16253.
35. Kvitek, L.; Vanickova, M.; Panacek, A.; Soukupov, J.; Dittrich, M.; Valentova, E.; Prucek, R.; Bancirova, M.; Milde, D.; Zboril, R. *J. Phys. Chem. C* **2009**, *113*, 4296–4300.
36. Reinsch, B. C.; Levard, C.; Li, Z.; Ma, R.; Wise, A.; Gregory, K. B.; Brown, J. G. E.; Lowry, G. V. *Environ. Sci. Technol.* **2012**, *46*, 6992–7000.
37. Panacek, A.; Prucek, R.; Safarova, D.; Dittrich, M.; Richtrova, J.; Benickova, K.; Zboril, R.; Kvitek, L. *Environ. Sci. Technol.* **2012**, *45*, 4974–4979.
38. García-Alonso, J.; Khan, F. R.; Misra, S. K.; Turmaine, M.; Smith, B. D.; Rainbow, P. S.; Luoma, S. N.; Valsami-Jones, E. *Environ. Sci. Technol.* **2012**, *45*, 4630–4636.
39. Xiu, Z. M.; Ma, J.; Alvarez, P. J. J. *Environ. Sci. Technol.* **2011**, *45*, 9003–9008.
40. Wigginton, N. S.; De Titta, A.; Piccapietra, F.; Dobias, J. A. N.; Nesatyy, V. J.; Suter, M. J. F.; Bernier-Latmani, R. *Environ. Sci. Technol.* **2010**, *44*, 2163–2168.
41. Kawata, K.; Osawa, M.; Okabe, S. *Environ. Sci. Technol.* **2009**, *43*, 6046–6051.
42. Lim, D.; Roh, J.-Y.; Eom, H.-J.; Choi, J.-Y.; Hyun, J.; Choi, J. *Environ. Toxicol. Chem.* **2012**, *31*, 585–592.
43. Eom, H.-J.; Choi, J. *Environ. Sci. Technol.* **2010**, *44*, 8337–8342.
44. Poynton, H. C.; Lazorchak, J. M.; Impellitteri, C. A.; Blalock, B. J.; Rogers, K.; Allen, H. J.; Loguinov, A.; Heckman, J. L.; Govindasmawy, S. *Environ. Sci. Technol.* **2012**, *46*, 6288–6296.

Chapter 11

Toxic PH_3 Catalytic Decomposition and High Purity Phosphorus Production by Amorphous Co-Based Alloy Nanomaterials

Xue-jiao Tang,[1] Zong-ming Xiu,[2] Chang-xiu Han,[*,3] and Bao-gui Zhang[1]

[1]College of Environmental Science and Engineering, Nankai University, Tianjin, 300071, P.R. China
[2]Department of Civil and Environmental Engineering, Rice University, Houston, Texas 77005, U.S.A.
[3]College of Environmental and Chemical Engineering, Tianjin Polytechnic University, Tianjin 300382, P.R. China
[*]E-mail: hanchangxiuhk@126.com. Phone: (86)13920957651.

PH_3 is a by-product of sodium hypophosphite industry, and is a high toxic waste gas for human and environment. Meanwhile, PH_3 can be decomposed to high purity yellow phosphorus (P_4), which is expensive and widely demanded in industries, such as the high-quality micro-electronic production. Therefore, catalytic decomposition of PH_3 into yellow phosphorus is of economic benefits and environmental significance. Co-based alloys nanomaterials in amorphous form were found to exhibit excellent catalytic activities for PH_3 decomposition in our investigation. In the paper, nano-sized CoB, CoP, CoBP amorphous alloys particles were prepared and their chemical compositions and structures were characterized by multiple analysis and characterization technologies, including ICP, DSC, SEM, TEM, X-ray diffraction, and BET analysis. The as-prepared Co-based amorphous alloy catalysts performed perfect catalytic activities for PH_3 decomposition under various temperatures. Depositing the CoP amorphous alloy on the support (TiO_2) significantly improved its thermal stability, surface area and anti-aggregation property, which were mainly attributed to the high dispersion of CoP alloy particles on the support. The catalytic activities of these as-prepared nano-catalysts were compared and they follow the sequence of CoB>CoBP>CoP/TiO_2>CoP. CoB exhibited the best PH_3 decomposition property, with 98.4% of PH_3 decomposed at 723 K. Overall, the Co-based nano-catalysts exhibited a unique property and a high potential in industrial application of PH_3 decomposition to yellow phosphorus, which has both important economic benefits and environment significance.

Introduction

Phosphine (PH_3) is a highly toxic gas with a permissible exposure level (PEL, OSHA air standard) of only 300 ppb, an LC_{50} of 20 ppm for rat at 1 hour exposure (*1*). Acute exposures to PH_3 cause respiratory tract irritation that attacks primarily the cardiovascular and respiratory systems causing peripheral vascular collapse, cardiac arrest and failure, and pulmonary edema (*1*). PH_3 is an exhausted gas from the sodium hypophosphite production industry. The precursor for sodium hypophosphite production is costly yellow phosphorus. Generally, only 75%-80% of yellow phosphorus could be converted to useful sodium hypophosphite, and the rest of them will be converted to PH_3 as waste gas (*2*). With the rapid progress of sodium hypophosphite industry in recent years, the exhausted PH_3 gas is becoming a serious problem for environmental safety and human health (*3*). For the sake of environmental and ecological safety, it is urgent to develop an economically effective way to control PH_3. The stringent environment regulation requires that PH_3 in the exhaust gas from chemical plants to be reduced to the lowest possible level (*4–6*).

The elimination of toxic chemicals (such as PH_3) that present in exhaust gas of chemical plant has become an important research topic (*7–11*). Usually the toxic PH_3 gas is trapped and converted to phosphoric acid or sodium hypophosphite or hypophinic acid by oxidation reactions (*12–14*). This is a practically feasible method, however, due to operational limitations, part of the PH_3 and phosphate acid mist can still leak into the environment. In the progress of sodium hypophosphite production, strict waste gas discharge requirement was demanded for the product quality and the processes were complicated and not easy to operate. Furthermore, phosphoric acid, sodium hypophosphite and hypophinic acid as end products have low economic values, therefore, these methods are not economically favorable. Another promising approach is to decompose PH_3 into high purity yellow phosphorus (P_4) and hydrogen gas (H_2) by special catalysts. H_2 is a clean energy source and high-purity yellow phosphorus is widely used in high-quality micro-electronic industries for super large scale integrate-circuit and liquid crystal display products. High-purity yellow phosphorus is expensive because of its complicated production process and it is only produced by several countries in the world. Therefore, the decomposition of PH_3 to high-purity yellow phosphorus has important economic benefits and environment significance.

It is well known that a high decomposition temperature of 1073K is required to break the high energy of P-H bonds in the absence of catalysts (*15*, *16*). Therefore, it's necessary to develop appropriate catalyst systems for complete PH_3 decomposition under lower temperature. In the past decades, the amorphous alloys have gained much attention as novel catalytic materials (*17–20*) due to their unique combination of small particle size with short-rang ordered and long-rang disordered structures (*21*). The presence of a large number of highly coordinated and unsaturated sites on the amorphous alloys enables surface adsorptions and surface reactions to occur more easily than on corresponding crystalline catalysts (*22*). Amorphous alloys have been widely applied in hydrogenation reactions due to their excellent properties in comparison with corresponding crystalline metal catalysts (*23–27*). Previous studies have shown that FeP amorphous alloys (*14*, *15*), CoP amorphous alloys (*28*) exhibited excellent catalytic property of decomposing PH_3 into pure phosphorous. In our previous report, a 99.8% PH_3 conversion rate was achieved at 360 °C by CoCe/CNTs (*29*), which is presumably due to the high dispersion of alloy nanoparticles on mesoporous structure of carbon nano-tubes (CNTs). However, the complex preparation processes and the high cost of CNTs inhibited its scale-up application. The group VIII metals are most widely used as active components in heterogeneous catalysis. For example, Co-based (*30–32*) and Ni-based (*33–36*) alloy materials have catalytic activities for hydrogenation reactions. In recent years, more investigations focused on Ni-based amorphous alloys (*37*, *38*). The catalytic activity sequence for hydrogenation reactions was NiB>NiBP>Raney>NiP. The nickel in the amorphous alloy NiP existed in electron-deficient state, while the nickel in NiB existed in electron-rich state. The electronic states of nickel in NiB and NiP resulted in the difference of their catalytic activities. The addition of P could enhance the thermal stability of NiB. Because Co has a higher activity to adsorb hydrogen, it is promising to investigate the Co-based amorphous alloy activity for PH_3 decomposition. In our research, Co-based amorphous alloys were found to be favorable and effective for PH_3 decomposition, and exhibited high catalytic activities even at temperature over crystallization.

In the work, nano-sized CoB, CoP, CoBP amorphous alloys were prepared by chemical reduction deposition method. The composition and structure of the alloys were characterized by multiple analysis and characterization technologies, including inductively coupled plasma (ICP), differential scanning calorimeter (DSC), scanning electron microscopy (SEM), transmission electron microscopy (TEM), X-ray diffraction (XRD), and BET surface area analysis. The catalytic properties of the nano-sized Co-based amorphous alloys particles in application of PH_3 decomposition were studied and compared based on the PH_3 conversion rate. The effects of inorganic element on the properties and catalytic activity of Co-based nanomaterials and the effect of TiO_2 support on the catalytic properties were also investigated. The catalysis mechanisms were addressed by analysis of changes of CoB before and after heat treatments.

Materials and Methods

Chemicals

The feeding gas (the mixture of 4.92% PH_3 and 95.08% N_2) was obtained from Beijng ZG Specialty Gases Science & Technology Co., Ltd. N_2 of 99.999% was obtained from Tianjin Sizhi Gases Co., Ltd. TiO_2 used as supports was a commercial product of Nanjing Titania Chemical Co., Ltd (Anatase structure>98.0%). $Co(NO_3)_2{\cdot}6H_3O$, KBH_4 and ethanol were obtained from Tianjin Chemistry Reagent Factory; NaOH, $Na_3C_6H_5O_7{\cdot}2H_2O$, $(NH_4)_2SO_4$ and $NaH_2PO_2{\cdot}H_2O$ were obtained from Tianjin Guangfu Finechemicals Co., Ltd; NaClO solution(free chlorine 0.6%) was obtained from Tianjin Baishi Chemical Engneering Co., Ltd. All chemicals used were reagent grade or better unless otherwise specified.

Preparation of Catalysts

The CoB amorphous samples were prepared by using the chemical reduction method. 100 ml 107.8 $g{\cdot}L^{-1}$ KBH_4 solution containing 0.8 g NaOH were added drop-wise into aqueous solution of 3.0 g Co^{2+} under vigorous stirring. The reaction was performed in an ice-water bath to prevent local overheating, which may cause the aggregation of small particles or the crystallization of the CoB amorphous alloy. The reaction mixture was stirred for 2 hours to ensure the complete reduction of Co^{2+}. The resulting black solid was washed thoroughly with distilled water to be neutral, and then washed further with ethanol and stored in ethanol for use.

The CoP and CoP/TiO_2 samples were prepared according to the following procedures: 2.0 g TiO_2 calcinated at 723 K for 4 hours were impregnated by a Co^{2+} aqueous solution for 4 hours and dried at 423 K for 1 hour. The obtained dry powders were triturated and then transferred into 250 ml solution containing 1.76 g Co^{2+}, 16.5g $Na_3C_6H_5O_7{\cdot}2H_2O$, 15.9g $(NH_4)2SO_4$ and 31.8 g $NaH_2PO_2{\cdot}H_2O$. The mixture was stirred at 363K under N_2 so that chemical reaction occurred. The reaction lasted about 1.5 hours until no significant bubbles were observed in the solution. The resulting CoP/TiO_2 samples were washed thoroughly with distilled water to be neutral. Then it was further washed with ethanol and finally kept in ethanol for use. The CoP amorphous catalyst was prepared in the same procedure without TiO_2.

To prepare the nano-sized CoBP samples. 30 ml KBH_4 solution of 1 $mol{\cdot}L^{-1}$ was added drop-wise into aqueous solution of 2.35 g Co^{2+} and 25.4 g $NaH_2PO_2{\cdot}H_2O$ under vigorous stirring within 5 min. The reaction was performed at 275 K under N_2. The mixture was stirred for 2 hours to ensure the complete reduction of the Co^{2+} in the solution. The resulting black solid was washed and stored similarly with distilled water and ethanol.

All the samples being kept in ethanol above would be dried in the N_2 flow (denoted as a fresh sample) before being used.

Characterization of Catalysts

The chemical compositions of the catalysts were analyzed by means of inductively coupled plasma (ICP, ICP-9000 (N+M), TJA Company, U.S.) and the surface area (S_{BET}) was determined by H_2 chemisorptions (Micromeritics ASAP 2010, U.S.). The morphological structure of the amorphous alloys was determined by X-ray powder diffraction (XRD, D/max-2500 with Cu $K\alpha$ radiation, 40 KV, 100 mA, Japan) and selective area electronic diffraction (SAED, JEM-2010). The thermal stability was determined by differential scanning calorimetric (DSC, NET2SCH DSC204, German) under N_2 (20 mL·min^{-1}) atmosphere at a heating rate of 10 K·min^{-1}. The surface morphology and particle sizes of the amorphous alloys were observed by scanning electron micrograph (SEM, JSM-6700F, JOLE company, Japan) and transmission electron micrograph (TEM, tecnai G220 S-Twin, FEI company).

PH_3 Decomposition Reactivity

PH_3 decomposition was chosen as the probe reaction to evaluate the catalytic activity of Co-based amorphous alloy catalysts, which was carried out in a quartz fixed-bed reactor depositing 0.10 g catalyst mixed with 0.40 g zeolite. The instrument system for testing activities of catalysts was shown in Figure 1.

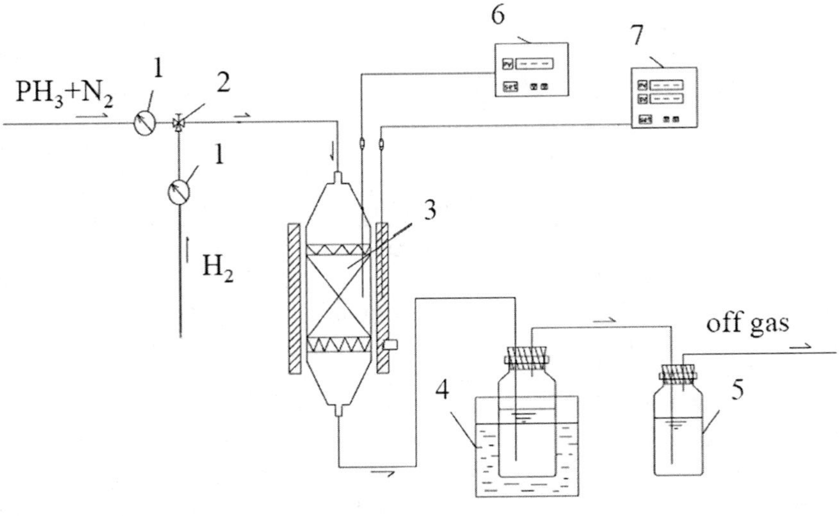

Figure 1. Flow chart of PH_3 decomposition by Co-based nanomaterials. (1) valve; (2) flow meter; (3) catalysis reactor; (4) phosphorus-collecting bottle; (5) Remaining PH_3 adsorption bottle; (6) PID temperature measurement; (7) PID temperature controller.

The resulting phosphorus gas was introduced into collecting bottle and agglomerated there. Due to the inflammable phosphorus exposure to air, the weight increase of collecting bottle was designed to the amount of phosphorus produced in testing. The remaining PH_3 was adsorbed and reduced by diluted NaClO solution. To ensure the reproducibility, the experiments were repeated for three times on the same batch of catalyst and the relative standard deviation (RSD) was found to be within acceptable limits (± 2%).

The conversion rate (η) of PH_3 in catalytic decompositions was calculated with the following equation:

$$\eta\,(\%)=\frac{(n_{PH3}^{total}-n_{PH3}^{remian})}{n_{PH3}^{total}}\times 100\%$$

The amount of remain PH_3 was calculated by determining the concentrations of NaClO before and after absorption. PH_3 was absorbed completely by NaClO and was oxidized to H_3PO_2. The reaction equation is : $PH_3+2NaClO=H_3PO_2+2NaCl$. The total amount of PH_3 can be measured by the flow meter. The measurement limits of the flow meter range from 25 $ml\cdot min^{-1}$ to 250 $ml\cdot min^{-1}$ and the turndown ratio is 10:1.

Results and Discussion

Characterization of As-Prepared Catalysts

SEM micrograph of CoB in Figure 2a demonstrated that the fresh CoB samples displayed a homogeneous morphology consisted of cotton-like nano-sized particles. The TEM micrographs of the fresh CoP and CoBP samples were shown in Figure 2b and Figure 2c, respectively. It can be seen that fresh CoP amorphous alloy was present in the form of nano-sized solid spherical particles with an average size around 60-140 nm. Nevertheless, the fresh CoBP amorphous alloy consisted of nanoparticles with core/shell structures in size no more than 20 nm. The surface areas (S_{BET}) of the catalysts prepared were listed in Table 1. CoBP had a larger surface area (51.3 $m^2\cdot g^{-1}$) than Co-B (37.4 $m^2\cdot g^{-1}$) and CoP (12.4 $m^2\cdot g^{-1}$). The advantageous morphology of CoBP might contribute to the higher catalytic activity for PH_3 decomposition than the other two.

The XRD patterns of different Co-based alloy nanoparticles, as shown in Figure 3, demonstrated that all the fresh CoB, CoP, CoBP samples were consistently present in a typical amorphous structure since only a broad peak around 2θ=45° was observed (*39–41*). The amorphous alloy is thermodynamically metastable. The DSC analysis, as shown in Figure 4, revealed the crystallization processes of the Co-based amorphous alloy nanoparticles with the increase of calcination temperature. The exothermic peaks of DSC curve were attributed to the crystallization of the amorphous alloys (*42*), and the peak temperature could roughly represent the crystallization temperature (Tc) of amorphous materials (*43*). In the DSC curve of CoB nanoparticles (Figure 4a), the small peaks at 519 K and 663 K were indicative of a rearrangement of the CoB amorphous alloy structure (*44*), while the large peak at 735 K corresponded to the occurrence of the crystallization of the CoB amorphous alloy. The DSC curve in Figure 4b demonstrated CoP and TiO_2 supported CoP amorphous nanoparticles (CoP/TiO_2) crystallized in two steps. The first crystallization of CoP performed at 540 K, much lower than that of CoB at 735 K, suggesting CoB amorphous alloy nanoparticles were much more stable than CoP. Both the original and final crystallization temperatures of CoP/TiO_2 were higher than thosc of CoP amorphous alloy indicating the promoting effect of the support (TiO_2) on the thermal stability of the amorphous structure. Li (*45*) believed that dispersion effect and heat sink effect of carrier as well as interaction between catalyst and carrier contributes to it. However, the crystallization temperature of CoP/TiO_2 (700 K) was still lower than that of fresh CoB (730 K). As shown in CoBP (Figure 4d), there is no significant exothermic peak of crystallization below temperature of 773 K. It is suggested that CoBP exhibited the most thermal stability among all the as-prepared nano-catalysts. P improved the thermal stability of Co-based amorphous nanoparticles due to the strong interactions between Co-P and B-P atoms. The chemical bonds of Co-P and B-P could form according to the quantum chemistry investigation for Ni-based and Co-based amorphous alloys (*46*, *47*).

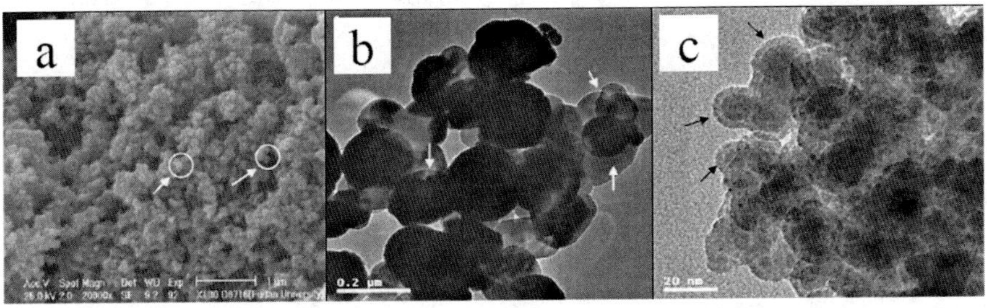

Figure 2. SEM photograph of the fresh CoB (a) and TEM photographs of the fresh CoP (b) and CoBP (c).

Table 1. Surface Area (S_{BET}) of Different Co-Based Nano-Catalysts

Catalysts	*CoB*	*CoP*	*CoP/TiO₂*	*CoBP*	*CoP-700*	*CoP/TiO₂-700*
S_{BET}/ $m^2 \cdot g^{-1}$	37.4	12.4	20.81	51.3	3.6	14.6

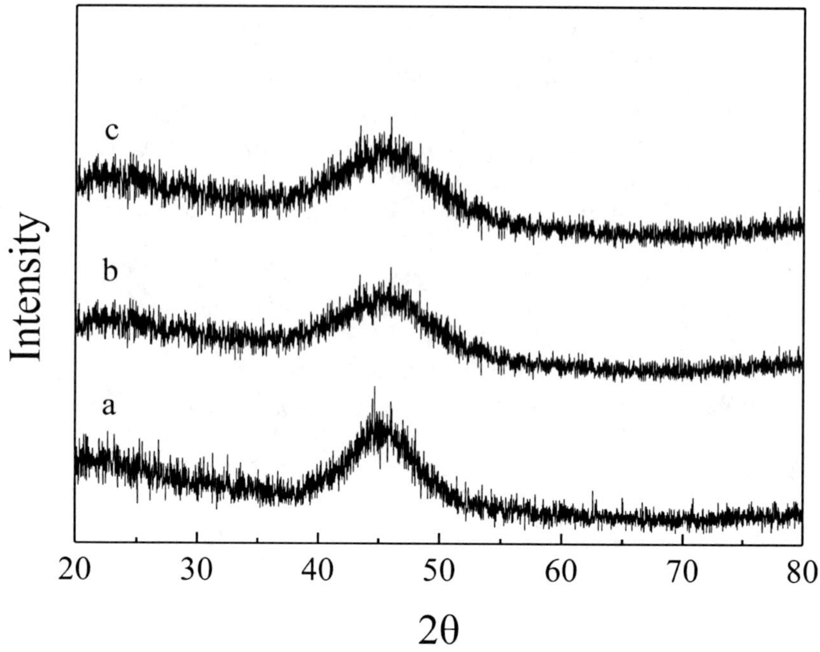

Figure 3. XRD patterns of the Co-based nano-catalysts (a. CoB; b. CoP; c. CoBP).

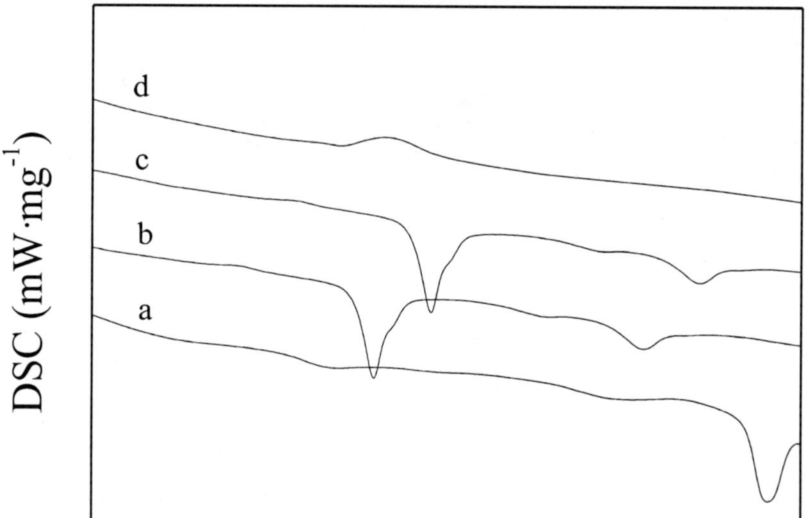

Figure 4. DSC curves of the Co-based nano-catalysts (a. CoB; b. CoP; c. CoP/TiO_2; d. CoBP).

Treating Temperature Affects the Structure and Property of CoB Nanoparticles

From the catalytic activity testing, it was found that the as-prepared nano-sized Co-based amorphous alloys catalysts were capable to reduce PH_3 decomposition temperature from over 1073 K to below 573 K. The PH_3 conversion rate over fresh CoB amorphous alloy could achieve 25.2% at 493 K, 90.2% at 693 K and even 98.4% at 723 K. Thc temperatures were all lower than its crystallization temperature of 735 K in its DSC curve. It is suggested that the nano-sized CoB alloy in amorphous form has effective catalytic activity for PH_3 decomposition.

To further investigate the effect of various temperatures on the structure of the nano-sized Co-based amorphous alloys, the fresh CoB nanoparticles were treated in H_2 flow for 2 hours at elevated temperatures from 493 K to 813 K, respectively. The results showed a good explanation for the performances and properties of catalysts under higher temperature. The XRD patterns and SEM photographs before and after treatment were recorded in Figure 5 and Figure 6, respectively. As seen in Figure 5a, there was no observed change in the XRD pattern of CoB sample treated at 493 K, indicating no change of CoB amorphous structure and this is in accordance with the results of DSC. The original broad peak disappeared gradually and several diffraction peaks appeared when CoB was treated under higher temperature. Both the strength and the numbers of the diffraction peaks increased with increasing temperature and reached maximum at 733 K. These results demonstrated that the CoB amorphous alloy began to rearrange at 623 K and reached completion of crystallization at 733 K, which was also consistent with the results of DSC where there was an exothermic peak at 733 K. Both metallic Co and Co_2B alloys were identified in XRD patterns during the rearrangement. The presence of metallic Co and crystalline Co_2B alloy species suggested that partial decomposition and rearrangement of CoB alloy occur at 623 K. The metallic Co and crystalline Co_2B alloy might be the catalytic centers for PH_3 decomposition before crystallization.

In Figure 6a, it was shown that CoB nanoparticles treated at 493 K were still presenting cotton-like appearance, similar to the fresh CoB nanoparticles (Figure 2a). With the increasing temperature, the CoB nanoparticles grew in size and coalesced to form larger clusters (Figure 6b). Big lumps appeared when the CoB samples were treated at 813 K (Figure 6c). Based on the XRD and DSC analysis, the SEM images in Figure 6 verified that the array of the metal atoms in CoB amorphous alloy nanoparticles unchanged at 493 K and got complete crystallization at 813 K.

In brief, it is suggested that the effective catalytic property of the nano-sized CoB alloy in amorphous form could be due to higher content of metallic Co and the separation of crystalline Co_2B from bulk to surface, which formed more catalytic active sites during the rearrangement processes, and this is in accordance with Wang's report (*48*). The cotton-like nano-sized particles promotes the active sites formation.

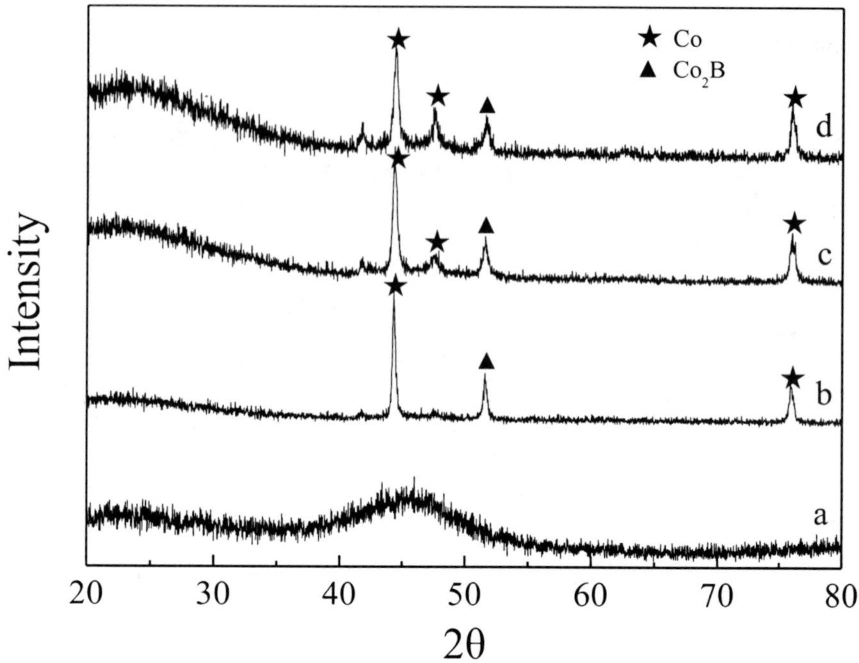

Figure 5. XRD patterns of treated CoB nano-catalysts under various temperatures (a. 493 K; b. 623 K; c. 733 K; d. 813 K).

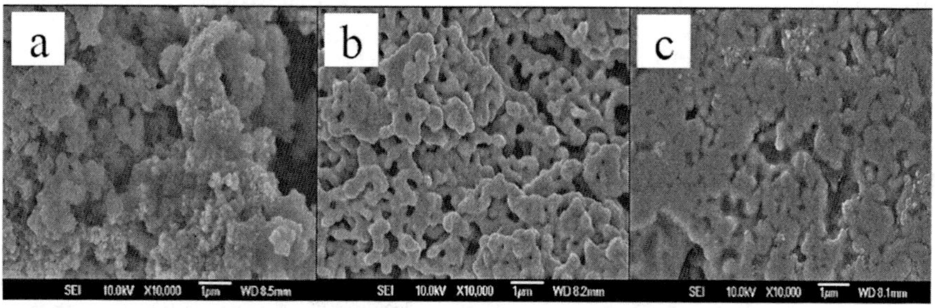

Figure 6. SEM photographs of treated CoB nano-catalysts under various temperatures (a. 493 K; b. 623 K; c. 813 K).

TiO_2 Support Improves the Catalytic Performance of CoP

The chemical compositions of the CoP and CoP/TiO_2 were obtained by ICP analysis. The molar ratio of P to Co was about 0.2 in both CoP and CoP/TiO_2 indicating TiO_2 support did not change the composition of the supported CoP alloy. To investigate the effect of TiO_2 support on the thermal stability of CoP, CoP and CoP/TiO_2 were treated under 700 K in N_2 flow for 2 hours. The treated catalysts were referred to CoP-700 and CoP/TiO_2-700, respectively. TEM micrographs of TiO_2 and the fresh and treated CoP and CoP/TiO_2 are shown in Figure 7. It can be seen that TiO_2 was composed of ultrafine particles with an average size of 80-120 nm. The fresh CoP amorphous alloy presented as ultrafine spherical particles with an average size of 60-140 nm. The particle of supported CoP had a smaller size with an average size of 50-110 nm. TEM micrograph of CoP-700, as shown in Figure 7d, demonstrated that the particle size of the treated CoP became larger. It is implied that the treatment at high temperature resulted in big lumps in CoP amorphous nanoparticles. Similar situation happed to TEM micrograph of CoP/TiO_2-700 (Figure 7e), but the extent was lower than the unsupported CoP due to its dispersion on TiO_2. As shown in Table 1, the fresh CoP/TiO_2 amorphous catalyst had larger surface area (20.81 $m^2 \cdot g^{-1}$) than fresh CoP (12.4 $m^2 \cdot g^{-1}$), and the fresh catalysts had larger area than the corresponding treated catalysts. After being treated, the surface areas of both CoP and CoP/TiO_2 decreased, but CoP decreased much more than CoP/TiO_2. It is verified that the introduction of TiO_2 not only increased the surface area but also enhanced the capacity of anti-aggregation. Briefly, the analysis of TEM, DSC and BET indicates that depositing the CoP amorphous alloy on the support TiO_2 can improve the thermal stability, surface area and anti-aggregation, which was mainly attributed to the high dispersion of CoP alloy particles on the support.

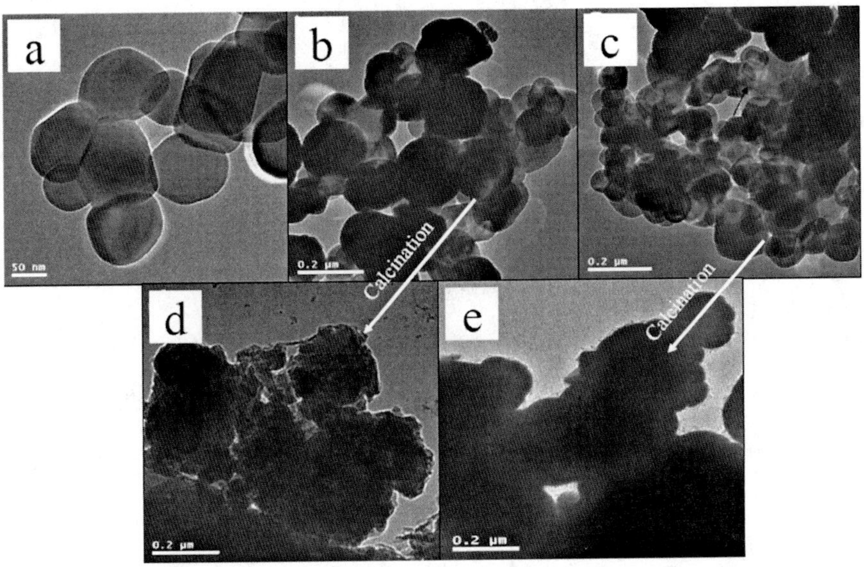

Figure 7. TEM photographs of TiO_2 (a), fresh CoP (b), fresh CoP/TiO_2 (c), CoP-700 (d), and CoP/TiO_2-700 (e).

Feeding Gas Flow Rate and N_2 to PH_3 Ratio Affect PH_3 Conversion Rate

To investigate the effects of testing parameters (the feeding gas flow rate and ratio of N_2 to PH_3) on PH_3 catalytic decomposition, PH_3 conversion rates with varied parameters were determined using CoP/TiO_2 as catalyst. Q_{PH3} and Q_{N2} were designated as the flow rate of PH_3 and N_2, respectively. Q_{N2+PH3} was designated as the total flow rate of PH_3 and N_2.

As shown in Figure 8, PH_3 conversion rate decreased from 97.8% to 83.8% with increasing Q_{PH3+N2} (from 150 ml·min^{-1} to 210 ml·min^{-1}, corresponding to Q_{PH3} from 7.1 ml·min^{-1} to 10 ml·min^{-1} with Q_{N2}:Q_{PH3} of 20:1). When the temperature of 733 K and Q_{N2}:Q_{PH3} of 20:1 were given, PH_3 conversion rate decreased with increasing Q_{N2+PH3} due to the decrease of PH_3 retention time in the reactor and its contact time with the catalysts. It is known that short retention time of gas and short contact time between gas and catalysts are not conducive to the catalysis reaction. Furthermore, during the same period (1 hour) larger Q_{N2+PH3} means more PH_3 need to be treated, but the catalytic capacity didn't changed. So lower concentration of PH_3 in total gas (C_{PH3}= Q_{PH3}/ Q_{N2+PH3}) benefits the PH_3 decomposition with the same catalyst dose. From Figure 8a, Q_{PH3} that is less than 7.8 ml·min^{-1} is appropriate for higher PH_3 decomposition rate (over 95%).

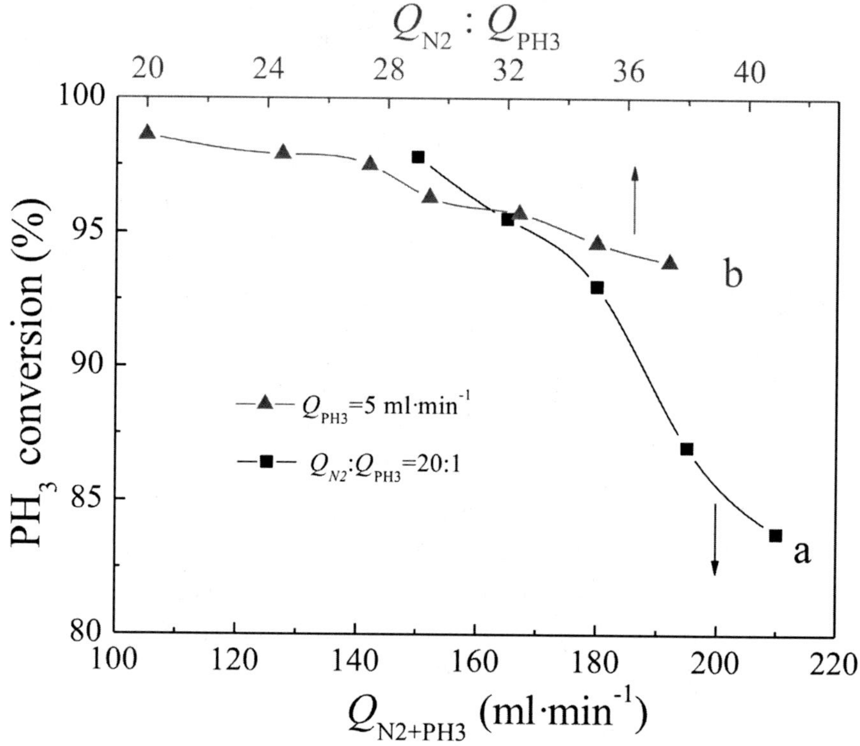

Figure 8. Effects of Q_{N2+PH3} and Q_{N2}:Q_{PH3} ratio on the conversion rate of PH_3 at 733 K for 1 h.

The relationship between Q_{N2}:Q_{PH3} and PH_3 conversion rate is shown in Figure 8b. It is showed that PH_3 conversion rate decreased with increasing Q_{N2}:Q_{PH3} with the given Q_{PH3} of 5 ml·min^{-1} at 733 K. It is suggested that high ratio of Q_{N2}:Q_{PH3} will inhibit PH_3 decomposition and a Q_{N2}:Q_{PH3} ratio of below 32 would be appropriate to perform over 95% PH_3 decomposition.

PH_3 Decomposition by Co-Based Amorphous Nanoparticles under Various Temperatures

Temperature plays a key role in the catalytic activity of Co-based amorphous alloy nanoparticles for PH_3 decomposition, as shown in Figure 9. PH_3 conversion rate of CoB and CoBP nano-catalysts increased with the increasing temperature from 693 K to 743 K. There was no observed decrease below 873 K for both CoB and CoBP (related data not shown). CoB nano-catalysts exhibited the best catalytic activity among the as-prepared catalysts at the corresponding temperature as shown in Figure 9. PH_3 conversion rate achieved 90.2% at 693 K, 99.8% at 743 K over CoB nano-catalysts, while achieved 83.5% at 693 K, 99.6% at 743 K over CoBP nano-catalysts. Figure 9 showed that CoP and CoP/TiO_2 presented a similar catalytic behavior but showed much lower catalytic activities than CoB or CoBP from 693 K to 743 K. They both achieved the maximum conversion rate of 89% at 713 K and the catalytic property decreased with increasing temperature. However, CoP/TiO_2 had a lower decline at higher temperature than unsupported CoP, which might be due to the TiO_2 support that enhanced the thermal stability and improved the dispersion of CoP. This result is in accordance with the DSC in Figure 4 and the surface areas analysis in Table 1.

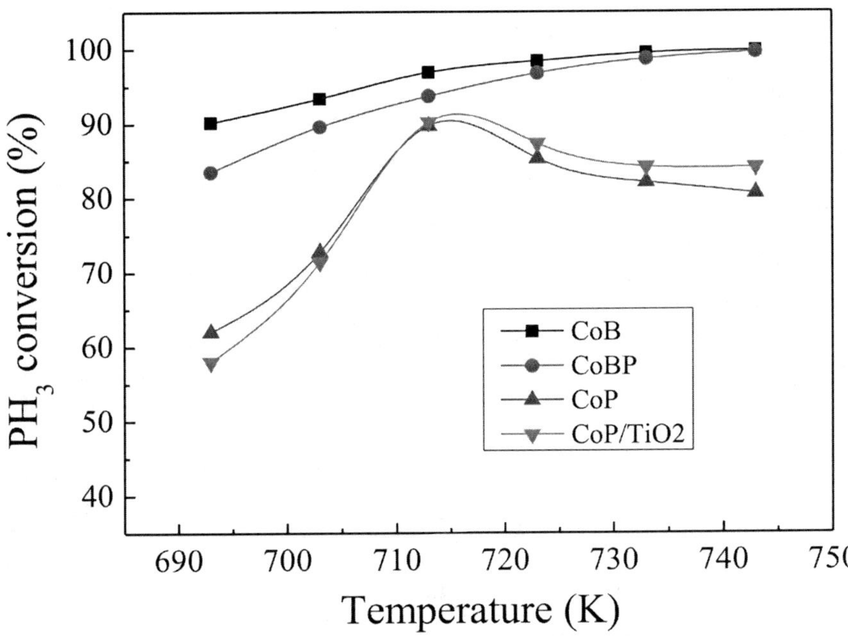

Figure 9. PH_3 conversion rate under various temperatures with Q_{PH3} of 4.2 ml·min^{-1} and Q_{N2}:Q_{PH3} of 20:1 for 1 h.

Besides the high intrinsic property, the dispersion of Co active sites on uniform nanoparticles and the strong electronic interaction between Co and H (which enhances the competitive adsorption of H on Co active sites against P-H groups of PH_3) are also responsible for their higher catalytic activity. Meanwhile, the stronger adsorption of hydrogen on Co active sites may also benefit PH_3 decomposition. The electronic states of Co species in catalyst resulted in the difference of their catalytic activities. Yang (*37*) and Li (*38*) believed that the nickel in the amorphous alloy NiP existed in electron-deficient state, while the nickel in NiB existed in electron-rich state. These could also explain the distinct behaviors of as-prepared Co-based nano-catalysts for PH_3 decomposition as shown in Figure 9 (with the sequence of CoB>CoBP>CoP/TiO_2>CoP, which is almost the same as the reported Ni-based catalysts (*37*, *38*)).

Catalytic Stability of CoB and Yellow Phosphorus Production

To demonstrate the commercialization feasibility, the catalytic stability of CoB nano-catalyst was investigated with Q_{PH3+N2} of 160 ml·min^{-1} in 48 hours. The time-on-stream performance of CoB nano-catalyst in PH_3 decomposition reaction was tested at 723 K. As shown in Figure 10, the CoB nano-catalysts exhibited stable and high catalytic activity during stability test. Almost 100% conversion was obtained and during the period of 48 hours testing, only a slight deactivation was observed.

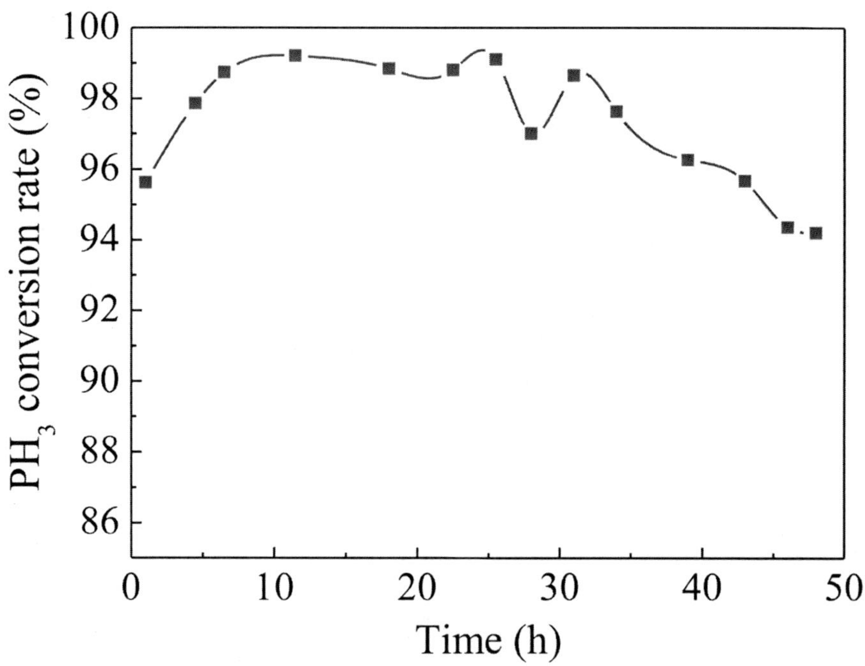

Figure 10. The catalytic stability of CoB nano-catalyst tested at 723 K with Q_{PH3+N2} of 160 ml·min^{-1} and Q_{N2}:Q_{PH3} of 20:1.

Figure 11 showed that the amount of phosphorus chemicals (including total PH_3, remained PH_3, P_4 produced and theoretical amount of P_4) varied with the reacting time of PH_3 decomposition (by CoB nano-catalysts). Theoretical amount of P_4 was designed to the amount of P_4 produced with the assumption that PH_3 was 100% decomposed to P_4 and H_2. The results indicated that CoB nano-catalysts still performed excellent catalytic property for PH_3 decomposition after 12 hours. PH_3 was nearly completely decomposed at all testing time point, indicated by the remaining PH_3. The amount of produced yellow phosphorus increased with increasing PH_3 amount and testing time. It was noticed that the yellow phosphorus determined was lower than the theoretical value, which could be due to the phosphorus loss (~30.0%) caused by agglomeration and absorption on inner surface of reactor and tubes.

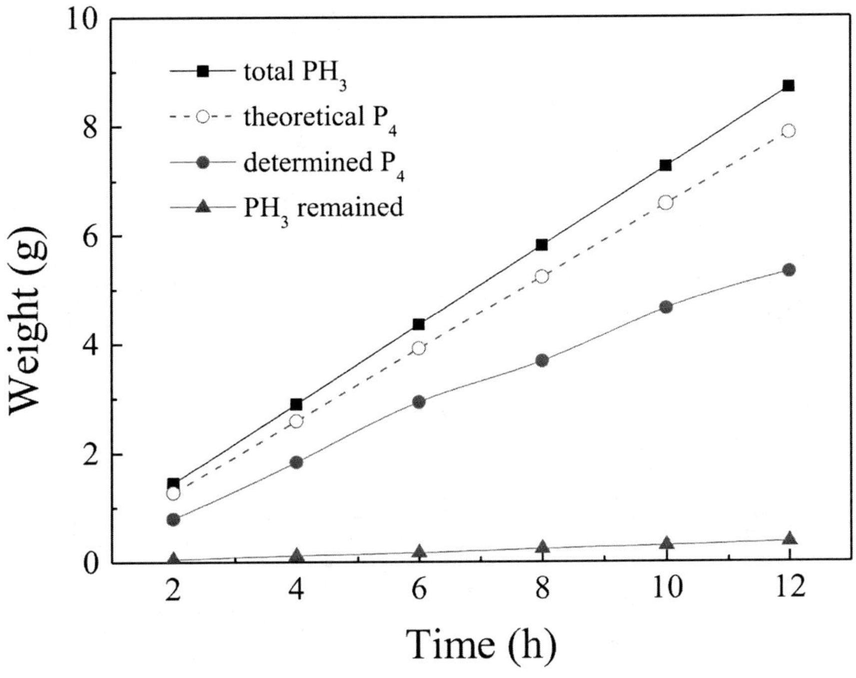

Figure 11. Phosphorous production amount at various time by CoB with Q_{PH3+N2} of 160 ml·min^{-1} and Q_{N2}:Q_{PH3} of 20:1 at 723 K.

Conclusions

CoB, CoP and CoBP amorphous alloys particles were prepared and their catalytic activities of PH_3 decomposition were compared, with the sequence of CoB>CoBP>CoP/TiO_2>CoP. Although P improved the thermal stability and the surface areas of Co-based amorphous nanoparticles, it inhibited the catalytic activity of PH_3 decomposition. CoB nano-catalysts exhibited the best catalytic activity among the as-prepared catalysts at the corresponding temperature, even though CoB presented a bigger particle size and a smaller surface area. PH_3 conversion achieved 90.2% at 693 K, 99.8% at 743 K over CoB nano-catalysts. This could be due to higher content of metallic Co and the separation of crystalline Co_2B from bulk to surface, which formed more catalytic active sites during the rearrangement processes. Overall, Co-B amorphous alloy nanoparticle appeared to be a promising catalyst for PH_3 decomposition to yellow phosphorus, which may not only treat the environmental contaminant, but also produce products of high economic value.

Acknowledgments

The authors gratefully acknowledge financial support by the special fund for basic research of central colleges and universities in China (65010451) and Tianjin National Science Foundation (043801311) of China.

References

1. Kamanina, N. V. *Features of Liquid Crystal Display Materials and Processes*; InTech: Croatia, 2011.
2. Wang, C. J.; Guo, A. H.; Wang, F. S.; Zhang, B. G. Treatmental technology of PH_3 wastegas in the process of industrial sodium hypophosphite production. *Tianjin Chem. Ind.* **2003**, *17*, 5, 37–38.
3. Han, X. Y.; Han, C. X.; Bi, C. L; Li, L.; Wang, X. Y.; Zhang, B. G. Improved risk assessment method in producing ultra-pure phosphorus by using phosphine vented from the industry. *J. Saf. Environ.* **2008**, *8* (4), 160–163.
4. Yu, Q.; Yi, H; Tang, X; Ning, P.; Wang, C. Progress on phosphine control technology. *Chin. Environ. Sci. Technol.* **2009**, *32* (10), 87–91.
5. Sun, T. H.; Zhou, Q. X.; Li, P. J. *Pollution Ecology*; Science Press: Beijing, 2001.
6. Zhou, Q. X.; Wei, S. H.; Zhang, Q. R. *Ecological Remediation*; China Environmental Science Press: Beijing, 2006.
7. Yu, Q. F.; Ning, P.; Yi, H. H.; Tang, X. L.; Li, M.; Yang, L. P. Effect of preparation conditions on the property Cu/AC adsorbents for phosphine adsorption. *Sep. Sci. Technol.* **2012**, *47* (3), 527–533.
8. Ren, B. N. Kinetics and thermodynamics of the phosphine adsorption on the modified activated carbon. *Front. Chem. Sci. Eng.* **2011**, *5* (2), 203–208.
9. Yi, H. H.; Yu, Q. F.; Tang, X. L.; Ning, P.; Yang, L. P.; Ye, Z. Q.; Song, J. H. Phosphine adsorption removal from yellow phosphorus tail gas over CuO–ZnO–La_2O_3/activated carbon. *Ind. Eng. Chem. Res.* **2011**, *50* (7), 3960–3965.
10. Hsu, J. N.; Bai, H.; Li, S. N.; Tsai, C. J. Copper loaded on sol-gel-derived alumina adsorbents for phosphine removal. *J. Air Waste Manage. Assoc.* **2010**, *60* (5), 629–635.
11. Yu, Q. F.; Tang, X. L.; Yi, H. H.; Ning, P.; Yang, L. P.; Yang, L. N.; Yu, L. L.; Li, H. Equilibrium and heat of adsorption of phosphine on $CaCl_2$-modified molecular sieve. *Asia-Pac. J. Chem. Eng.* **2009**, *4*, 612–617.
12. Wisnouskas, J. S.; Ho, R. Process for Preparing Phosphorus Acid from Industrial Waste Materials. U.S.Patent 4380531, 1983.

13. Arzoumanidis, G. G.; Darragh, K. V. Process for Producing Hypophosphorous Acid and Non-Transition Metal Hypophosphites. U.S. Patent 4265866, 1981.
14. Brent, E.; Frank, B.; Frederick, J. Exhaust gas incineration and the combustion of arsine and phosphine. *Solid State Technol.* **1990**, *33* (1), 89–92.
15. Matsubara, H.; Tabei, S.; Ichimura, S.; Iso, A. Production of Elemental Phosphorus. Japan Patent 01-313309, 1989.
16. Han, C. X.; Ren, J. L.; Bi, C. L.; Zhang, B. G.; Liu, S. X. Preparation of Fe(Pd)P alloy and its catalysis on decomposition of PH_3. *Chin. J. Mol. Catal.* **2009**, 418–421.
17. Liang, Q.; Zhang, A.; Li, L. Hydrogenation of phenylpyruvic acid to phenylalanine catalyzed by Ni-B/SiO2. *J. Nat. Gas Chem.* **2008**, *17*, 303–308.
18. Van Wonterghen, J.; RuP, S. M.; Koch, C, J. W. Formation of ultra-fine amorphous allys particles by reduction in aqueous solution. *Nature* **1986**, *322*, 622–623.
19. Rajesh, B.; Sasirkha, N.; Lee, S. P.; Kuo, H. Y.; Chen, Y. W. Investigation of Fe-P-B ultralfine amorphous nanomaterials: Influence of synthesis parameters on physicochemical and catalytic properties. *J. Mol. Catal. A: Chem.* **2008**, *289*, 69–75.
20. Ma, Y.; Li, W.; Zhang, M.; Zhou, Y.; Tao, K. Preparation and catalytic properties of amorphous alloys in hydrogenation of sulfolene. *Appl. Catal., A* **2003**, *243*, 215–223.
21. Song, L.; Li, W.; Wang, G.; Zhang, M; Tao, K. A new route to prepare supported nickel phosphide/silica-alumina hydrotreating catalysts from amorphous alloys. *Catal. Today* **2007**, *125*, 137–142.
22. Rajesh, B.; Sasirekha, N.; Chen, Y. W. Physicochemical and catalytic properties of Fe-P ultralfine amorphous catalysts. *J. Mol. Catal. A: Chem.* **2007**, *275*, 174–182.
23. Chen, Y. Chemical preparation and characterization of metalmetalloid ultrafine amorphous alloy particles. *Catal. Today* **1998**, *44*, 3–15.
24. Deng, J. F.; Li, H. X.; Wang, W. J. Progress in design of new amorphous alloy catalysts. *Catal. Today* **1999**, *51*, 113–125.
25. Wang, W. J.; Qiao, M. H.; Li, H. X.; Dai, W. L.; Deng, J. F. Study on the deactivation of amorphous NiB/SiO_2 catalyst during the selective hydrogenation of cyclopentadiene to cyclopentene. *Appl. Catal., A* **1998**, *168*, 151–157.
26. Li, H.; Yang, P. F.; Chu, D. S.; Li, H. X. Selective maltose hydrogenation to maltitol on a ternary Co–P–B amorphous catalyst and the synergistic effects of alloying B and P. *Appl. Catal., A* **2007**, *325*, 34–40.
27. Tong, D. G.; Chu, W.; Luo, Y. Y.; Chen, H.; Ji, X. Y. Preparation and characterization of amorphous Co-B catalysts with mesoporous structure. *J. Mol. Catal. A: Chem.* **2007**, *269*, 149–157.
28. Han, C. X.; Lin, X. M.; Zhang, B. G. Effect of support TiO_2 on the characteristics of Co-P amorphous alloy. *Acta Chim. Sin.* **2007**, *9*, 793–797.

29. Li, L.; Han, C.; Han, X.; Zhou, Y.; Yang, L.; Zhang, B.; Hu, J. Catalytic decomposition of toxic chemicals over metal-promoted carbon nanotubes. *Environ. Sci. Technol.* **2011**, *45*, 726–731.
30. Pei, Y.; Fang, J.; Hu, H. R.; Li, H. X.; Fan, K. N.; Li, H. X.; Qiao, M. H. Selective hydrogenation of crotonaldehyde over amorphous Co-B catalysts: the effect of Zn. *Acta Chim. Sin.* **2005**, *63* (4), 289–294.
31. Li, H. X.; Li, H.; Zhang, J.; Dai, W. L.; Qiao, M. H. Ultrasound-assisted preparation of a highly active and selective Co-B amorphous alloy catalyst in uniform spherical nanoparticles. *J. Catal.* **2007**, *246*, 301–307.
32. Li, H. X.; Wu, Y. D.; Luo, H. S.; Wang, M.; Xu, Y. Liquid phase hydrogenation of acetonitrile to enthylamine over the Co-B amorphous alloy catalyst. *J. Catal.* **2003**, *214*, 15–25.
33. Wang, L. J.; Zhang, M. H.; Li, Wei; Tao, K. Y. Preparation of supported NiB amorphous alloy catalysts by powder electroless plating. *Chin. J. Catal.* **2005**, *26*, 91–92.
34. Li, H. X.; Wang, W. J.; Li, H.; Deng, J. F. Crystallization deactivation of NiP/SiO_2 amorphous catalyst and stabilizing effect of silica support on the Ni-P amorphous structure. *J. Catal.* **2000**, *194*, 211–221.
35. Guo, Z.; Keng, K. G.; Sha, W. Crystallization and phase transformation behavior of electroless nickel phosphorus plating during continuous heating. *J. Alloy. Comp.* **2003**, *358*, 112–119.
36. Lee, S. P.; Chen, Y. W. Nitrobenzene hydrogenation on Ni-P, Ni-B and Ni ultrafine material. *J. Mol. Catal. A: Chem.* **2000**, *152*, 213–223.
37. Yang, J. M.; Lu, J.; An, Z. W. Hydrogenation of 4-ethylbenzoic acid over the amouphous alloy nickel ultrafine particles. *Chin. J. Mol. Catal.* **2002**, *5*, 364–368.
38. Li, L.; Liang, Q. F.; Zhang, A. Q. Preparation of phenylalanine from phenylpyruvic acid by hydrogenation over Ni-B amouphous alloy. *Chin. J. Catal.* **2007**, *28* (12), 1031–1033.
39. Yokoyama, A.; Komiyama, H.; Inoue, H.; Masumoto, T.; Kimura, H. M. The hydrogenation of carbon monoxide by amorphous ribbons. *J. Catal.* **1981**, *68*, 355–361.
40. Li, H. X.; Chen, H. Y.; Dong, S. Z.; Yang, J. S.; Deng, J. F. Study on the crystallization process of Ni-P amorphous alloy. *Appl. Surf. Sci.* **1998**, *125*, 115–119.
41. Yamashita, H.; Yoshikawa, M.; Funabiki, T.; Yoshida, S. Catalysis by amorphous metal alloys. Part 2: Effects of oxygen pretreatment on the catalytic activity of amorphous and crystallised Ni–P alloys. *J. Chem. Soc., Faraday Trans. 1* **1985**, *81*, 2485–2493.
42. Wang, W. J.; Qiao, M. H.; Li, H. X.; Deng, J. F. Amorphous NiP/SiO_2 aerogel: Its preparation, its high thermal stability and its activity during the selective hydrogenation of cyclopentadiene to cyclopetene. *Appl. Catal., A* **1998**, *166*, 243–247.
43. Deng, J. F.; Li, H. X.; Wang, W. J. Progress in design of new amorphous alloy catalysts. *Catal. Today* **1999**, *51*, 113–115.

44. Yamashita, H.; Funabiki, T.; Yoshida, S. Structural modification towards metastable states and catalytic activity of an amorphous Ni-B alloy. *J. Chem. Soc., Chem. Commun.* **1984**, 868–869.
45. Li, H. X.; Li, Hui; Wang, W. J.; Deng, J. F. Crystallization deactivation of CoP-SiO_2 amorphous catalyst and the stabilizing effect of silica support on the Co-P amorphous structure. *J. Catal.* **2000**, *194*, 211–221.
46. Cao, L. J.; Fang, Z. G.; Du, B.; Tian, X. Y.; Zhu, X. M. DFT study on local structure and stability of amorphous $Co_{80}P_{20}$ alloy. *Chem. World* **2011**, *52* (1), 5–9.
47. Fang, Z. G.; Shen, B. R.; Fan, K. N.; Deng, J. F. DFT study of electron transfer between P and Ni in Ni-P amorphous alloy. *Acta Chim. Sin.* **1999**, *57*, 1246–1251.
48. Wang, Y. D.; Ai, X. P.; Yang, H. X. Electrochemical hydrogen storage behaviors of ultrafine amorphous Co-B alloy particles. *Chem. Mater.* **2004**, *16*, 5194–5197.

Chapter 12

Chapter Green Nanotechnology: Development of Nanomaterials for Environmental and Energy Applications

Changseok Han,[1] Joel Andersen,[1] Suresh C. Pillai,[2] Rachel Fagan,[2] Polycarpos Falaras,[3] J. Anthony Byrne,[4] Patrick S. M. Dunlop,[4] Hyeok Choi,[5] Wenjun Jiang,[6] Kevin O'Shea,[6] and Dionysios D. Dionysiou*,[1,7]

[1]Environmental Engineering and Science Program, University of Cincinnati, Cincinnati, Ohio 45221-0012, U.S.A.
[2]Centre for Research in Engineering Surface Technology (CREST) FOCAS Institute, Dublin Institute of Technology, Kevin St., Dublin 8, Ireland
[3]Division of Physical Chemistry, Institute of Advanced Materials, Physicochemical Processes, Nanotechnology and Microsystems (IAMPPNM), National Center for Scientific Research "Demokritos", 15310 Aghia Parakevi Attikis, Athens, Greece
[4]Nanotechnology and Integrated BioEngineering Centre, School of Engineering, University of Ulster, Northern Ireland, BT37 0QB, United Kingdom
[5]Department of Civil Engineering, The University of Texas at Arlington, Arlington, Texas 76019-0308, U.S.A.
[6]Department of Chemistry and Biochemistry, Florida International University, University Park, Miami, Florida 33199, U.S.A.
[7]Nireas-International Water Research Centre, University of Cyprus, 20537 Nicosia, Cyprus
*E-mail: dionysios.d.dionysiou@uc.edu

This book chapter discusses the syntheses of various nanomaterials, for green nanotechnology applications in detail. Special attention is given to the development of emerging areas, such as environmental as well as energy materials. Various approaches for preparing nanostructured photocatalysts, such as titanium dioxide, zinc oxide, iron oxide, and metal sulfides, different conventional methods and novel methods, including sol-gel methods, hydrothermal methods, microwave-assisted methods and sonochemical methods are introduced. The use of nanomaterials as photocatalysts, supporting materials for solar cells, and disinfectants is reported for environmental remediation and energy applications. Advanced applications of nanomaterials for water detoxification, air purification, and the inactivation of pathogenic microorganisms in water as well as dye-sensitized solar cells is also discussed. The enhancement of selectivity of photocatalysis, especially TiO_2 systems, for the destruction of target contaminants in water is comprehensively presented. Finally, the role of reactive oxygen species (ROS), such as hydroxyl radical ($^{\bullet}OH$), superoxide anion radical ($O_2^{\bullet-}$), singlet oxygen ($^1O_2$) and hydrogen peroxide (H_2O_2), in semiconductor photocatalysis is introduced and various experimental techniques to detect ROS are also discussed.

1. Introduction

Environmental pollution due to anthropogenic activities has become a serious problem for developing and developed countries in the world. Pollution has affected various plants and animals in the ecosystem, as well as humans around the world. Therefore, there is a need for techniques to remediate the polluted environment. Photocatalytic oxidation using semiconductors as catalysts has become an attractive candidate as a green technology for environmental applications, such as water cleaning (*1*), air purification (*2*), and soil remediation (*3*) ever since Fujishima and Honda (*4*) discovered the photoinduced water cleavage of TiO_2. Photocatalysis has been employed to degrade toxic organic compounds, dyes, pesticides, volatile organic compounds, harmful gases, and bacteria (*3, 5–9*) in different media. Therefore, photocatalysis has been intensively studied over the last two decades as a clean and green process to decompose pollutants in the environment. Recently, nanomaterials have received significant attention for environmental and energy applications because their size, shape, crystallinity, and surface can provide unique physical and chemical properties, such as high surface area, well-defined structure, high dispersibility, and high reactivity (*10, 11*).

Intensive research efforts have been focused on the development of novel nanostructured materials. Many nanostructured materials, such as titanium dioxide, zinc oxide, iron oxide, metal sulfide, have been synthesized with different techniques, including sol-gel methods, hydrothermal methods, and chemical deposition methods (*12–15*). Green nanotechnology has the potential to help us to obtain desirable materials with low toxicity and cost, high chemical and thermal stability, and high degradation activity for environmental remediation (*12, 13*). This chapter will discuss the syntheses of various nanomaterials and their environmental and energy applications. In addition, selectivity issues of TiO_2 photocatalysis and reactive oxygen species during photocatalysis will be discussed.

2. Preparation of Nanomaterials for the Green Nanotechnology

There are many approaches to creating nanostructured materials. With respect to materials such as titanium dioxide, zinc oxide, and iron oxide, conventional approaches include sol-gel (*16*), chemical vapor deposition (CVD) (*17*), pyrolysis (*18*) and hydrothermal treatment (*19*, *20*). The sol-gel approach will be emphasized in detail, but brief introductions to the other conventional methods will be provided, and new developments will be addressed later.

All the synthesis methods have the end goal of creating a high-purity product. To reach this end, however, each method exploits different aspects of chemistry. For example, the CVD method for creating ceramics involves the transport of gaseous metal halides with gaseous oxygen, carbon, nitrogen, or boron depending upon the desired product (*21*, *22*). This transport is done above a heated substrate upon which the solid material can deposit. One alternative to CVD is spray pyrolysis. In spray pyrolysis, solutions containing precursors (e.g., $TiCl_4$ for titanium or NH_4F for nitrogen/fluorine) are atomized by a nebulizer. The droplets undergo pyrolysis while passing through a high-temperature container. Collection of the generated powders can be done on a ceramic filter at the container's terminal. (*23*). Another high-temperature approach is called hydrothermal treatment. In hydrothermal treatment, synthesis of a new material is performed under elevated temperature and pressure conditions in a sealed vessel (*24*). In this case, formation behavior occurs according to a liquid nucleation model. One final synthesis approach, the sol-gel approach, allows for great customizability in the final titania product.

The sol-gel process was thoroughly reviewed by Hench and West in 1990 (*25*). Two major benefits of the sol-gel process include increased purity and homogeneity in comparison to conventional synthesis methods for ceramics. Hench and West describe a "sol" as a liquid containing dispersed colloids; however, it is pertinent to include nanoparticles in addition to colloids. "Gel" implies a rigid network of interconnected pores (size < 1 μm). The sol-gel is approach is thus a wet-chemistry based approach. In preparing the sol-gel, precursors are selected and supplied to control the various desired product parameters (surface area, roughness, dopants, etc.). The final solid product can be obtained by subsequently calcining in a furnace. During this calcination process, volatile species in the sol-gel are removed, leaving the desired product in good purity.

2.1. TiO_2 Photocatalyst

As described above, various methods are capable of synthesizing various ceramic products. One commonly synthesized ceramic product of interest is titanium dioxide. This semiconductor is a photocatalyst with a band gap of 3.2 eV (*4*, *26*). Examples of applications include air and water remediation, anti-fogging glass, cancer treatment, and solar cells (*27*, *28*). Effectiveness of a given application may be optimized by the synthesis method chosen. Examples of parameters of titanium dioxide that can be optimized for various applications include, but are not limited to: crystal phase, band gap, surface area, porosity, roughness, and particle-size distribution. The sol-gel approach has enough flexibility to allow for many parameters to be altered simultaneously.

To begin, photocatalytic effectiveness is strongly correlated with adsorptive capabilities of the catalyst for a given contaminant. The adsorptive efficacy of a surface is a function of film properties such as surface area, porosity, and roughness, amongst other factors. TiO_2 surface area may be increased during the sol-gel process by incorporating a surfactant (*29*, *30*), which can simultaneously tailor pore size and porosity to meet various needs. This tailoring occurs at a sufficient surfactant concentration, the critical micelle concentration (CMC). At the CMC, the surfactant will form micelles in solution, which act as a scaffolding of sorts around which the titania framework forms. The framework stays in place during and after calcination even as the surfactant is removed by heat. The inclusion of a surfactant also offers an opportunity for doping.

When a photocatalyst is doped with impurities, the effective band gap can be decreased. As mentioned earlier, conventional TiO_2 has a band gap of 3.2 eV. This requires an incident photon whose energy corresponds to ultraviolet light. UV light only accounts about 5% of the solar spectrum; therefore, it is beneficial to decrease the effective band gap and extend the activity to a large portion of the solar spectrum (*30*). Pelaez *et al.* used a nonionic fluorosurfactant (Zonyl FS-300), which not only controlled adsorptive qualities, but also lent fluorine atoms to the final structure (*30*). In this study, a non-surfactant doping method was included as well, ethylenediamine, which acted as a nitrogen source. The final product was N, F-codoped titania, with a band gap of 2.75 eV. Other non-metal dopants have been employed such as sulfur (*31*) and carbon (*32*). Metallic dopants, both transition metals and rare earth metals, have been used with success as well, including Fe (*33*), Cr, Mn, and Co (*34*), and rare earth metals (La, Ce, Er, Pr, Gd, Nd, Sm) (*35*). Figure 1 shows the effect of doping on the photoexcitation mechanism of TiO_2. An additional factor that has influence on band gap and activity is the crystal phase.

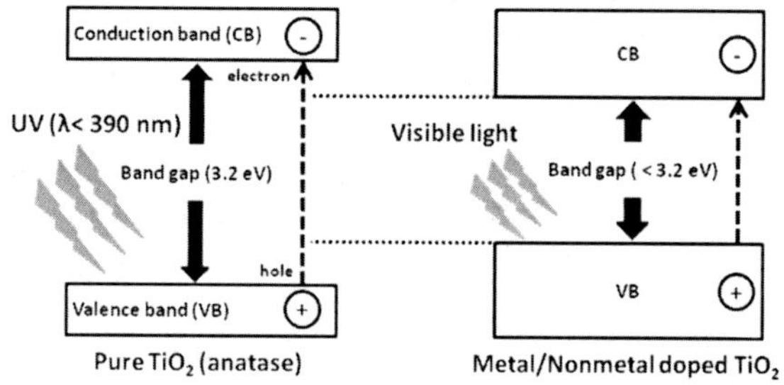

Figure 1. Schematic of photoexcitation mechanism of TiO_2.

Titania has several possible crystal phases: anatase, rutile, and brookite. If a titania sample is not one of the above phases, it is amorphous. Although rutile titania has a smaller band gap (3.0 eV), anatase titania offers higher adsorptive capability for organic compounds and also has lower recombination rates (*36*). Therefore, anatase is commonly considered the preferred phase. Crystal phase can be altered in a calcination step. Rutile is generally the most common phase at high temperatures (T > ~1000 K), and at lower temperatures, crystallinity will depend strongly on not only temperature, but also particle size (*37*).

Particle size in prepared samples generally is not a well-defined value, but rather it is a distribution of values. However, work to develop relatively monodisperse titania has been successful through the addition of salts and polymers (*38*). Han *et al.* (*39*) employed a simple approach that involved altering the concentration of calcium chloride in solution. With this method, particle size was inversely proportional to calcium chloride concentration. Such a method allows for nanotoxicology researchers to do toxicity studies on novel nanoparticles while controlling for size effects.

Beyond particle size, band gap, crystal phase, surface area, and similar structural properties, research has also explored the possibility of decorating titania. Decorating, not be confused with doping, involves the addition of a metal, such as silver (*40*), gold (*41*), copper (*42*), and platinum (*43*) to the external surface of the titania. Using silver as an example, Srisitthiratkul *et al.* (*40*) decorated particles by dispersing titania particles in solution (1g L^{-1}) with sonication for 10 minutes and subsequently adding 0.1 g L^{-1} silver nitrate with stirring for another 10 minutes. The benefit of decorating with silver comes from silver's antimicrobial activity. They observed antimicrobial behavior by Ag-decorated TiO_2 in the absence of light, whereas control TiO_2 demonstrated no such behavior.

2.2. Metal Oxide and Metal Sulfide Photocatalysts

Other metal oxide and metal sulfide nanomaterials, such as ZnO, Fe_2O_3, CdS, and ZnS, have been employed as photocatalysts for the degradation of organic contaminants (*12*, *44–46*). To synthesize those nanomaterials, various methods, such as chemical bath deposition, hydrothermal treatment, ball milling, and sol-gel method, has been employed. A brief synopsis of several synthesis methods was provided in the previous section 1.

Zinc oxide (direct band gap of 3.37 eV) has been intensively studied over the past few decades due to its broad applications in solar cells, sensors, light-emitting diodes, and photocatalyst (*44*, *47–49*). ZnO nanoparticles were synthesized by a ball milling method using a conventional ball mill with chemical reactions, which were activated mechanically during the grinding. The average particle size of synthesized ZnO nanoparticles ranged from 20 to 30 nm (*50*). Ao *et al.* (*51*) and Shen *et al.* (*52*) reported that particle size depended on the milling time and the temperature of heat treatment. The particle size decreased with increasing the milling time while it increased at higher heat treatment temperature. In order to improve the physical and chemical properties of ZnO and enhance the photocatalytic activity of ZnO, much research effort has been focused on the control of materials' morphology. Various nanostructured ZnO nanomaterials, such as nanowire, nanorods, nanotubes, and nanodisks, have been synthesized (*44*, *53–56*). ZnO nanowires for photocatalytic application were synthesized by a modified carbothermal reduction method at 1150 °C (*44*). A large amount of well-crystallized nanowires with an average diameter of 33 nm were obtained. Eighty percent of source ZnO powders converted to ZnO nanowires within 5 min. In addition to nanowires, the synthesis of ZnO nanotubes, nanorods, and nanodisks has been reported in order to enhance the photocatalytic reactivity. Lu *et al.* (*53*) obtained ZnO nanotubes by thermal oxidation of Zn nanowires. In the furnace chamber, Zn nanowires were first deposited on Si substrates, and then the nanowires were oxidized at different temperatures ranging from 400 to 700 °C under a pressure of 20 Pa. At 400 °C, well-defined, crack-free ZnO nanotubes with 60-130 nm in outer diameter and 30.7 $m^2\ g^{-1}$ BET surface area were synthesized. Different ZnO nanotubes were synthesized by a chemical bath deposition method (*54*). Firstly, ZnO nanoparticles prepared by a sol-gel method were immobilized on fluorine doped tin oxide glass substrate. The solvent on the substrate was removed by heat treatment. Then, the substrate was suspended to grow ZnO nanotubes in a solution of 0.1 M zinc nitrate hexahydrate and 0.1 M methenamine at 75 °C for 3h. The substrate was rinsed with deionized water, immersed into 0.3 M KOH solution at 80 °C for 1h following drying at 100 °C for 1h. Highly vertically-aligned, hexagonal ZnO nanotube arrays with ~ 650 nm in diameter were obtained. In addition, the synthesis of ZnO nanorods and nanodisks by a sol-gel method has been reported (*55*, *56*). To obtain ZnO nanorods, NaOH, zinc acetate, manganese acetate, and sodium dodecyl sulfate were used. ZnO nanorods with 7-10 nm in diameter and 200-300 nm in length were grown in the presence of sodium dodecyl sulfate (SDS) as soft template at room temperature (*55*). The diameter of nanorods decreased with an increase in the molar ratio of Mn/OH in the solution. However, the diameter increased when the ratio reached a certain level. The nanorods were not obtained with only manganese acetate or SDS in the solution. Also, the formation of nanorods was only influenced by Mn^{2+} in the solution. No formation of nanorods was observed by replacing Mn^{2+} with other cations, such as Cr^{2+}, Cu^{2+}, Co^{2+}, Ca^{2+}, Cd^{2+}, Ni^{2+}, Pb^{2+}, Fe^{3+}, Al^{3+}, Er^{3+}, or Sn^{4+}. Zhai *et al.* (*56*) reported the synthesis of hierarchical ZnO nanodisks through a chemical hydrolysis method in a dimethyl sulfoxide (DMSO)-H_2O system. To obtain ZnO nanodisks, the concentration of DMSO in the solution was a critical parameter. Nanodisks with a uniform size of *ca.* 200-300 nm were formed with 90% DMSO. The nanodisks were composed of many 20-40 nm nanocrystallines. Hexagonal nanorings (~ 300 nm, 40-60 nm one side diameter), nanoparticles (200-350 nm in diameter), and microtyres (1.1 μm in diameter) were formed with 60, 70, and 80% DMSO, respectively. The BET surface area of nanodisks, nanorings, nanoparticles, and microtyres was 37.3, 21.9, 26.1, and 12.8, respectively.

Iron oxides also have been widely used as photocatalysts to remove organic pollutants (*57–60*). Hematite (α-Fe_2O_3) is usually used as a photocatalyst due to the visible light absorption capacity (2.1 eV bandgap) (*60*, *61*). α-Fe_2O_3 was obtained using starch as a capping reagent (*61*). Ferric hydroxide was synthesized by adding triethyl amine after reaction with $FeCl_3$ and starch solution. After centrifugation and calcination, very uniform α-Fe_2O_3 nanoparticles with a narrow size distribution (average particle size of 30 nm) were synthesized. BET surface area increased from 24 to 126 $m^2\ g^{-1}$ with increasing precursor Fe^{3+} concentration. Particle size and shape were influenced by pH of solution. Zhou *et al.* (*60*) reported the synthesis of hematite by thermal dehydration. α-Fe_2O_3 nanorods with 40 nm length, 11 nm width, and 93 m^2 g^{-1} BET surface area were obtained by heating α-FeOOH precursors at 300 °C in air for 1 h.

In addition, metal sulfides have been extensively investigated for photocatalytic degradation of environment pollutants. Recently, CdS, ZnS and their suspensions have been intensively studied (*14*, *15*, *62–64*). CdS nanocrystalline particles are considered attractive candidates for visible light-induced photocatalysis because they can produce reactive oxygen species, such as superoxide ($O_2^{\bullet -}$) and hydroxyl ($OH^{\bullet}$) radicals in aqueous solution (*62*). Bao *et al.* (*14*) synthesized nanoporous CdS, including nanosheets and hollow nanorods, by a two-step aqueous route. Nanoporous nanosheets with size less than 60 nm and thickness of ~9 nm and hollow nanorods with 30 nm length and 7-14 nm in outer diameter were obtained by an initial precipitation of nanoporous $Cd(OH)_2$ intermediates following S^{2-}/OH^- ion exchange. Nanopores' diameter and BET surface area of nanosheets were ~ 3 nm and 112.8 m^2 g^{-1}, respectively. Yang *et al.* (*63*) synthesized visible light-activated, Zn-doped CdS (Zn-CdS) nanostructures by hydrothermal treatment. To obtain Zn-CdS, an aqueous solution containing $Cd(CH_3COO)_2$, $Zn(CH_3COO)_2$, $(NH_2)_2CS$ (thiourea), and deionized water was autoclaved at 120-180 °C for several hours. BET surface area and bandgap increased with increasing molar ratio of Zn/Cd. Zn doping enhanced photocatalytic activity and improved the stability of CdS.

In comparison with CdS, ZnS has relatively large bandgap (3.6 eV), which is a limitation to utilize the visible light portion of the solar spectrum. Therefore, UV light sources were employed to activate ZnS photocatalysts. Hu *et al.* (*15*) synthesized ZnS nanoporous nanoparticles by a solution-phase thermal deposition route in the presence of poly(*N*-vinyl-2-pyrrolidone). The spherical monodisperse nanoparticles with 60 nm diameters and large BET surface area (156.1 m^2 g^{-1}) were obtained. Ren *et al.* (*64*) prepared ZnS microspheres consisting of interwoven nano-sheets by hydrothermal conditions using NaOH. The primary particle size (3 nm to bulk) and nanostructure (compact microspheres to nanosheets) were significantly influenced by NaOH concentration.

2.3. Novel Methods for Preparation of Nanomaterials through Green Technologies

2.3.1. Microwave-Assisted Processing

The microwave oven was invented by Dr. Percy Spencer after the Second World War from the radar technology developed for the war. The development of a magnetron to generate microwaves of a definite frequency was the most significant step in this invention. Domestic microwave ovens appeared in the consumer markets in 1950. However, microwave technology was not employed in inorganic material synthesis until the late 1990s. It was revealed in 1996 by Komerneni and co-workers that, under similar processing circumstances, preparation of hematite by a microwave hydrothermal reaction is 36 times faster than the conventional hydrothermal synthesis (*65*). The use of the microwave in chemistry has been seen to improve yields, selectivity and allows conversions that would otherwise be impossible. Since then microwave-assisted synthesis is becoming a widespread method to synthesize organic and inorganic nanostructured materials. The major advantage of this technology is that no further annealing is required for developing crystalline materials. Microwave processing is found to be significantly faster compared with the synthesis using a conventional chamber furnace. Researchers at the Dublin Institute of Technology, Ireland have recently developed (Patent publication number WO/2009/113045 A2) a microwave-assisted synthesis of mesoporous titanium dioxide with high surface area (250 m^2 g^{-1}), small particle size (15 nm) and excellent visible-light activity (up to four times higher than commercial samples) (*66*). Research into microwave-assisted preparation of nanomaterials has been studied for its effectiveness against more traditional methods. It has been seen that the majority of the other preparation methods such as sol-gel methods require high temperatures (450-1400 °C) and a longer processing time, typically six to eight hours, to obtain crystalline particles, whereas with the addition of the microwave in the preparation, temperatures as low below 250 °C can be used to produce quality crystalline TiO_2 (*67*). The efficiency of hydrothermal processing in the preparation of particles can be seen in the reduction of time taken to prepare highly crystalline particles as it allows for rapid heating and extremely rapid kinetics of crystallization. Compared to sol-gel processing, the microwave-assisted technique offers an energy efficient, eco-friendly, rapid and convenient method of synthesizing doped and undoped metal oxides such as TiO_2 (*68*), CuO nanoparticles (*69*) and copper sulfate nanorods (*70*). An industrially viable method for the preparation of mesoporous titania nanomaterials for NanoChromic display device applications by the use of microwave processing was reported (*66*). Spherical aggregates of nanocrystalline titania materials were rapidly prepared using titanium butoxide, deionized water and alcohol at comparatively low microwave power intensity (300 W) for 2 min irradiation (Figure 2). The samples prepared via microwave process using an 2.45 GHz industrial microwave oven, CEM Mars 5 (Figure 3) showed a surface area up to 240 m^2 g^{-1}, which was significantly higher than similar traditional sol–gel or commercial samples (40 m^2 g^{-1}).

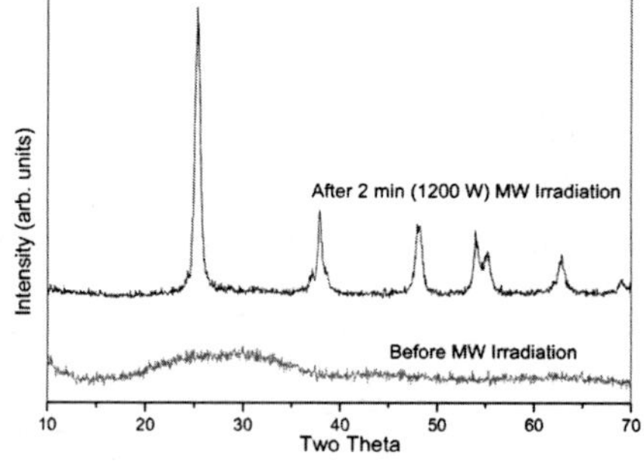

Figure 2. XRD pattern of titania precursor sample before and after microwave irradiation.

Figure 3. CEM MARS-5 2.45 GHz Microwave processing facility housed at CREST, Dublin Institute of Technology.

The development of materials that can utilize visible light (>400 nm) is essential for the effective exploitation of solar energy-driven photocatalysis. Photocatalysis is a reaction that uses light to activate a substance that modifies the rate of a chemical reaction without being involved itself and the photocatalyst is the substance which can modify the rate of chemical reaction using light irradiation. Like above, microwave irradiation can be used in the preparation of visible-light active (VLA) materials which can be used to break down contaminants of interest. Photocatalytic anatase TiO_2 has been formed by using a mixed solution of tetrabutyl titanate, hydrogen fluoride (HF) solution and additional water at 200 °C for 30 min in a rapid synthesis using microwave-assisted hydrothermal treatment (*71*). TiO_2 nanostructures of various morphologies have been prepared using microwave irradiation as its synthesizing technique. These different patterns are seen to all have different photocatalytic responses when tested for their degradation abilities against methylene blue under ultraviolet light. They also are seen to have a superior photocatalytic rate when compared to the industry standard, Degussa P25 TiO_2 (*72*). A simple microwave-irradiation method has been seen in the making of silica-coated ZnO nanoparticles. These were characterized by X-ray photoelectron spectroscopy (XPS), Fourier transform infrared spectroscopy (FT-IR), high-resolution transmission electron microscopy (HR-TEM), CHN elemental analysis and zeta potential measurements for the presence of Zn, O and Si. These particles have certain photocatalytic properties, where the silica effectively inhibits the photocatalytic activity of the ZnO nanoparticles (*73*). The use of microwave irradiation has the ability to effectively reduce time and energy used in the making of many different nanomaterials in the area of nanoscience.

2.3.2. Sonochemical Methods

Advanced research into the chemical effects of ultrasound radiation has been a rapidly growing research area for the last two decades. An ultrasonic reaction occurs when ultra-sound waves are irradiated in a liquid phase. During such an occurence, intense local heating and high pressure are produced (hot spots). These local hot spots initiate and propagate high-energy chemical reactions. There are potentially many benefits to using ultrasonic technology as a strategic synthetic method for making nanomaterials at a low temperature. A number of investigations using ultrasound radiation to synthesize ceramic materials have been explored with potential uses in materials science. For example, the materials group at Trinity College Dublin, prepared superparamagnetic γ-Fe_2O_3 nanoparticles by using ultrasonic irradiation of an Fe(II) alkoxide derivative (*74, 75*). There have been many other studies carried out in the field of sonochemistry in relation to nanoparticles. Sonochemical decomposition of $Fe(CO)_5$ was carried out in the presence of various surfactants, resulting in the production of stable colloids of undecenoate, dodecyl sulfonate, and octyl phosphonate coated Fe_2O_3 nanoparticles with a diameter of 5-16 nm (*76*). The use of sonochemistry in relation to the synthesis of nanomaterials has been seen to have several effects on the particles produced. One such effect is the different morphologies it can produce and its ability to control the shape of chemicals such as $PbWO_4$ nanostructures (*77*), Zn(II)bis(1-(2-pyridylazo)-2-naphthol) ($Zn(PAN)_2$) complex nanorods (*78*), porous Cu_2O nanospheres (*79*), and single-crystalline PbF_2 nanorods (*80*). Increasing ultrasound irradiation power can also be seen to decrease the particle size of nanoparticles such as ZnO (*81*). Nanosize ZnO was similarly prepared using a simple synthesis method using zinc acetate as a precursor and 1, 3-propanediol as a solvent using ultrasound sonication (*82*). Previously, a novel sonochemical process in which the hydrolysis of tetraisopropyl titanate, a titanium alkoxide, under high-intensity ultrasonic irradiation (20 KHz, 100 W cm^{-2}) at 90 °C for 3 hr was used to directly prepare anatase nanocrystalline TiO_2 (*83*). The above results show that sonochemical processing has the ability to produce functional materials at a low temperature in a shorter time compared to conventional synthetic methods.

3. Applications of Nanomaterials for the Green Nanotechnology

3.1. Water, Air, and Soil Remediation and Energy Application

Most formal definitions of nanotechnology revolve around the study and control of phenomena and materials at length scales below 100 nm. In October 2011 the European Commission adopted the definition of a nanomaterial as a natural, incidental or manufactured material containing particles, in an unbound state or as an aggregate or as an agglomerate and where, for 50% or more of the particles in the number size distribution, one or more external dimensions is in the size range 1 nm - 100 nm (*84*). Today, nanotechnology efforts are centered at controlling matter at near-atomic scales to produce unique or enhanced materials, products and devices, optimizing the fabrication processes, proposing innovative solutions for environmental remediation/protection, enhancing the perspectives for optimum exploitation of renewables and contributing to economic development. In addition, there is an important trend of increasing societal concerns about the safety aspects of nanomaterials and rational use of related products. Aimed at answering the above questions and building on existing work on green chemistry and engineering, green nanotechnology is almost as broad as nanotechnology, aiming at developing technology (made in a green way) that can tackle environmental and energy challenges, for example with new solar energy and water treatment technologies (*85*).

3.1.1. Advanced Oxidation Processes Involving Innovative Nanocatalysts for Environmental Repair

Nanotechnology can guarantee cleaner drinking water providing important achievements that constitute the basis for a sustainable and cost-effective, solar energy technology for water treatment (*86*). Very recently, special interest was expressed in exploiting solar energy and recent advances in nano-engineered titania photocatalysts and membranes for the destruction of extremely hazardous compounds in water (*87*), including novel emerging pollutants (pharmaceuticals, toxins, hormones) of high environmental and health impact (*88–90*). Thus, the synthesis of highly reactive TiO_2 photocatalysts has been reported by means of an efficient sol-gel method combined with the removal of larger aggregates, proper aeration and optimization of the calcination temperature (*88*). In addition, an innovative sol-gel synthesis based on the self-assembling template method has been also applied to synthesize mesoporous anion doped TiO_2 with N-F, S and C hetero-atoms using suitable surfactants and reagents, to improve simultaneously the structural, morphological and electronic properties of TiO_2 nanomaterials and achieve anion doping of titania with high visible light photo-induced reactivity (*31, 88, 89*). Co-doping of TiO_2 by N and F anions was achieved by a modified sol–gel method using a nonionic fluorosurfactant, as pore template material and fluorine source combined with ethylenediamine as nitrogen source (*91*). Moreover, the synthesis of m-modified TiO_2 was optimized using combustion sol-gel synthesis employing urea as nitrogen source (*92*). The evaluation of the photocatalytic activity of the above nanostructured catalysts for water detoxification from cyanotoxin compounds released by cyanobacteria blooms (e.g. microcystin-LR/MC-LR, one of the most common and toxic congeners) as well as taste and odor pollutants released by cyanobacteria, geosmin (GSM) and 2-methylisoborneol (2-MIB) was performed using UVA, solar and visible light (*93*). This work was further expanded to water detoxification by endocrine disrupting compounds (EDCs) such as bisphenol A and 17β-estradiol (*94*). Finally, composite carbon nanotubes/titania nanostructures were developed on well-aligned carbon nanotubes (CNTs) produced by means of chemical vapor deposition (CVD) using surfactant-stabilized Fe_3O_4 particles and anodized alumina disks as templates. All these innovative approaches have been implemented by integrating extensive R&D efforts spanning a wide range of interdisciplinary activities from the development of novel nanomaterials and membranes to reactor engineering and analytical methods for water pollutants of emerging concern as well as pilot plant scale studies and cost analysis for water purification (*95–97*).

Besides the establishment and validation of methods for detection of emerging contaminants, nanotechnology also permits the development of smart instruments (sensors). Thus, an innovative multi-walled carbon nanotube (MWCNT) electrochemical biosensor was developed (*98*) for monitoring MC-LR in drinking water supplies. This novel approach was devised relying on the combination of the work carried out on the functionalization of carbon nanotubes with the MC-LR analytical determination.

Using innovative sol-gel chemistry and electrochemistry, UV and visible-light responding nanostructured titania based catalysts can be developed that are highly-efficient in photoinduced multi-dynamic processes. This comprises advanced oxidation processes with applications outside the field of water detoxification including self-cleaning photoinduced superhydrophilicity (*99*) and self-sterilizing (*100*) as well as atmospheric depollution from volatile organic compounds (VOCs) and nitrogen oxides (*101*). Thus, optimized N, F-TiO_2 films were exploited as immobilized photocatalysts for air purification and specifically in the photocatalytic oxidation of NOx gas pollutants under UV and visible light, exhibiting very efficient photocatalytic activity under visible-light excitation (*102*). A maximum percentage of NO removal equal to 24.2 %, one of the best reported up to now for visible-light photocatalysis, was reported together with excellent stability and reproducibility in successive light on-off experiments. Moreover, very promising photocatalytic materials (ECT-1023t, GOT composites and N-TiO_2) embedded into polymeric aerogel type hollow fibers (*95*) were applied for NOx and volatile organic compounds (VOCs) abatement in air in a continuous-flow photocatalytic air treatment process in which the gas stream was forced to flow through the pore structure of light transparent fibers in contact with the photocatalysts.

Besides water and air remediation, advanced oxidation processes and especially photocatalysis are also capable of organic pollutants removal in soil. In a recent study (*103*), a pulsed-discharge plasma-TiO_2 catalytic (PDPTC) technique was proposed to investigate the remediation of p-nitrophenol (PNP) contaminated soil, by combining TiO_2 photocatalysis and nonthermal discharge plasma, where catalysis of TiO_2 is driven by the pulsed discharge plasma. In addition, removal of alkylphenols from polluted sites using surfactant-assisted soil washing and photocatalysis was observed, in the presence of TiO_2 dispersions irradiated with simulated solar light (*104*).

3.1.2. Combating Climate Change

Although there is no single solution to the challenge of reducing greenhouse gas emissions, nanotechnology can help to combat climate change by improving efficiency and bringing new energy sources (developing low-carbon forms of energy) to the market. This represents an enormous opportunity for the international community. A recent report lists five areas where nanotechnology can make a difference: fuel additives to increase the efficiency of diesel engines; photovoltaic technology for solar cells; the hydrogen economy and fuel cells; batteries and supercapacitors for energy storage; and improved insulation for houses and offices (*105*).

Nanotechnology offers the promise of better solar cells. In fact, based on a process mimicking natural photosynthesis in plants that separates the reaction centers of light absorption and charge transport/storage processes, easily processable and cost-effective solar cells (dye-sensitized solar cells-DSCs) can be developed, based on the dye-sensitization approach (*106*, *107*). In this device, the active photoelectrode consists of a thin layer of mesoporous nanostructured wide band-gap semiconductor (TiO_2 has been almost exclusively used as a semiconductor in high performance DSCs) deposited onto a conducting substrate and a monolayer of a visible light-absorbing dye chemically adsorbed onto the nanoparticles of the semiconductor. This new type of nanotechnology-based third generation photovoltaic (hybrid organic-inorganic) currently presents power conversion efficiencies (η) of the order of 12% (*108*, *109*). Coordinated efforts including the participation of partners from both academia and private companies are now under intensive investigation at a global level, in order to endow the device with a solid character as well as with enhanced life time and stability, features that make this technology a very promising alternative to classical solid state PVs (*110–112*).

3.2. Photocatalytic Disinfection of Water

Water is an important natural resource and safe drinking water is vital for human existence and good quality of life. Clean water resources are becoming depleted due to population growth, over-use of resources and climate change. Since the adoption of the Millennium Development Goals (MDG), the WHO/UNICEF Joint Monitoring Programme for Water Supply and Sanitation has reported on progress towards achieving Target 7c: "reducing by half the proportion of people without sustainable access to safe drinking water and basic sanitation" (*112*). It was reported that, as of 2010, the target for drinking water has been met; however, 780 million people are still without access to an improved drinking water source. Many more are forced to rely on sources that are microbiologically unsafe, leading to a higher risk of contracting waterborne diseases, including typhoid, hepatitis A and E, polio and cholera (*113–116*). Although the MDG drinking-water target refers to sustainable access to safe drinking water, the MDG indicator – "use of an improved drinking water source" – does not include a measurement of either drinking water safety or sustainable access. This means that accurate estimates of the proportion of the global population with sustainable access to safe drinking water are likely to be significantly lower than estimates of those reportedly using improved drinking water sources. It is estimated that, at the current rate of progress, 672 million people will not use improved drinking water sources in 2015. It is likely that many hundreds of millions more will still lack sustainable access to safe drinking water.

Piped-in water supplies are a long term goal and interventions to improve water supplies at the source (point of distribution) have long been recognized as effective in preventing waterborne disease. Recent reviews have shown household-based (point-of-use) interventions to be significantly more effective than those at the source for the reduction of diarrhoeal diseases in developing regions (possibly due to contamination of water between collection and use). As a result, there is increasing interest in such household-based interventions that can deliver the health gains of safe drinking water at lower cost (*116*). Household water treatment and safe storage (HWTS) is one option for improving the quality of water for consumption within the home, especially where water handling and storage is necessary and recontamination is a real risk between the point of collection and point of use. Limitted access to nearby sources, unreliable piped supplies and reliance on rainwater are all factors that make household storage a necessity. Living conditions in many humanitarian crises also call for effective HWTS. The practice of household water treatment and safe storage can help improve water quality at the point of consumption, especially when drinking water sources are distant, unreliable or unsafe. However, HWTS is a stop-gap measure only and does not replace the obligation of a service provider to supply access to safe drinking water. Household water treatment (HWT) methods include boiling, filtration, adding chlorine or bleach, and solar disinfection.

In 2008, Clasen and Haller reported on the cost and cost effectiveness of household based interventions to prevent diarrhoea (*117*). They compared: chlorination using sodium hypochlorite following the "Safe Water System" (SWS) developed and promoted by the US Centers for Disease Control and Prevention (CDC); gravity filtration using either commercial "candle" style gravity filters or locally fabricated pot-style filters developed by Potters for Peace; solar disinfection following the "SODIS" method in which clear 2 L PET bottles are filled with raw water and then exposed to sunlight for 6 - 48 h (Figure 4); and flocculation disinfection using Procter & Gambles PUR® sachets, which combine an iron-based flocculant with a chlorine-based disinfectant to treat water in 10 L batches. They concluded that household-based chlorination was the most cost-effective. Solar disinfection was only slightly less cost-effective, owing to its almost identical cost but lower overall effectiveness. Given that household-based chlorination requires the distribution of sodium hypochlorite, solar disinfection has a major advantage in terms of non-reliance on chemical distribution.

Sunlight is freely available on Earth and the combined effects of heat and UV from the sun can inactivate pathogenic organisms present in water. Of course, there are a number of parameters which affect the efficacy of the solar disinfection (SODIS) process, including the solar intensity, temperature, and the level and nature of the contamination (some pathogens are more resistant to SODIS than others). One approach to SODIS enhancement is the use of heterogeneous photocatalysis.

In 1985, Matsunaga *et al.* reported the inactivation of bacteria using TiO_2 photocatalysis (*118*). To date, there have been many papers published presenting results on the photocatalytic inactivation of microorganisms including bacteria, viruses, protozoa, fungi and algae. In 1999, Blake *et al.* published an extensive review of the microorganisms reported to be inactivated by photocatalysis (*119*). More recently, McCullagh *et al.* reviewed the application of photocatalysis for the disinfection of water contaminated with pathogenic microorganisms (*120*). Malato *et al.* have published an extensive review on the decontamination and disinfection of water by solar photocatalysis (*121*) and, in 2010, Dalrymple *et al.* reviewed the modeling and mechanisms of photocatalytic disinfection (*122*).

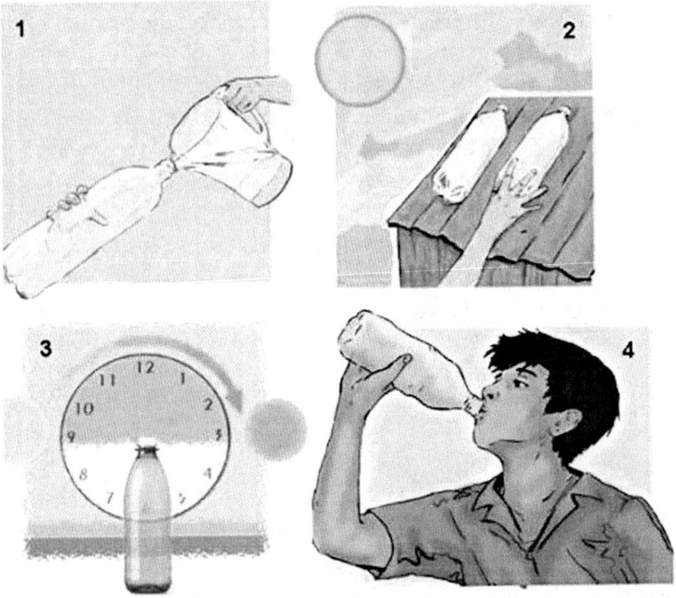

Figure 4. Simple protocol for the solar disinfection of water SODIS process. (courtesy of Dr. Kevin McGuigan, Royal College of Surgeons in Ireland, drawn by Maria Boyle).

In most studies, the hydroxyl radical is suggested to be the primary species responsible for microorganism inactivation. Some researchers have examined the role of other reactive oxygen species, such as H_2O_2 and $O_2^{\bullet-}$ in disinfection (*123*). Reactive oxygen species cause fatal damage to microorganisms by disruption of the cell membrane and/or by attacking DNA and RNA (*119*). Other modes of action include damage to the respiratory system within the cells (*124*). Some researchers have attributed cell death to lipid peroxidation of bacterial cell membrane (*125–127*). The peroxidation of the unsaturated phospholipids that are contained in the bacterial cell membrane causes loss of respiratory activity (*124*) and/or leads to a loss of fluidity and increased ion permeability (*126*). This is suggested to be the main reason for cell death. Other researchers suggested that the cell membrane damage can open the way for further oxidative attack of internal cellular components, ultimately resulting in cell death (*128*).

The photocatalyst (most commonly TiO_2) can be used as an aqueous suspension or it may be immobilized on a supporting substrate. Most studies report that suspension reactors are more efficient due to large surface area available for the reaction and no mass transfer limitations (*129*, *130*). The main drawback of using nanoparticles in suspension is the requirement for post-treatment separation and recycling of the catalyst, making the treatment more complex and expensive. Treatment systems utilizing immobilized TiO_2 have gained more attention. There are a range of methods available for the preparation of immobilized photocatalyst films on supporting substrates (*129*).

It is clear from the literature that photocatalytic disinfection is more effective than UVA treatment alone (*131–134*). However, scale-up and testing under real sun conditions may show different results depending on the reactor configuration and on the model organism being tested. For example, Alrousan *et al.* tested solar photocatalytic (SPC-DIS) and solar disinfection (SODIS) of water at pilot scale using different reactor configurations with and without immobilized TiO_2 (Evonik Aeroxide P25). The model organism used was *E. coli.* The use of compound parabolic collectors improved the SODIS and SPC-DIS of water, however, the improvement was less significant compared to the improvements reported previously for SODIS in static batch reactors. Kinetic fitting yields a log-linear component (1st order rate constant). The following order was found for k where coated refers to TiO_2 coating and the equals sign indicates no significant difference; uncoated external – coated internal ≥ double coated tube ≥ uncoated double tube. It is known that *E. coli* is inactivated by SODIS and it may be a 'soft' target for comparing the effectiveness of SODIS vs SPC-DIS. Nevertheless, photocatalysis presents advantages in terms of the non-recovery of inactivated organisms and the inactivation of SODIS resistance organisms (*135*).

Conventional TiO_2 is a UV absorber and will only utilize the around 4% of the solar spectrum. Much research effort has been placed on increasing the solar efficiency of TiO_2 by doping to shift the optical absorption towards the visible region of the electromagnetic spectrum. Different approaches have been attempted including doping the TiO_2 with metal ions. According to the literature, one of the more promising approaches to achieve visible light activity is doping with non-metal elements as reported in the previous section. However, the number of publications concerning the photocatalytic activity of these materials for the inactivation of microorganisms is limited. The UV activity of undoped TiO_2 may be greater than the visible light activity of a doped material. Therefore, for solar applications, the efficiency should be tested under simulated solar irradiation or under real sun conditions. Rengifo-Herrera and Pulgarin reported on the photocatalytic activity of N, S co-doped and N-doped commercial anatase (Tayca TKP 102) TiO_2 powders towards phenol oxidation and *E. coli* inactivation (*136*). However, these novel materials did not present any enhancement as compared to Degussa P-25 (now Evonik Aeroxide P25) under simulated solar irradiation. They suggest that while the N or N, S co-doped TiO_2 may show a visible light response, the localized states responsible for the visible light absorption do not play an important role in the photocatalytic activity.

More research is required to determine if visible-light active materials can deliver an increase in the efficiency of photocatalysis under solar irradiation. Improvements in photocatalytic reactor design could make this technology available at low cost for the solar disinfection of water in developing regions.

4. Selective Photocatalysis for the Decomposition of Target Contaminants

4.1. Selectivity Issue in TiO_2 Photocatalysis

Photocatalysis has become a major discipline and research area due to the mutual enrichment of researchers in many different fields. Since the photocatalytic process is governed by a free radical mechanism, it is characterized with non-selective attack on organic chemicals in water. Decomposition rate of many organic molecules was reported to be more or less similar (*137*). The non-selectivity of the process sounds great since re-designing photocatalysts is not required for the decomposition of a certain contaminant. However, the poor selectivity also implies that the oxidation does not differentiate between highly toxic contaminants and less toxic contaminants in water (*138*). When a new TiO_2 photocatalyst is synthesized, its reactivity is evaluated with a probe chemical in pure water to compare it with a benchmark catalyst (e.g., Degussa P-25 TiO_2 nanoparticles) and mechanisms for the enhanced reactivity are also investigated (*139*). This is a common practice in the area. However, an important issue has been invoked: would the catalyst work for the decomposition of such a probe chemical in natural water matrix containing co-existing chemicals, in particular natural organic matter (NOM)? Typical source water contains low concentrations of highly toxic organic substances (e.g., anthropogenic chemicals) together with high concentrations of less toxic naturally present organic matter (i.e., NOM). As shown in Figure 5, the non-selective nature of TiO_2 photocatalysis comes under criticism when the process is applied for water treatment to finally polish the effluent from treatment plants. Target contaminants have to compete with abundant NOM for catalytic sites and hydroxyl radicals (*140*). The nonspecific sorption of NOM inhibits the sorption of target chemicals and also contributes to fouling of the catalyst surface. These significantly decrease the oxidation rate of target chemicals.

4.2. Current Strategies for Selectivity Enhancement

Prevention of the nonspecific sorption and oxidation of coexisting organic chemicals at the TiO_2 surface (e.g., NOM, organic interferents, proteins, microorganisms, and any organic chemicals other than target chemicals) is of interest (*141–147*). Operational parameters, including temperature, pH, residence time, solvent, and UV wavelength, can change the degradation ratio of more than two competing chemicals (*138*). Methods for synthesizing new TiO_2 materials with controlled physical and chemical properties (e.g., particle size, crystal size and phase, and surface structure) can also provide many pathways to the preferential oxidation (*138*). Photocatalysts can also be specially designed for selectivity. This includes doping with noble metals, coating with a specific material, impregnation onto inert organic/inorganic domains, and chemical imprinting (*138*, *144–147*). Mixing activated carbon and TiO_2 particles was found to increase the degradation rate of phenol (*148*). Coating photocatalysts with a material that has specific affinity towards a certain contaminant (or a group of contaminants) can decompose the target contaminant selectively. In most cases, coating materials are molecules which have hydrophobic moieties to accommodate sorption of hydrophobic contaminants (*149*). There is a sorption-and-shuttle approach, where sorption of contaminants on organic or inorganic domains located closely to TiO_2 is followed by their diffusion from the adsorptive sites to the photocatalytic sites (*146*). Predesigned immobile organic molecular recognition sites can also be introduced to inert domains in the vicinity of photocatalysts to facilitate the selective adsorption (*150*). Another new method is to imprint cavities of target molecules on the TiO_2 surface (*151*). However, all of the approaches are either too general or too specific to practically apply to water treatment (note Figure 5). For example, when TiO_2 is modified to increase hydrophobicity, it adsorbs and decomposes preferentially a group of hydrophobic contaminants. When TiO_2 is functionalized with a chemical marker towards a specific contaminant, it exhibits an ultimate selectivity only to the contaminant. When a target changes, the chemical marker should be replaced accordingly. The chemical functionalization is case-specific and it is also not an easy task. Only few studies aiming at achieving preferential photocatalytic degradation in mixed streams, have been reported.

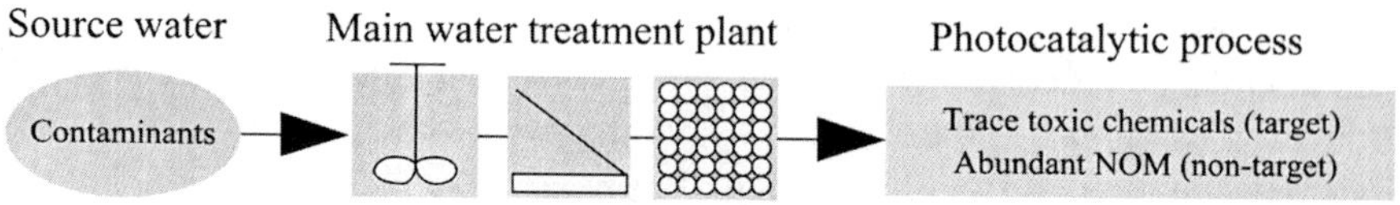

Figure 5. Application of TiO_2 photocatalysis to water treatment to polish the effluent from a main water treatment facility composed of coagulation/flocculation-sedimentation-filtration. Since conventional water treatment plants do not have a capability to remove toxic organic chemicals, the photocatalytic process, as a tertiary treatment option, should focus on the destruction of toxic organic chemicals remaining in the effluent.

4.3. Prospects and Research Direction

Selective decomposition of target chemicals can be achieved by controlling their preferential transport to the TiO_2 surface, considering the heterogeneous nature of the catalytic process. Selection by size has been researched by developing innovative photocatalytic materials with well-defined specific porous structures (*147*, *152–154*). As an example, incorporation of TiO_2 particles into sheet silicates of clay was suggested (*152*). Only molecules that are smaller than the distance between adjacent sheet layers are allowed to contact with the incorporated TiO_2 (*153*). Similarly, TiO_2 nanoparticles are grafted onto the surface of mesoporous silicates (*154*). They can also be encapsulated with a porous SiO_2 shell, where molecules larger than the pore size of a SiO_2 shell (e.g., macromolecules and NOM) are excluded from the physical contact with core TiO_2 particles (*155*). In many cases, however, the encapsulated TiO_2 catalysts exhibit low reactivity due to limitations in light penetration and mass transfer. Detailed research studies are needed to find ways of enhancing both the selectivity and the reactivity of such composite materials, including precise control of the mesoporous and microporous structure of a shell, fabrication of an ultrathin and uniform layer with high mechanical stability, encapsulation of individual TiO_2 particle with a shell, and introduction of a void space between core particles and a shell layer. As discussed, common approaches for the size exclusion have required introduction of a secondary material with a well-defined porous structure. Selection of the second material might have been based on the easiness of processing and controlling its porous structure. This can be explained by the wide use of silicates in many applications (*147*, *154*). Meanwhile, there has been no attempt to use porous TiO_2 photocatalysts for the size exclusion. This is because synthesis and control of the porous structure of TiO_2 is not as easy as that of SiO_2. A pure TiO_2 catalyst with an engineered porous structure, when properly designed, is expected to work for the size exclusion, without introduction of any other complex physical and chemical modifications or aids. This would comply with the green engineering and chemistry principle. An engineered TiO_2 photocatalyst with controlled properties through innovative material design and processing can serve as an excellent green material for the selective destruction of target contaminants in water.

5. Reactive Oxygen Species (ROS) in Semiconductor Photocatalysis

5.1. Generation of ROS

Semiconductor photocatalysis has shown tremendous promise for the environmental remediation of an extensive number of pollutants and toxins in water and air streams. Photoexcitation of a semiconductor material can promote an electron from the valence band (VB) to the conduction band (CB) resulting in an electron-hole pair. In competition with recombination the electron can act as a reducing entity and the hole as an oxidizing entity. The VB and CB positions are critical to the effectiveness and economic feasibility of semiconductor photocatalytic materials for water purification. The energy difference between the VB and CB dictates the wavelength of irradiation required for photoexcitation. Large band gaps require higher energy UV irradiation, while semiconductor materials with smaller band gaps can be photoexcited with lower energy visible light irradiation. Photoexcitation with visible light has a significant economic advantage because of the high cost associated with generating the UV light required for activating materials with large band gaps. Doping of TiO_2 materials can extend the absorbance into the visible light range, but the semiconductor photocatalytic reaction pathways initiated by UV and visible light can involve significantly different processes (*156*). Another critical factor is the oxidation and reduction potential of the hole and electron (band position). The VB and CB positions as well as band gaps for a number of common semiconductor materials are summarized in Figure 6 (*157*).

A variety of ROS can be produced during semiconductor photocatalysis. ROS are central to the semiconductor photocatalytic and advanced oxidative water treatment technologies (*158*, *159*). Hydroxyl radical ($^{\bullet}OH$), superoxide anion radical ($O_2^{\bullet -}$), singlet oxygen ($^1O_2$) and hydrogen peroxide (H_2O_2), can be produced during semiconductor photocatalysis but the production and role of the different ROS are dependent on the photocatalyst, reaction conditions and target compound. UV TiO_2 photocatalysis has been extensively studied and hydroxyl radical is generally believed to be responsible for the primary degradation of target pollutants. Visible light initiated (VLA) photocatalysis typically does not possess the oxidation potential required for the direct formation of hydroxyl radical and thus superoxide anion radical and singlet oxygen appear to play important roles in the degradation processes associated with VLA photocatalysis. Ultimately, the VB and CB positions are critical to the effective production of ROS and the economic feasibility of semiconductor photocatalytic materials for water purification.

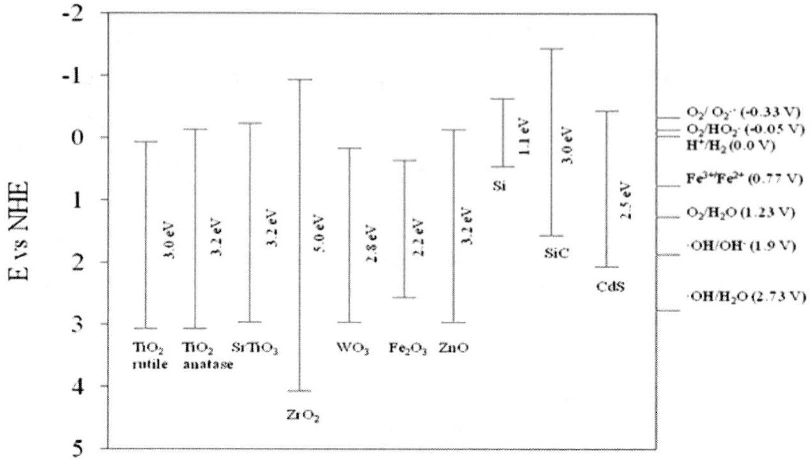

Figure 6. VB and CB positions, band gaps for a number of common semiconductor materials at pH 0. The energy scale is referenced by normal hydrogen electrode (NHE).

5.2. Roles of ROS during Photocatalysis

The actual ROS produced during semiconductor photocatalysis depends on the band edge position of semiconductors. Among the ROS, $^{\bullet}OH$ is the most powerful oxidant and is primarily responsible for the degradation processes during UV photocatalysis in aqueous solution (*160*, *161*), and advanced oxidation technologies (*162*, *163*). Hydroxyl radical reacts by the addition to double and triple bonds, and aromatic rings, hydrogen-atom abstraction from C(sp3)-H bonds, and electron transfer with most organic compounds and many inorganic substrates often at nearly diffusion-controlled rates (*164*). Singlet oxygen, superoxide anion radical, and hydrogen peroxide have lower oxidation potentials than hydroxyl radical, however these species can still lead to the remediation of a variety of pollutants and toxins. The lower oxidation of these species can lead to better selectivity, which can be advantageous for treatment of solutions containing innocuous substrates such as dissolved organic matter, which can compete for and quench hydroxyl radical. H_2O_2 and singlet oxygen are generally formed as secondary oxidants but their specific impact in the UV and VLA semiconductor photocatalysis are still not clear. Singlet oxygen can be produced through a photosensitized process and from the reduction of superoxide anion radical, while hydrogen peroxide can be generated from disproportionation of superoxide anion radical and combination of two hydroxyl radicals. Hydroperoxide is commonly employed for remediation through UV/H_2O_2 disinfection processes, Fenton-type reactions and Haber-Weiss reactions. $O_2^{\bullet-}$ can act as a oxidizing or reducing agent depending on the reactivity of the target substrate. The role of $O_2^{\bullet-}$ especially in VLA photocatalysis is not well understood (*165*).

5.3. Measurement of ROS

Accurate determination of ROS can be challenging because of their high reactivities, short lifetimes and the requirement for specialize equipment. Direct measurement of $^{\bullet}OH$ is not feasible given the short wavelength absorbance and short lifetime. Common methods for detection and measurement of hydroxyl radical include monitoring of initial reaction products, use of trapping agents and competition (*166*). The formation of hydroxyl radical products with absorbance from ~260-700 nm can be directly monitored by transient absorption spectroscopy and absolute rate constants can be extracted by varying the concentration of substrate. Trapping experiments and competition kinetics employ probe compounds with known hydroxyl radical selectivity and reactivities. A number of simple compounds, including t-butanol and formic acid, have been added as hydroxyl radical inhibitors during photocatalysis to access the role of hydroxyl radical in the observed degradation of a target compound. A decrease in the degradation upon addition of a hydroxyl radical inhibitor implies hydroxyl radical plays an important role in the degradation process. While inhibitors are effective for a qualitative assessment, trapping of hydroxyl radical can provide an accurate determination of the concentration and production rate of hydroxyl radical. Hydroxyl radical spin traps produce a spin adducts which can be measured by electron spin resonance (ESR) spectroscopy (*167*, *168*). Terephthalic acid and coumarin react with hydroxyl radical to yield adducts with characteristic fluorescence (*169*, *170*). These fluorescence probes are easy to use and excellent for quantification of hydroxyl radical in a variety of systems.

Singlet oxygen is in an excited state unlike the other ROS mentioned earlier. While it requires specialized equipment it is possible to measure the presence of singlet oxygen directly based on the phosphorescence at 1270 nm of the $^1\Delta_g$ state (*171*). Another challenge in measuring singlet oxygen during photocatalysis is its short lifetime in aqueous media. 1,4-diazabicyclo[2.2.2]octane (DABCO), azide, and furfuryl alcohol are used to probe singlet oxygen reactions. Addition of these probes to the reaction solution will quench singlet oxygen and inhibit reactions with target substrates. The generation rate and steady-state concentration of $^1O_2$ can be measured by using furfuryl alcohol as a probe (*172*). Thus, simple addition of a singlet oxygen quencher can provide qualitative evidence for the involvement of singlet oxygen. Another way to probe the role of $^1O_2$ during photocatalysis is to run the experiment in water and in heavy water (D_2O). The lifetime of singlet oxygen in D_2O is ~ 20 times longer than in H_2O, thus singlet oxygen mediated processes are dramatically enhanced in D_2O compared to H_2O. Singlet oxygen production has also been monitored by ESR spectroscopy using a trapping agent (*173*).

Simple reliable colorimetric methods based on peroxidic-type reactions are commonly used for the detection of H_2O_2, i.e., the I_3^- method is based on the spectrophotometric determination of formed I_3^-, when H_2O_2 is mixed with a concentrated I^- solution. The dimerization of p-hydroxyphenyl acetic acid and enzymatic-colorimetric method are also employed for H_2O_2 measurement (*174*). Catalase has also been used to assess the involvement of hydrogen peroxide in oxidative and biological systems. The involvement of superoxide anion radical can be assessed using competition kinetics employing benzoquinone and superoxide dismutase. Electron spin resonance spectroscopy using a spin trap (*175*), a luminal chemiluminescent probe compound (*176*), the combination of reduction of 2,3-bis(2-methoxy-4-nitro-5-sulfophenyl)-2H-tetrazolium-5-carboxanilide and superoxide dismutase method (*177*) have been used to monitor the production of superoxide anion radical. While there is an extensive number of techniques for measuring and assessing the production of ROS, semiconductor photocatalysis involves heterogeneous systems, which can complicate measurements and limit the use of spectroscopic methods. Accurately determining the production and understanding the interplay among ROS generated under specific semiconductor photocatalytic conditions is critical for practical water treatment applications.

6. Conclusions

In conclusion, this book chapter summarizes the preparation of various nanostructured photocatalysts, such as titanium dioxide, zinc oxide, iron oxide, and metal sulfides, by conventional synthesis methods including sol-gel methods, hydrothermal methods, and ball milling methods. Novel processing technologies, such as microwave-assisted and sonochemical methods, were also described for the synthesis of different nanomaterials. By these two latter methods, desirable functional nanomaterials can be produced at lower temperature and in shorter time compared to conventional synthetic methods. With respect to the applications of nanomaterials for environmental remediation, water detoxification, air purification, and photocatalytic disinfection, as well as selectivity of TiO_2 for the destruction of target compounds, were discussed. The nanostructured catalysts efficiently degrade toxic compounds released by cyanobacteria blooms and EDCs in water. NOx and VOCs in air are efficiently removed by photocatalysis under UV or visible irradiation. For photocatalytic disinfection, semiconductor photocatalysts effectively enhances SODIS at the pilot scale and under natural sunlight conditions. In addition, the nanocrystalline titania materials are excellent candidates to develop efficient DSCs for direct solar energy conversion to electricity. The enhancement of the selectivity of TiO_2 photocatalysis through nanoscale modification and functionalization of TiO_2 photocatalysts was also described. An engineered TiO_2 photocatalyst with controlled properties through innovative material design and processing can serve as an excellent green material for the selective destruction of target contaminants in water. Finally, the formation and roles of ROS during the semiconductor photocatalysis were discussed. The production and role of the different ROS are dependent on the photocatalyst, reaction conditions and target compounds. It is critical to accurately determine the production of ROS and understand their interaction during specific semiconductor photocatalysis for practical water treatment applications.

Acknowledgments

All authors contributed equally to this work.

D. D. Dionysiou, S. C. Pillai, J. A. Byrne, P. S. M. Dunlop, and K. O'Shea wish to acknowledge financial support under the US-Ireland R&D Partnership Initiative from the Department of Employment and Learning Northern Ireland (DELNI), the Science Foundation Ireland (SFI-grant number 10/US/I1822) and the US National Science Foundation-CBET (Award 1033317). D. D. Dionysiou would like to acknowledge the Cyprus Research Promotion Foundation through Desmi 2009-2010 which was co-funded by the Republic of Cyprus and the European Regional Development Fund of the EU under contract number NEA IPODOMI/STRATH/0308/09. P. Falaras acknowledges funding by the European Community's Seventh Framework Programme (FP7/2007-2013) under grant agreement of the CLEAN WATER (n° 227017) and SANS (n° 246124) projects.

References

1. Vilar, V. J. P.; Gomes, A. I. E.; Ramos, V. M.; Maldonado, M. I.; Boaventur, R. A. R. *Photochem. Photobiol. Sci.* **2009**, *8*, 691–698.
2. Pichat, P. *Appl. Catal., B* **2010**, *99*, 428–434.
3. Higarashi, M. M.; Jardim, W. F. *Catal. Today* **2002**, *76*, 201–207.
4. Fujishima, A.; Honda, K. *Nature* **1972**, *238*, 37–38.
5. Zoua, L.; Luo, Y.; Hooper, M.; Hu, E. *Chem. Eng. Process.* **2006**, *45*, 959–964.
6. Dalton, J. S.; Janes, P. A.; Jones, N. G.; Nicholson, J. A.; Hallam, K. R.; Allen, G. C. *Environ. Pollut.* **2002**, *120*, 415–422.
7. Konstantinou, I. K.; Albanis, T. A. *Appl. Catal., B* **2004**, *49*, 1–14.
8. Hajkova, P.; Spatenka, P.; Horsky, J.; Horska, I.; Kolouch, A. *Plasma Process. Polym.* **2007**, *4*, S397–S401.
9. Jardim, W. F.; Moraes, S. G.; Takiyama, M. M. K. *Water Res.* **1997**, *31*, 1728–1732.
10. Dimitrov, D. *Colloids Surf., A* **2006**, *282-283*, 8–10.
11. Dastjeridi, R.; Montazer, M. A. *Colloids Surf., B* **2010**, *79*, 5–18.
12. Xu, P.; Zeng, G. M.; Huang, D. L.; Feng, C. L.; Hu, S.; Zhao, M. H.; Lai, C.; Wei, Z.; Huang, C.; Xie, G. X.; Liu, Z. F. *Sci. Total Environ.* **2012**, *424*, 1–10.
13. Comenares, J. C.; Luque, R.; Campelo, J. M.; Colmenares, F.; Karpiński, Z.; Romero, A. A. *Materials* **2009**, *2*, 2228–2258.
14. Bao, N.; Shen, L.; Takata, T.; Domen, K. *Chem. Mater.* **2008**, *20*, 110–117.
15. Hu, J.-S.; Ren, L.-L.; Guo, Y.-G.; Liang, H.-P.; Cao, A.-M.; Wan, L.-J.; Bai, C.-L. *Angew. Chem., Int. Ed.* **2005**, *44*, 1269–1273.
16. Gonzalez-Oliver, C.; Jamesl, P.; Rawson, H. *J. Non-Cryst. Solids.* **1982**, *48*, 129–152.
17. Sladek, K. *Ind. Eng. Chem. Prod. Res. Dev.* **1972**, *11*, 92–96.
18. Mooney, J.; Radding, S. *Annu. Rev. Mater. Sci.* **1982**, *12*, 81–101.
19. Kuznetzon, V. *J. Cryst. Growth* **1968**, *3–4* (C), 405–410.

20. Lindh, A. *Lithos* **1972**, *4*, 325–343.
21. Choy, K. *Prog. Mater. Sci.* **2003**, *48*, 57–170.
22. Marcilly, C.; Courty, P.; Delmon, B. *J. Am. Ceram. Soc.*, *53*, 56–57.
23. Li, D.; Haneda, H.; Hishita, S.; Ohashi, N. *Chem. Mater.* **2005**, *17*, 2588–2595.
24. Feng, S.; Xu, R. *Acc. Chem. Res.* **2001**, *34*, 239–247.
25. Hench, L.; West, J. *Chem. Rev.* **1990**, *90*, 33–72.
26. Umebayashi, T.; Yamaki, T.; Itoh, H.; Asai, K. *Appl. Phy. Lett.* **2002**, *81*, 454–456.
27. Fujishima, A.; Rao, T.; Tryk, D. *J. Photochem. Photobiol., C* **2000**, *1*, 1–21.
28. O'Regan, B.; Gratzel, M. *Nature* **1991**, *353*, 737–740.
29. Kresge, C.; Leonowicz, M.; Roth, W.; Vartuli, J.; Beck, J. *Nature* **1992**, *359*, 710–712.
30. Pelaez, M.; de la Cruz, A.; Elias, S.; Polycarpos, F.; Dionysiou, D. *Catal. Today* **2009**, *144*, 19–25.
31. Han, C.; Pelaez, M.; Likodimos, V.; Kontos, A.; Falaras, P.; O'Shea, K.; Dionysiou, D. *Appl. Catal., B* **2011**, *107*, 77–87.
32. Barnes, P.; Randeniya, L.; Murphy, A.; Gwan, P; Plumb, I.; Glasscock, J.; Grey, I.; Li, C. *Dev. Chem. Eng. Miner. Process.* **2006**, *14*, 51–70.
33. Eder, D.; Motta, M.; Windle, A. *Nanotechnology* **2009**, 20, 055602.
34. Dvoranova, D.; Brezova, V.; Mazur, M.; Malati, M. *Appl. Catal., B* **2002**, *37*, 91–105.
35. Xu, a.; Gao, Y.; Liu, H. *J. Catal.* **2002**, *207*, 151–157.
36. Hurum, D.; Agrios, A.; Gray, K. *J. Phys. Chem. B* **2003**, *107*, 4545–4549.
37. Chen, X.; Nao, S. *Chem. Rev.* **2007**, *107*, 2891–2595.
38. Eiden-Assmann, S.; Widoniak, J.; Maret, G. *Chem. Mater.* **2004**, 6–11.
39. Han, C.; Luque, R.; Dionysiou, D. *Chem. Commun.* **2012**, *48*, 1860–1862.
40. Srisitthiratkul, C.; Pongsorrarith, V.; Instasanta, N. *Appl. Surf. Sci.*, *257*, 8850–8856.
41. Dokou, E.; Farneth, W.; Barteau, M. *Stud. Surf. Sci. Catal.* **2000**, *130*, 3167–3172.
42. Tan, J.; Fernandez, Y.; Liu, D.; Maroto-Valer, M.; Bian, J.; Zhang, X. *Chem. Phys. Lett.* **2012**, *531*, 149–154.
43. Kang, Q.; Yang, L.; Cai, Q. *Bioelectrochemistry* **2008**, *74*, 62–65.
44. Ma, C.; Zhou, Z.; Wei, H.; Yang, Z.; Wang, Z.; Zhang, Y. *Nanoscale Res. Lett.* **2011**, *6*, 536.
45. Ge, L.; Zuo, F.; Liu, J.; Ma, Q.; Wang, C.; Xun, D.; Bartels, L.; Feng, P. *J. Phys. Chem. C* **2102**, *116*, 13708–13714.
46. Tsuji, I.; Kato, H.; Kobayashi, H.; Hudo, A. *J. Am. Chem. Soc.* **2004**, *126*, 13406–13413.
47. Zhang, Q. F.; Chou, T. P.; Russo, B.; Jenekhe, S. A.; Cao, G. Z. *Adv. Funct. Mater.* **2008**, *18*, 1654–1660.
48. Umar, A.; Rahman, M. M.; Hahn, Y. B. *Talanta* **2009**, *78*, 284–289.
49. Bao, J.; Zimmler, M. A.; Capasso, F. *Nano Lett.* **2006**, *6*, 1719–1722.
50. Aghababazadeh, R.; Mazinani, B.; Mirhabibi, A.; Tamizifar, M. *J. Phys.: Conf. Ser.* **2006**, *26*, 312–314.

51. Ao, W.; Li, J.; Yang, H.; Zeng, X.; Ma, X. *Powder Technol.* **2006**, *168*, 148–151.
52. Shen, L.; Bao, N.; Yanagisawa, K.; Domen, K.; Gupta, A.; Grimes, C. A. *Nanotechnology* **2006**, *17*, 5117–5123.
53. Lu, H. B.; Li, H.; Liao, L.; Tian, Y.; Shuai, M.; Li, J. C.; Hu, M. F.; Fu, Q.; Zhu, B. P. *Nanotechnology* **2008**, *19*, 045604, 7pp.
54. Chu, D.; Masuda, Y.; Ohji, T.; Kato, K. *Langmuir* **2010**, *26*, 2811–2815.
55. Chung, Y.-A.; Chang, Y.-C.; Lu, M.-Y.; Wang, C.-Y.; Chen, L.-J. *J. Electrochem. Soc.* **2009**, *156*, F75–F79.
56. Zhai, T.; Xie, S.; Zhao, Y.; Sun, X.; Lu, X.; Yu, M.; Xu, M.; Xiao, F.; Tong, Y. *CrystEngComm.* **2012**, *14*, 1850–1855.
57. Xu, P.; Zeng, G. M.; Huang, D. L.; Feng, C. L.; Hu, S.; Zhao, M. H.; Lai, C.; Wei, Z.; Huang, C.; Xie, G. X.; Liu, Z. F. *Sci. Total Environ.* **2012**, *424*, 1–10.
58. Liu, S.-Q. *Environ. Chem. Lett.* **2012**, *10*, 209–216.
59. Kato, R.; Komatsu, T. *J. Inorg. Organomet. Polym.* **2012**, *22*, DOI:10.1007/s10904-012-9725-5.
60. Zhou, X.; Yang, H.; Wang, C.; Mao, X.; Wang, Y.; Yang, Y.; Liu, G. *J. Phys. Chem. C* **2010**, *114*, 17051–17061.
61. Chowdhury, P. S.; Arya, P. R.; Raha, K. *Synth. React. Inorg. Met.* **2007**, *37*, 447–451.
62. Ipe, B. I.; Lehnig, M.; Niemeyer, C. M. *Small* **2005**, *1*, 706–709.
63. Yang, F.; Yan, N.-N.; Huang, S.; Sun, Q.; Zhang, L.-Z.; Yu, Y. *J. Phys. Chem. C* **2102**, *116*, 9078–9084.
64. Ren, G.; Lin, Z.; Gilbert, B.; Zhang, J.; Huang, F.; Liang, J. *Chem. Mater.* **2008**, *20*, 2438–2443.
65. Komarneni, S.; Menon, V. C.; Li, Q. H *Ceram. Trans.* **1996**, *62*, 1042–1122.
66. Periyat, P.; Leyland, N. l.; McCormack, D. E.; Colreavy, J.; Corr, D.; Pillai, S. C *J. Mater. Chem.* **2010**, *20*, 3650–3655.
67. Komarneni, S.; Rajha, R. K; Katsuki, H. *Mater. Chem. Phys.* **1999**, *60*, 50–54.
68. Wilson, G. J.; Will, G. D.; Frost, R. L.; Montgomery, S. A. *J. Mater. Chem.* **2002**, *12*, 1787–1791.
69. Wang, H.; Xu, J. Z.; Zhu, J. J.; Chen, H. Y. *J. Cryst. Growth* **2002**, *244*, 88–94.
70. Liao, X. H.; Chen, N. Y.; Xu, S.; Yang, S. B.; Zhu, J. J. *J. Cryst. Growth* **2003**, *252*, 593–59.
71. Zheng, Y.; Lv, K.; Wang, Z.; Deng, K.; Li, M. *J. Mol. Catal. A: Chem.* **2012**, *356*, 137–143.
72. Suprabha, T.; Roy, H. G.; Thomas, J.; Kumar, K. P.; Mathew, S. *Nanoscale Res. Lett.* **2009**, *4*, 144–152.
73. Siddiquey, I. A.; Furusawa, T.; Sato, M.; Suzuki, M. *Mater. Res. Bull.* **2008**, *43*, 3416–3424.
74. Biddlecombe, G. B.; Gun'ko, Y. K.; Kelly, J. M.; Pillai, S. C.; Coey, J. M. D.; Venkatesan, M.; Douvalis, A. P. *J. Mater. Chem.* **2011**, *11*, 2937–2939.
75. Gun'ko, Y. K.; Pillai, S. C.; McInerney, D. *J. Mater. Sci.: Mater. Electro.* **2001**, *12*, 299–302.

76. Shafi, K. V. P . M.; Ulman, A.; Yan, X.; Yang, N. L.; Estourne‘s, C.; White, H.; Rafailovich, M. *Langmuir* **2001**, *17*, 5093–5097.
77. Geng, J.; Lv, Y.; Lu, D. J.; Zhu, J. J. *Nanotechnology* **2006**, *17*, 2614–2620.
78. Pan, H. C.; Liang, F. P.; Mao, C. J.; Zhu, J. J. *Nanotechnology* **2007**, *18*, 195606, 6pp.
79. Xu, L.; Jiang, L. P.; Zhu, J. J. *Nanotechnology* **2009**, *20*, 045605, 6pp.
80. Xu, K.; Mao, C. J.; Geng, J.; Zhu, J. J. *Nanotechnology* **2007**, *18*, 315604, 5pp.
81. Khanjani, S.; Morsali, A.; Joo, S. W. *Ultrason. Sonochem.* **2013**, *20*, 734–739.
82. Bhatte, K. D.; Sawant, D. N.; Pinjari, D. V.; Pandit, A. B.; Bhanage, B. M. *Mater. Lett.* **2012**, *77*, 93–95.
83. Guo, W.; Lin, Z.; Wang, X.; Song, G. *Microelectron. Eng.* **2003**, *66*, 95–101.
84. Definition of a Nanomaterial. European Commission. http://ec.europa.eu/environment/chemicals/nanotech/index.htm#definition.
85. Editorial. *Nat. Nanotechnol.* **2007**, *2*, 257.
86. Litter, M. I.; Choi, W.; Dionysiou, D. D.; Falaras, P.; Hiskia, A.; Li Puma, G.; Pradeep, T.; Zhao, J. *J. Hazard. Mater.* **2012**, *211-12*, 1–2, Editorial.
87. Likodimos, V.; Dionysiou, D. D.; Falaras, P. *Rev. Environ. Sci. Biotechnol.* **2010**, *9*, 87–94.
88. Pastrana-Martinez, L. M.; Faria, J. L.; Dona-Rodriguez, J. M.; Fernandez-Rodriguez, C.; Silva, A. M. T. *Appl. Catal., B* **2012**, *113– 114*, 221–227.
89. Liu, G.; Han, C.; Pelaez, M.; Zhu, D.; Liao, S.; Likodimos, V.; Ioannidis, N.; Kontos, A. G.; Falaras, P.; Dunlop, P. S. M.; Byrne, J. A.; Dionysiou, D. D. *Nanotechnology* **2012**, *23*, 294003, 10pp.
90. Pastrana-Martínez, L. M.; Morales-Torres, S.; Likodimos, V.; Figueiredo, J. L.; Faria, J. L.; Falaras, P.; Silva, A. M. T. *Appl. Catal., B* **2012**, *123– 124*, 241–256.
91. Pelaez, M.; Falaras, P.; Likodimos, V.; Kontos, A. G.; de la Cruz, A. A.; O’Shea, K.; Dionysiou, D. D. *Appl. Catal., B* **2010**, *99*, 378–387.
92. Moustakas, N. G.; Kontos, A. G.; Likodimos, V.; Katsaros, F.; Boukos, N.; Tsoutsou, D.; Dimoulas, A.; Romanos, G. E.; Dionysiou, D. D.; Falaras, P. *Appl. Catal., B* **2013**, *130−131*, 14–24.
93. Triantis, T. M.; Fotiou, T.; Kaloudis, T.; Kontos, A. G.; Falaras, P.; Dionysiou, D. D.; Pelaez, M.; Hiskia, A. *J. Hazard. Mater.* **2012**, *211*, 196–202.
94. Mboula, V. M.; Hequet, V.; Gru, Y.; Colin, R.; Andres, Y. *J. Hazard. Mater.* **2012**, *209*, 355–364.
95. Papageorgiou, S. K.; Katsaros, F. K.; Favvas, E. P.; Romanos, G.; Athanasekou, C. P.; Beltsios, K. G.; Tzialla, O. I.; Falaras, P. *Water Res.* **2012**, *46*, 1858–1872.
96. Romanos, G. E.; Athanasekou, C. P.; Katsaros, F. K.; Kanellopoulos, N. K.; Dionysiou, D. D.; Likodimos, V.; Falaras, P. *J. Hazard. Mater.* **2012**, *211–212*, 304–316.
97. Athanasekou, C. P.; Romanos, G. E.; Katsaros, F. K.; Kordatos, K.; Likodimos, V.; Falaras, P. *J. Membr. Sci.* **2012**, *392–393*, 192–203.

98. Han, C.; Doepke, A.; Cho, W.; Likodimos, V.; de la Cruz, A. A.; Back, T.; Heineman, W. R.; Halsall, H. B.; Shanov, V. N.; Schulz, M. J.; Falaras, P.; Dionysiou, D. D. *Adv. Func. Mater.* **2012**, DOI: 10.1002/adfm.201201920.
99. Kontos, A. I.; Kontos, A. G.; Tsoukleris, D. S.; Vlachos, G. D.; Falaras, P. *Thin Solid Films* **2007**, *515*, 7370–7375.
100. Chorianopoulos, N. G.; Tsoukleris, D. S.; Panagou, E. Z.; Falaras, P.; Nychas, G.-J. E. *Food Microbiol.* **2011**, *28*, 164–170.
101. Kontos, A. G.; Katsanaki, A.; Likodimos, V.; Maggos, T.; Kim, D.; Vasilakos, C.; Dionysiou, D. D.; Schmuki, P.; Falaras, P. *Chem. Eng. J.* **2012**, *179*, 151–157.
102. Pelaez, M.; Nolan, N. T.; Pillai, S. C.; Seery, M. K.; Falaras, P.; Kontos, A. G.; Dunlop, P. S. M.; Hamilton, J. W. J.; Byrne, J. A.; O'Shea, K.; Entezari, M. H.; Dionysiou, D. D. *Appl. Catal.. B* **2012**, *125*, 331–349.
103. Wang, T. C.; Lu, N.; Li, J.; Wu, Y. *Environ. Sci. Technol.* **2011**, *45*, 9301–9307.
104. Davezza, M.; Fabbri, D.; Prevot, A. B.; Pramauro, E. *Environ. Sci. Pollut. Res.* **2011**, *18*, 783–789.
105. Editorial. *Nat. Nanotechnol.* **2007**, *2*, 325.
106. O'Regan; Grätzel, M. *Nature* **1991**, *353*, 737.
107. Hagfeldt, A.; Grätzel, M. *Chem. Rev.* **1995**, *95*, 49–68.
108. Hagfeldt, A.; Boschloo, G.; Sun, L.; Kloo, L.; Pettersson, H. *Chem. Rev.* **2010**, *110*, 6595.
109. Green, M. A.; Emery, K.; Hishikawa, Y.; Warta, W.; Dunlop, E. D. *Prog. Photovoltaics* **2011**, *19*, 565–572.
110. Vougioukalakis, G. C.; Philippopoulos, A. I.; Stergiopoulos, T.; Falaras, P. *Coord. Chem. Rev.* **2011**, *255*, 2602–2621.
111. Stergiopoulos, T.; Kontos, A.; Likodimos, V.; Perganti, D.; Falaras, P. *J. Phys. Chem. C* **2011**, *115*, 10236–10244.
112. Stergiopoulos, T.; Rozi, E.; Hahn, R.; Schmuki, P.; Falaras, P. *Adv. Energy Mater.* **2011**, *1*, 569–572.
113. *Progress on Sanitation and Drinking Water: 2012 Update*; World Health Organization, UNICEF: Geneva, Switzerland, 2012.
114. Clasen, T.; Edmondson, P. *Int. J. Hyg. Environ. Health* **2006**, *209*, 173–181.
115. *Economic and Health Effects of Increasing Coverage of Low Cost Household Drinking Water Supply and Sanitation Interventions to Countries Off-Track To Meet MDG Target 10*; World Health Organization: Geneva, Switzerland, 2007.
116. Burch, J. D.; Thomas, K. E. *Sol. Energy* **1998**, *64*, 87–97.
117. Clasen, T. F.; Haller, L. *Water Quality Interventions To Prevent Diarrhoea: Cost and Cost-Effectiveness*; World Health Organization: Geneva, Switzerland, 2008.
118. Matsunaga, T.; Tomoda, R.; Nakajima, T.; Wake, H. *FEMS Microbiol. Lett.* **1985**, *29*, 211–214.
119. Blake, D. M.; Maness, P. C.; Huang, Z.; Wolfrum, E. J.; Huang, J.; Jacoby, W. A. *Sep. Purif. Method.* **1999**, *28*, 1–50.
120. McCullagh, C; Robertson, J. M.; Bahnemann, D. W.; Robertson, P. J. K. *Res. Chem. Intermed.* **2007**, *33*, 359–375.

121. Malato, S.; Fernandez-Ibanez, P.; Maldonado, M. I.; Blanco, J.; Gernjak, W. *Catal. Today* **2009**, *147*, 1–59.
122. Dalrymple, O. K.; Stefanakos, E.; Trotz, M. A.; Goswami, D. Y. *Appl. Catal., B* **2010**, *98*, 27–38.
123. Cho, M.; Chung, H.; Choi, W.; Yoon, J. *Water Res.* **2004**, *38*, 1069–1077.
124. Rincon, A. G.; Pulgarin, C.; Adler, N.; Peringer, P. *J. Photochem. Photobiol., A* **2001**, *139*, 233–241.
125. Huang, Z.; Maness, P.; Blake, D. M.; Wolfrum, E. J.; Smolinski, S. L.; Jacoby, W. A. *J. Photochem. Photobiol., A* **2000**, *130*, 163–170.
126. Wainwright, M. *Int. J. Antimicrob. Agents* **2000**, *16*, 381–394.
127. Sunada, K.; Watanabe, T.; Hashimoto, K. *J. Photochem. Photobiol., A* **2003**, *156*, 227–233.
128. Rincon, A.; Pulgarin, C. *Sol. Energy* **2004**, *77*, 635–648.
129. Byrne, J. A.; Eggins, B. R.; Brown, N. M. D.; McKinney, B.; Rouse, M. *Appl. Catal., B* **1998**, *17*, 25–36.
130. McMurray, T. A.; Byrne, J. A.; Dunlop, P. S. M.; Winkelman, J. G. M.; Eggins, B. R.; McAdams, E. T. *Appl. Catal,. A* **2004**, *262*, 105–110.
131. Alrousan, D. M. A.; Dunlop, P. S. M.; McMurray, T. A.; Byrne, J. A. *Water Res.* **2009**, *43*, 47–54.
132. Dunlop, P. S. M.; Byrne, J. A.; Manga, N.; Eggins, B. R. *J. Photochem. Photobiol., A* **2002**, *148*, 355–363.
133. Dunlop, P. S. M.; McMurray, T. A.; Hamilton, J. W. J.; Byrne, J. A. *J. Photochem. Photobiol., A* **2008**, *196*, 113–119.
134. Sunnotel, O.; Verdoold, R.; Dunlop, P. S. M.; Snelling, W. J.; Lowery, C. J.; Dooley, J. S. G.; Moore, J. E.; Byrne, J. A. *J. Water Health* **2010**, *8*, 83–91.
135. Alrousan, D. M. A.; Polo-López, M. I.; Dunlop, P. S. M.; Fernández-Ibáñez, P.; Byrne, J. A. *Appl. Catal., B* **2012**, *128*, 126–134.
136. Rengifo-Herrera, J. A.; Pulgarin, C. *Sol. Energy* **2010**, *84*, 37–43.
137. Matthews, R. W. *Water Res.* **1986**, *20*, 569–578.
138. Paz, Y. *C. R. Chim.* **2006**, *9*, 774–787.
139. Choi, H.; Kim, Y.; Varma, R. S.; Dionysiou, D. D. *Chem. Mater.* **2006**, *18*, 5377–5384.
140. Doll, T. E.; Frimmel, F. H. *Water Res.* **2005**, *39*, 403–411.
141. Yu, J. G.; Yu, J. C.; Leung, M. K. P.; Ho, W. K.; Cheng, B.; Zhao, X. J.; Zhao, J. C. *J. Catal.* **2003**, *217*, 69–78.
142. Shiraishi, Y.; Saito, N.; Hirai, T. *J. Am. Chem. Soc.* **2005**, *127*, 12820–12822.
143. Morishita, M.; Shiraishi, Y.; Hirai, T. *J. Phys. Chem. B* **2006**, *110*, 17898–17905.
144. Ghosh-Mukerji, S.; Haick, H.; Paz, Y. *J. Photochem. Photobiol., A* **2003**, *160*, 77–85.
145. Sharabi, D.; Paz, Y. *Appl. Catal., B* **2010**, *95*, 169–178.
146. Paz, Y. *Solid State Phenom.* **2010**, *162*, 135–162.
147. Nussbaum, M.; Paz, Y. *Phys. Chem. Chem. Phys.* **2012**, *14*, 3392–3399.
148. Matos, J.; Laine, J.; Herrmann, J. M. *Appl. Catal., B* **1998**, *18*, 281–291.
149. Inumaru, K.; Murashima, M.; Kasahara, T.; Yamanaka, S. *Appl. Catal., B* **2004**, *52*, 275–280.

150. Ghosh-Mukerji, S.; Haick, H.; Schvartzman, M.; Paz, Y. *J. Am. Chem. Soc.* **2001**, *123*, 10776–10777.
151. Ramstörm, O.; Ansell, R. J. *Chirality* **1998**, *10*, 195–209.
152. Yoneyama, H.; Haga, S.; Yamanaka, S. *J. Phys. Chem.* **1989**, *93*, 4833–4837.
153. Shimizu, K. I.; Kaneko, T.; Fujishima, T.; Kodama, T.; Yoshida, H.; Kitayama, Y. *Appl. Catal., A* **2002**, *225*, 185–195.
154. Aronson, B. J.; Blanford, C. F.; Stein, A. *Chem. Mater.* **1997**, *9*, 2842–2851.
155. Kato, R.; Shimura, N.; Ogawa, M. *Chem. Lett.* **2008**, *37*, 76–77.
156. Choi, H.; Antoniou, M. G.; Pelaez, M.; de la Cruz, A. A.; Shoemaker, J. A.; Dionysiou, D. D. *Environ. Sci. Technol.* **2007**, *41* (21), 7530–7535.
157. Choi, W. *Catal. Surv. Asia* **2006**, *10* (1), 16–28.
158. Serpone, N; Emeline, A. V. *J. Phys. Chem. Lett.* **2012**, *3*, 673–677.
159. Kim, D. K.; O'Shea, K. E.; Cooper, W. J. *Ultrason. Sonochem.* **2012**, *19*, 959–968.
160. Linsebigler, A. L.; Lu, G.; Yates, J. T. *Chem. Rev.* **1995**, *95*, 735–758.
161. Sun, J.; Qiao, L.; Sun, S.; Wang, G. *J. Hazard. Mater.* **2008**, *155*, 312–319.
162. O'Shea, K. E.; Dionysiou, D. D. *J. Phys. Chem. Lett.* **2012**, *3* (15), 2112–2113.
163. Cooper, W. J.; Cramer, C. J.; Martin, N. H.; Mezyk, S. P.; O'Shea, K. E.; von Sonntag, C. *Chem. Rev.* **2009**, *109* (3), 1302–1345.
164. Buxton, G. V.; Greenstock, C. L.; Helman, W. P.; Ross, A. B. *J. Phys. Chem. Ref. Data* **1988**, *17*, 513–886.
165. Li, W.; Li, D.; Lin, Y.; Wang, P.; Chen, W.; Fu, X.; Shao, Y. *J. Phys. Chem. C* **2012**, *116*, 3552–3560.
166. Song, W.; Xu, T.; Cooper, W. J.; Dionysiou, D. D.; de la Cruz, A, A.; O'Shea, K. E. *Envrion. Sci. Technol.* **2009**, *43* (5), 1487–1492.
167. Schwarz, P. F.; Turro, N. J.; Bossmann, S. H.; Braun, A. M.; Wahab, A. A. A.; Dürr, H. *J. Phys. Chem. B* **1997**, *101*, 7127–7134.
168. Fu, H.; Zhang, L.; Zhang, S.; Zhu, Y. *J. Phys. Chem. B* **2006**, *110* (7), 3061–3065.
169. Louit, G.; Foley, S.; Cabillic, J.; Coffigny, H.; Taran, F.; Valleix, A.; Renault, J. P.; Pin, S. *Radiat. Phys. Chem.* **2005**, *72*, 119–124.
170. Ishibashi, K.; Fujishima, A.; Watanabe, T.; Hashimoto, K. *J. Photochem. Photobiol., A.* **2000**, *134*, 139–142.
171. Nosaka, Y.; Daimon, T.; Nosaka, A. Y.; Murakami, Y. *Phys. Chem. Chem. Phys.* **2004**, *6*, 2917–2918.
172. Haag, W. R.; Hoigne, J. *Envrion. Sci. Technol.* **1986**, *20*, 341–348.
173. Konaka, R.; Kasahara, E.; Dunlap, W. C.; Yamamoto, Y.; Chien, K. C.; Inoue, M. *Free Radical Biol. Med.* **1999**, *27*, 294–300.
174. Miller, W. L.; Kester, D. R. *Anal. Chem.* **1988**, *60* (24), 2711–2715.
175. Diaz-Uribe, C. E.; Daza, M. C.; Martínez, F.; Páez-Mozo, E. A.; Guedes, C. L. B.; Mauro, E. D. *J. Photochem. Photobiol., A.* **2010**, *215*, 172–178.
176. Hirakawa, T.; Nakaoka, Y.; Nishino, J.; Nosaka, Y. *J. Phys. Chem. B.* **1999**, *103*, 4399–4403.
177. Auffan, M.; Pedeutour, M.; Rose, J.; Masion, A.; Ziarelli, F.; Borschneck, D.; Chaneac, C.; Botta, C.; Chaurand, P.; Labille, J.; Bottero, J. *Environ. Sci. Technol.* **2010**, *44*, 2689–2694.

Chapter 13

Dye-Sensitized Photocatalyst: A Breakthrough in Green Energy and Environmental Detoxification

Pankaj Chowdhury, Hassan Gomaa, and Ajay K. Ray*

Department of Chemical and Biochemical Engineering, University of Western Ontario, London, ON N6A 5B9, Canada
***E-mail: aray@eng.uwo.ca. Fax: +1 519 661 3498. Tel: +1 519 661 2111 x81279.**

Sensitization of semiconductor material is well recognized in the field of photography and photo-electrochemistry. Recently, dye sensitization technique has found its application in solar cells. Dye sensitization can further be applied for water treatment and sacrificial hydrogen generation using photocatalysis. Success of the process depends on the choice of suitable dye, semiconductor material, electron donor, and light sources. Attachment of dye molecule on photocatalyst surface leads to subsequent electron transfer into the conduction band of semiconductor, therefore strongly bound anchoring groups are preferred. Fixation of dye molecule on semiconductor surface also improves the electron transfer process. Incorporation of noble metals on semiconductor surface enhances their photocatalytic activity by reducing the electron/hole recombination rate. Dye-sensitized photocatalyst is applied for degradation of a wide range of compounds such as i) aliphatic compounds (carbon tetrachloride, trichloroethylene, hydrazine, and pesticides), ii) aromatic compounds (non-sensitizing dye, phenol, chlorophenol, and benzyl alcohol) in aqueous medium. Hydrogen generation is also possible in visible light with dye-sensitized photocatalyst in presence of sacrificial reagents. Ruthenium based dyes in solar cells and dye-sensitized photocatalysis are the best reported so far, however, researchers are gradually switching towards inexpensive and environment-friendly organic dyes and/or natural dyes from vegetable sources. Eosin Y is an organic dye which has been widely used for hydrogen generation, reportedly providing quantum yields between 9-19 %.

Keywords: Dye; Sensitization; Platinum; Wastewater; Hydrogen

Introduction

Photocatalysis is a well known technology where light energy is utilized to excite the semiconductor material producing electron/hole pair which ultimately involves in the detoxification of pollutants and/or water splitting producing hydrogen. Pollutant degradation under UV light is already proven to be an effective method as reported by several researchers. At the same time, photocatalytic water splitting showed remarkable progress in the last few decades under UV light which started with the pioneer work of Fujishim and Honda (*1*). The basic difficulty with UV light was related to its narrow wavelength ranges (4 % of the solar spectrum) which really drive the researcher to utilize the more profuse visible light (46 % of solar spectrum) for the excitation of semiconductor.

There are two significant impediments to visible light induced photocatalysis. First is the rapid recombination of electron/hole (e^-/h^+) pair. The second is poor activation of semiconductor photocatalyst. Moreover, there is an additional problem of rapid backward reaction during water splitting. Efforts have been made to improve the photocatalytic activity and visible light response. This includes modification of semiconductor photocatalyst to expand their photo-response to visible region in several ways such as doping with cation/anion, sensitizing with dye, coupling with another small band gap semiconductor and implantation of metal ion (*2*). Another approach, which attracts extensive industrial interest, is the field of dye-sensitization with potential applications in fields such as photography, photochromatic devices and photolithography (*3*).

Dye-sensitization of semiconductor is also a very popular concept that has been successfully utilized in solar cell technology for generating electricity. The same principle can be applied to split the water molecule under visible light to produce hydrogen as well as to degrade toxic pollutants.

This review will first describe the basic phenomena for dye-sensitization and a few brief methodologies used for different semiconductor materials such as ZnO, TiO_2, Cu_2O etc. It will also discuss different methods for deposition of novel metals as co-catalyst on semiconductor surface. Finally dye-sensitization method will be elaborated with different semiconductors for hydrogen production and degradation of different organic pollutants.

Dye Sensitization

Theory of Dye Sensitization

The process of expanding the sensitivity of transparent semiconductor materials to the visible spectra is known as spectral sensitization. When the sensitization of a large band gap semiconductor to the visible region is achieved with a dye at the molecular level, it is called dye-sensitization (*4*). The operating mechanism of dye in dye-sensitized semiconductor is similar to that in the field of photography (*5*). Dye is adsorbed chemically onto the semiconductor surface by conventional adsorption process and the chemisorbed dye molecules play the role of spectral sensitizer which upon excitation with visible light inject electron into the conduction band of the semiconductor (*5*). This phenomenon is also known as anodic sensitization. On the contrary, cathodic sensitization occurs when adsorbed dye molecule inject holes into valance band of the semiconductor (*6*). This review only considers anodic sensitization of semiconductor.

Some early studies of spectral sensitization of semiconductors were performed with ZnO and CdS. The first report discussed sensitization of pressed ZnO powder by adsorbed rhodamine B, eosin, erythrosine, and cyanine dyes (*3*, *7*). But most of these early studies focused on organic dyes related to the photographic industry. Gleria and Memming (*3*) carried out the first experiment of sensitization using ruthenium metal based polypyridine complex dyes on SnO_2 electrodes.

Gratzel and Kalyansundaram demonstrated the dye sensitization principles of an n-type semiconductor electrode. In the first step, the organic or inorganic dye molecule adsorb photon and results in the formation of excited state (equation 1). Then the excited dye molecule inject electron into the conduction band of the semiconductor (equation 2). The oxidized dye (S^+) is consequently reduced to ground state (S) by an electron donor. An electron donor can be a molecule or a mediating redox couple in a regenerative cell. The injected electron travels through the semiconductor and then through the external circuit to the counter electrode where the reduction of the oxidized donor takes place. So, the photon of irradiated light transports an electron through the semiconductor and the external circuit, and this results in conversion of light to electrical energy (*3*).

$$\mathrm{S} + h\vartheta \rightarrow S^* \tag{1}$$

$$S^* \rightarrow S^+ + e^-_{CB}\ (semiconductor) \tag{2}$$

The efficiency of the sensitization process can be expressed in terms of incident photon-to-current conversion efficiency (IPCE).

$$\mathrm{ICPE} = \left[\frac{\text{number of electrons produced by light in the external circuit}}{\text{number of incident photons (monochromatic)}}\right] \tag{3}$$

The IPCE value depends on the i) chemical nature of the dye, ii) mediating redox electrolyte, and iii) nature of the semiconductor and the interface with the electrolyte (*3*). The excited states of most of the metal complexes and organic dyes have short life span. So the efficiency of charge injection process depends on those dyes molecules that are in close proximity of the semiconductor electrodes (*3*).

The dye molecule should possess few basic properties regarding surface anchoring group, energy levels and ground state redox potential, to achieve a successful electron injection (*5*). Anchoring groups, such as phosphonates or carboxylates form strong covalent bonds with the semiconductor surface and that covalent bond ultimately increases the strength of the electronic coupling between the molecular orbitals of the dye and the semiconductor levels leading to fast electron injection rates (*8*). Carboxylic and phosphonic acid groups show very good electron transfer process (80-90 %) as mentioned for some ruthenium complexes (*4*). Different binding or interaction modes of TiO_2 surface and –COOH group is shown in Figure 1 as described by Galoppini (*8*). The anchoring group can be arranged in the following order of reactivity: $-P(O)(OH)_2$ > -COOH > -COOR (R=alkyl group) > -COX (X=Cl, Br etc) > $-COONH_2$ ~ COOM (M=Na, K etc). There are few more functional groups such as silanes, ethers, acetylacetone and salicylates that can also form bonds with metal oxide (semiconductor) by reacting with surface hydroxyl groups (*8*).

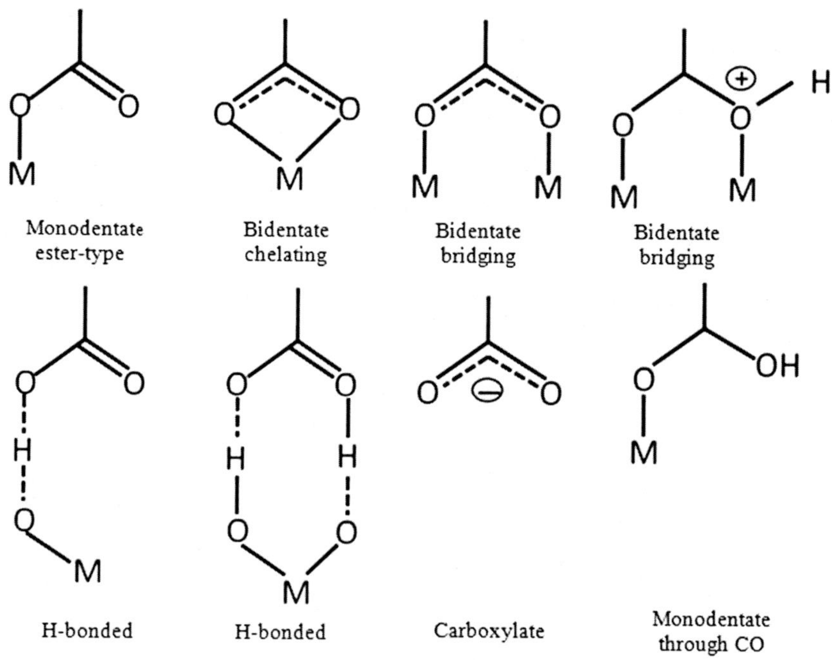

Figure 1. Possible binding modes of TiO_2 surface and –COOH. (Reproduced with permission from reference (8) Copyright 2004 Elsevier).

Dye molecules in the bulk phase of the electrolyte solution undergo severe charge collection problem similar to that in photogalvanic cells. To overcome this problem dye molecules are fixed on the semiconductor surface either by chemisorptions or by some form of derivatization. Several approaches were adopted to attach dye molecule on semiconductor surface since the mid seventies (*3*). Matsumura et al (*9*) investigated the possibility of chelate formation on the semiconductor surface. Fixation of xanthene dye to the surface was attempted either by etherification or by using a Si-CH_2-CH_2 -CH_2 -NH-CO-dye linkage (*3*). Visible light sensitization of ZnO was observed by certain azo dyes which form 1:1 Zn complex at the surface of ZnO powders. In another study, colloidal TiO_2 sensitization was achieved by 8-hydroquinoline to generate hydrogen from water invisible light irradiation. 8-hydroquinoline experienced a chelation reaction with colloidal TiO_2 that results in the formation of an active yellow complex during sensitization (*3*, *10*).

The type of sensitization (anodic or cathodic) mainly depends on the band position of the semiconductor and energy level of the dye. In anodic sensitization the energy level of the dye molecule at excited state should be comparable with the lower level of semiconductor conduction band for better electron injection (*5*). Here an electron excited from the lower singlet level (S_0) to upper excited singlet level (S_1) is transferred to conduction band (Figure 2). Moreover a higher value of ground state redox potential of the dye would be beneficial for rapid regeneration of exhausted dye in presence of suitable electrolyte such as I^-/IO_3^-. The best reported value of ground state oxidation potential is 0.5 V vs SCE (Standard Calomel Electrode) (*4*).

The electron injection mechanism was explained by several authors but the most recent work by Willig's group enlightened the interfacial electron transfer (ET) mechanism in a proper way. They demonstrated the ET dynamics in dye-TiO_2 system and found that on leaving the dye, the electron was not trapped on the surface but entered bulk TiO_2. Their observation with femtosecond laser spectroscopy was not in agreement with the usual model of electron transfer (*11*).

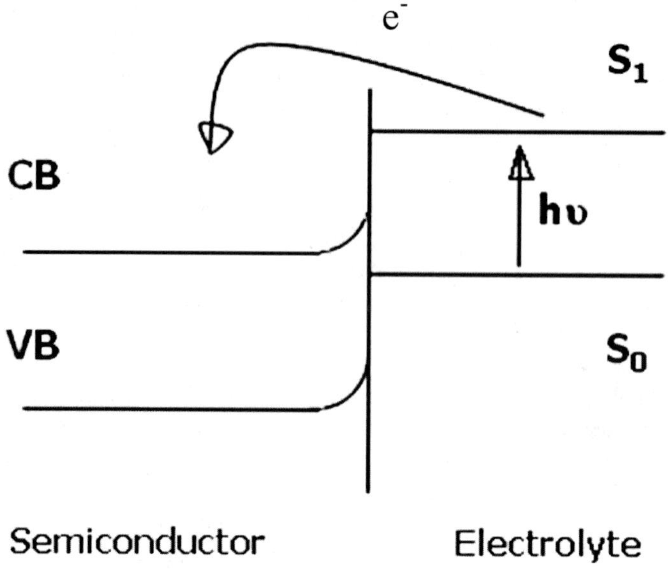

Figure 2. Energy level diagram of anodic dye sensitization process.

Dye Sensitization Methodology

In dye sensitization the dye is adsorbed on to the semiconductor surface having a high surface area and forms a dye-sensitized semiconductor film which adsorbs the visible light efficiently. In common practice the semiconductor material (powder or immobilized form) is dipped into the dye solution for specified time interval under dark. Then the semiconductor material is separated from dye solution and washed with solvent (alcohol or water) to remove unadsorbed or loosely bound dye molecules from semiconductor surface. The final dye-sensitized material is obtained only after drying at some specific temperature in oven. Although the process is quite straight forward, wide variations are expected in sensitization with the change in dye type, dye concentration, semiconductor type, semiconductor surface morphology, solution pH and incorporation of electron donor into the system. Thus, the visible light absorption efficiency of dye-sensitized semiconductor is a function of all the above parameters.

Selection of Dye

A group of semiconductor sensitizers such as porphyrins, coumarin, phalocyanines and carboxilate derivatives of anthracine are used by several authors. Among these photosensitizers, transition metal based sensitizers have been proved to be the best (*12*). Transition metals such as Ru (II), Fe (II) and Os (II) form d^6 complex and undergo intense charge transfer absorption across the entire visible range (*12*). In dye sensitized solar cells mainly Ru (II) polypyridine complexes are used. In the field of dye-sensitized solar cell Gratzel reported the best sensitizer, cis-[Ru(dcbH$_2$)$_2$(NCS)$_2$] (N3) where he applied thiocyanate (NCS$^-$) as an ancillary ligand. Subsequently, several investigators attempted to substitute the NCS$^-$ ligands which led to a variety of new photosensitizers, but the efficiency did not improve (*5*).

The ruthenium polypyridyl complex contains heavy metals and thus it is not environment friendly. Moreover, the process to synthesize the complex is complicated and costly. Recently few authors reported the use of natural dyes as an alternative for dye sensitization. The natural dyes are generally found in fruits, flowers and leaves of plants. A list of several natural dyes is presented in Table 1.

Selection of Semiconductor Material

A range of semiconductors such as ZnO, TiO_2, $SrTiO_3$, SnO_2 and Cu_2O have been studied with different dyes (*16*, *17*). TiO_2 became the semiconductor of choice because of several advantages such as i) low cost, ii) wide availability, iii) non-toxic nature, iv) high stability against photo-corrosion, and v) superior electronic energy band structure (*5*, *18*). The anatase form of TiO_2 (E_g=3.2 eV) has been found to be the most active for photocatalysis. TiO_2 sensitization have also been studied with different dyes such as $Ru(bpy)_3^{2+}$, RuL_3^{2+} (L=2,2′-bipyridine-4,4′-dicarboxilate) and metal-quinolinol (*16*). Composite semiconductors such as ZnO/SnO_2, SnO_2/MgO and Cds/MgO showed almost equal efficiencies as achieved with TiO_2 in solar cells (*19*).

Table 1. Natural Extracts Used for Dye Sensitization

Extract source	*References*
Rosella	Wongcharee et al. (*13*)
Blue Pea	Wongcharee et al. (*13*)
Jaboticaba's skin	Polo and Iha (*14*)
Chaste tree fruit	Gracia et al. (*15*)
Mulberry	Gracia et al. (*15*)
Cabbage-palm fruit	Polo et al. (*12*)
Java Palm	Polo et al. (*12*)
Pomegranate seeds	Polo et al. (*12*)

Semiconductor surface morphology and particle size are very important as these ultimately affect the e^-/h^+ recombination and photocatalytic reaction rate (*19*). Nanoporous semiconductor film can be a solution to improve the process performance. There are several other methods to prepare the semiconductor material such as high temperature solidification methods from the elements in bulk form, or vapor phase and vacuum methods in the form of thin films (*20*).

High temperature processes can be replaced with cathodic electrodeposition to produce thin film of semiconductor in more economically efficient manner (*21*). One step electrodeposition of ZnO/dye hybrid thin film was reported by Yoshida et al. (*22*), where the morphology, the crystal size, and orientation of ZnO were significantly modified by the added dyes such as eosin Y, tetrabomophenol blue, and tetrasulfonated metallophthalocyanines. Sol-gel method is another widely used method for the preparation of nanocystalline TiO_2 and ZnO (*23–27*).

Incorporation of Noble Metal Co-Catalyst on to Semiconductor

In the last few years interest has been shown in improving activity of TiO_2 by noble metal doping. Metal-ion-doped TiO_2 was studied to improve the photocatalytic activity under UV radiation (*28–30*). Choi et al. (*31*) reported increased UV photocatalytic degradation of chloroform with Fe^{3+}, Ru^{3+}, V^{4+}, Mo^{5+}, Os^{3+}, Re^{5+} and Rh^{3+} ion doped TiO_2. At the same time they observed decreased photoactivity in case of CO_3+ and Al^{3+} ion doped TiO_2.

Gas phase oxidation rate of ethanol (*32*), acetone (*33*), and acetaldehyde (*34*), was increased with TiO_2/Pt catalyst. It has been reported that in liquid phase, addition of Pt and other noble metals to TiO_2 improve photocatalytic performance for the degradation of different pollutants (*35–37*). Platinum deposits on semiconductor helps in separating photogenerated e^-/h^+ pair (*32*). Modification of TiO_2 surface by platinum is an effective method to increase its photoactivity for water and wastewater decontamination (*38*). The generated e^-/h^+ pairs, through band gap excitation, undergo recombination in a short duration of time, and this recombination rate is a crucial factor to determine the photocatalytic activity of TiO_2. Usually e^-/h^+ recombination rate exceeds photosensitization rate, so the overall process is often not very efficient with respect to photons (*39*). Platinum metal consists of a Fermi levels lower than TiO_2 conduction band and could function as an electron trap to speed up the discharge of electrons from TiO_2. This increases the lifespan of e^-/h^+ pairs and thereby improves the photocatalytic activity of TiO_2 (*40*).

Platinum Deposition Methods

Several methods are available for platinum deposition on catalyst surface, such as photo-deposition (*41–43*), impregnation (*37*, *44*, *45*), chemical vapor deposition (*46*), chemical reduction of Pt salts (*47*, *48*), and atomic layer deposition (*40*, *49*, *50*). They yield different degree of surface modification and subsequent catalytic efficiency. A brief summary of platinum deposition methods is illustrated in the following section.

Photodeposition Method

TiO_2 powder is suspended in a deaerated solution and dispersed by sonication. The solution contains a platinum precursor (e.g. H_2PtCl_6, K_2PtCl_6, $H_2Pt(OH)_6$, $Pt(NH_3)_2(NO_2)_2$ etc.), water and a sacrificial organic reagent such as methanol, ethanol, propan-2-ol, 2-methylpropan-2-ol, and acetic acid (*41*). The solution pH is adjusted to 3.0 before the reaction. The suspension is stirred and irradiated with either UV or solar light generated from Hg vapor lamp or Xe arc lamp. After 1-2 h irradiation, the color of the suspension changes from white to black owing to Pt deposition. Completion of photodeposition can be ascertained by an analysis of Pt species in the solution. In photodeposition method the platinum precursor $Pt^{(IV)}Cl_6^{2-}$ is reduced by the photo generated electron at the valance band of TiO_2, and the sacrificial organic reagent act as hole scavenger. According to Li et al. (*42*) $Pt^{(IV)}Cl_6^{2-}$ is first adsorbed on TiO_2 surface, then reduced into $Pt^{(II)}Cl_4^{2-}$ and Pt^0 or $(Pt^0)_m$ (Pt metal cluster). They confirmed the presence of Pt, $Pt(OH)_2$, and PtO_2 on the surface of TiO_2. Yang et al. (*43*) also reported a mixture of $Pt^{(II)}$ and Pt^0 state even after 24 h of photoreduction, which could suggest the simultaneous agglomeration of Pt atoms and cathode-like reduction during the particle growth of Pt metal. They also suggested a particle growth mechanism of Pt on the basis of EXAFS and XPS results. Nakamatsu et al. (*41*) reported the diameter of photodeposited platinum metal on TiO_2 particles in the range of 0.3-0.8 μm. XRD measurement showed that most of the TiO_2 particles were rutile and some were anatase (anatase (101)/rutile (110) peak ratio=0.16). TEM and SEM analysis were carried out to identify the size distribution and orientation of the platinum deposits on TiO_2. Dispersion of platinum deposit was dependent on sacrificial agent; the particle size decreased in the following order: acetic acid (100 nm) 2-methylpropane-2-ol (100 nm) > methanol (30 nm) propane-2-ol (30 nm) > ethanol (5 nm).

Impregnation

In impregnation process, metal complexes are dissolved in aqueous solution to allow the contact with a porous oxide catalyst support such as TiO_2. In the aqueous solution the noble metal adsorbs onto the high surface area catalyst support. The catalyst slurry is then filtered if large excess solution has been employed, or just evaporated to dryness. Then it is treated further to transform the metal from its precursor state into its active form (*44*). Platinum modified catalysts are synthesized by impregnation of TiO_2 (anatase) with aqueous solution of platinum precursor. The mixture is evaporated to dryness and then calcined at 450°C for few hours. Then the deposited platinum salt is reduced to lower oxidation state in hydrogen atmosphere at 250°C. Kryukova et al. (*37*) reported 0.5-2 wt % platinum loading on TiO_2 surface, with specific surface area of around 170 $m^2\ g^{-1}$ regardless of Pt loadings. Bavykin et al. (*45*) reported the impregnation of platinum metal on TiO_2 nanotube. In order to produce TiO_2/Pt, the nanotube titanium dioxide was mixed with aqueous H_2PtCl_6 solution. The process was very sensitive to humidity, so the reaction mixture was placed into a controlled humidity chamber for 7 days. H_2PtCl_6 reduction was carried out in a quartz U-tube with hydrogen at room temperature. This method allowed the deposition of nanowires of Pt metal inside nanotubes but resulted in very uneven distribution of platinum.

Chemical Reduction of Pt Salts

In this method inorganic or organic reducing agents are used to reduce platinum precursors. The use of formaldehyde and ethylene glycol (EG) as potential reducing agents were reported by Li et al. (*47*), to deposit Pt on multiwalled carbon nanotube (MWNT) surface. Aqueous solution of H_2PtCl_6 was employed as platinum precursor. The MWNT/Pt catalyst with a metal loading of 10 wt % was obtained. In case of formaldehyde reduction, Pt particles have a wide particle-size distribution ranging from 2 to 9 nm with a mean particle size of 3.4 nm. On the contrary, for EG reduction Pt particles had narrow size dispersion ranging from 2 to 5 nm with its peak centered at 2.6 nm. This observation explained a better interaction of platinum precursor with EG solvent to produce a high homogeneous dispersion of spherical Pt metal particles with a narrow particle size distribution. They also mentioned about two main factors, surface modification of MWNTs, and water content in EG solvent which actually control the particle size and distribution of Pt particles deposited on the MWNT support.

$NaBH_4$ can also be used as reducing agent to prepare Pt deposited surface. Mei et al. (*48*) mentioned about the immobilization of platinum nanoparticles on spherical polyelectrolyte brushes. After reduction of H_2PtCl_6 by $NaBH_4$ nanosized particles were formed. TEM analysis confirmed the formation of Pt nanoparticles with 2 nm diameter.

Chemical Vapor Deposition

There are wide applications of chemical vapor deposition (CVD) of noble metals in electronics, protective coating and catalyst industries. High volatility, thermal stability and clean decomposition of suitable precursors results in the effectiveness of CVD. Metal organic complexes are commonly used as precursors of noble metals in CVD. CVD method is applied to produce fine particles of platinum, which can be used to prepare catalysts. There are several precursors available for platinum CVD, such as $Pt(acac)_2$, $Pt(CO)_2Cl_2$, (MeCp)Pt, $Pt(allyl)_2$, $(Cp)PtMe_3$, and $(MeCp)PtMe_3$. The efficiency of CVD can be further improved by applying plasma or laser as an assisting energy source to CVD (*46*).

Atomic Layer Deposition

It is a thin film growth technology which provides outstanding conformal and uniform growth with good control over both composition and thickness. It also follows sequential self-limiting surface reaction steps for metal deposition in an atomic layer-by-layer approach (*50*, *51*).

According to Floro et al. (*49*) the nucleation of precursors on surface sites in the early stages of atomic layer deposition (ALD) growth can be based on the Volmer–Weber growth mechanism. Volmer–Weber growth mode includes the following microstructural stages i) nucleation of discrete islands, ii) island growth, iii) island impingement and coalescence, iv) percolation of the island array, and v) channel filling to eventually form a continuous thin film. Variations in deposition temperatures can result in varying particle densities. Therefore, changes in the frequency of ALD cycles and deposition temperatures can control the loading and dispersion of platinum. Such characteristics enable ALD to produce both high-quality films and nanoparticles on the surfaces of flat or particulate substrates.

Zhou et al. (*40*) used fluidized bed reactor (FBR) for ALD of Pt nanoparticles on the surfaces of primary TiO_2. They also described ALD as the best method in term of precise control the Pt deposition, to uniformly modify surfaces with small, highly dispersed Pt nanoparticles.

Application of Dye-Sensitized Photocatalyst for Water and Wastewater Detoxification

Maximum allowable exposure and concentration limits of some of organic pollutants are listed in Table 2.

Table 2. Exposure and Regulatory Limits of Some Organic Pollutants. (adapted from ATSDR, 1997, 1999, 2003, 2005, 2008)

Compounds	*Use/Exposure*	*Health effect*	*Maximum allowable contaminant level*
Hydrazine	i) rocket fuels, ii) chemical manufacturing, iii) boiler water treatment	may cause nervous system effect, liver and kidney damage, human carcinogen	0.03-0.06 ppm
Trichloroethylene	i) metal degreasing agent, ii) common ingredient in cleaning agents, paint and adhesive, varnishes and ink	cause liver and lung damage, abnormal heartbeat, coma and possible death, human carcinogen	5 ppb
Carbon tetrachloride	i) used in the production of refrigeration fluid and propellants for aerosol cans, ii) as a cleaning fluid and degreasing agent, in fire extinguishers, and in spot removers	can damage the liver, kidney and nervous system, human carcinogen	5 ppb
Phenol	i) used in the manufacture of phenolic resins nylon and other synthetic fiber, ii) used as a disinfectant and antiseptic iii) it is found in aqueous effluent from industries such as petroleum refining, steel production, coal gasification, textile, tannery, pulp and paper, pesticides, pharmaceuticals and food processing	high amount can produce skin burns, liver damage, dark urine, and even death, not a human carcinogen	2 ppm
Chlorophenols	i) used in pesticides and antiseptics, ii) produced in pulp and paper industry in bleaching process with chlorine, iii) formed as a result of chlorination of humic matter during the chlorination of drinking water	high levels can cause damage to liver and immune system	0.04-0.05 ppm

Degradation Processes of Aliphatic Compounds

Carbon Tetrachloride (CCl_4) Degradation

The sensitized degradation of CCl_4 in water was successfully demonstrated by several authors under visible light illumination (λ>420 nm) (*52–55*). Ruthenium based dyes have very good potential to sensitize TiO_2 photocatalyst as well as non ionic surfactant (Brij-35). Cho et al. (*54*) studied the effect of different parameters such as solution pH, dissolved oxygen and number of dye layer on TiO_2 surface as a function of CCl_4 dechlorination rate. Tris (4,4-dicarboxy-2,2-bipyridyl) ruthenium (II) complex dye was used as sensitizer under visible light with 2-propanol as an electron donor. Adsorption of the sensitizer on TiO_2 surface showed strong pH dependency. Thus the pyrolysis rate was influenced greatly by pH, and at pH~3.0, a maximum degradation rate was achieved. The photolysis rate of CCl_4 showed a maximum at a sensitizer surface coverage of 30 % monolayer.

Dissolved oxygen showed negative effect towards dechlorination as oxygen competed for conduction band electrons. Similar result was observed by Cho et al. (*52*) where they used [Ru^{II} $(bpy)_3$] photosensitizer with nonionic surfactant. The optimum concentration of CCl_4, surfactant, and sensitizer for maximum CCl_4 dechlorination rate were reported as 30 mM, 0.4 g L^{-1} and 5 mM respectively. With the ruthenium sensitizer alone in the absence of the surfactant, the dechlorination rate was negligible as the photoinduced electron transfer from the excited sensitizer to CCl_4 occurred only in the presence of the surfactant which concentrates both reactants within a micelle.

Fung et al. (*53*) anchored a Ru(II) photosensitizer onto the surface of anatase TiO_2 particles via in situ silylation. The silyl linkage provided huge stability over a wide pH range for aqueous media and in organic solvents. The electronic coupling between photosensitizer and TiO_2 was achieved by the silyl anchoring group. Upon photoexcitation, this coupling interceded electron injection from photosensitizer to the TiO_2 conduction band. Photodegradation of CCl_4 in prescence of TiO_2-[Ru^{II}(py-pz-Si)$_3$] and electron donor was possible under both anaerobic and aerobic conditions. CCl_4 degradation rate followed Langmuir-Hinshelwood type relation which confirmed the adsorption of CCl_4 on TiO_2 surface during degradation.

Trichloroethylene (TCE) Degradation

Alexander et al. examined the photo oxidation of TCE with natural anthocyanine dye sensitized-buoyant TiO_2 coated microspheres. The naural dyes were obtained from commercially available frozen fruits. The dye-sensitized system provided excellent remediation with 93 % of the TCE removal in 12 h. One of the shortcomings of the photo-oxidative remediation technique is the considerable decline in pH associated with conversion of TCE to HCl, which may not be a desirable process to some sensitive aquatic environments (*56*).

Pesticides Degradation

Muszkat et al. suggested accelerated photocatalytic oxidation of pesticides in water by dye sensitization. They studied two widely applied herbicides, bromacil (5-bromo-3-sec-butyl-6-methyluracil) and metribuzin (4-amino-6-tert-butyl-4,5-dihydro-3-methylthio-1,2,4-triazine-5-one). In absence of TiO_2, different reaction patterns of photo-oxidation were shown by the above mentioned pesticides. Oxygen had a distinct effect on the rate of photo-oxidation of metribuzin while the influence of H_2O_2 was quite moderate. On the contrary, in presence of H_2O_2, photo-oxidation of bromacil improved a lot although oxygen showed nominal effect. In dye-sensitized system, bromacil (BR) degradation was observed in presence of small amount of methylene blue (MB) and red 22 (R) dyes in UV and natural sun light respectively. Degradation of BR in UV light was slightly higher than that of sun light. The photo-oxidation mechanism of the reaction seems to happen in presence of dual oxidant. First one was hydroxyl radicals produced by band gap excitation of semiconductor and subsequent formation of electron/hole pair. Second one was singlet oxygen formed through dye-sensitization process (*57*).

Hydrazine Degradation

Chatterjee, studied the reduction of hydrazine to ammonia via dye-sensitized photocatalysis with Pt/TiO_2-[Ru^{III}(EDTA)(H_2O)] - system. The reaction mechanism was explained by the formation of [Ru^{III}(EDTA)(N_2H_5)] species (adsorbed on TiO_2 surface) which experienced a two-electron transfer reduction followed by cleavage of the N-N bond of coordinated hydrazine as described below:

$$TiO_2 \xrightarrow{360\ nm} TiO_2\ (e_{CB}^- + h_{VB}^+) \quad (i)$$

$$[Ru^{II}(EDTA)(H_2O)]^- + N_2H_5^+ \rightarrow [Ru^{II}(EDTA)(N_2H_5)] + H_2O \quad (ii)$$

$$[Ru^{II}(EDTA)(N_2H_5)] + 2e_{CB}^- \rightarrow [Ru^{I}(EDTA)(N_2H_5)]^{2-} \quad (iii)$$

$$[Ru^{I}(EDTA)(N_2H_5)]^{2-} \rightarrow [Ru^{III}(EDTA)(N_2H_5)]^{2-} + NH_3 \quad (iv)$$

$$[Ru^{III}(EDTA)(N_2H_5)]^{2-} \xrightarrow{H^+/H_2O} [Ru^{III}(EDTA)(H_2O)]^- + NH_3 \quad (v)$$

The rate controlling step of the photocatalytic process was the surface chemical step (electron transfer) which could be coupled with adsorption of reactants and desorption of ammonia molecule (*58*).

Dye Degradation

A non-sensitizing dye (e.g. acid blue1 (AB1)) in an effluent can be degraded by just mixing with another effluent containing sensitizing dye (eosin Y or thionine) and subjecting the mixture of effluent to visible light and aqueous suspensions of TiO_2. However, the sensitizing dye also degrade gradually due to self-sensitized degradation mechanism (*59*).

Different dye degradation such as bromothymol blue, acid blue 1, methylene blue, eosin Y, thionine, and chrysoidine Y were reported in recent literatures (*59–63*). Degradation rates are strongly influenced by several factors such as solution pH, catalyst concentration, substrate concentration, type of electron acceptor and semiconductor type. Degussa TiO_2 P25 showed highest efficiency compared to Hombikat UV100, PC 500, TTP and ZnO (*60, 61*). Solution pH is an important parameter in the photocatalytic reactions, since it influences the surface charge of the catalyst and therefore the adsorption behavior of the pollutants. The efficiency of bromothymol blue degradation was found to be more or less similar in pH range of 4-9. Formation of reaction intermediate was also confirmed by new absorption peak at pH 9. At pH 2, slightly lower mineralization rates were observed (*60*). In case of chrysoidine Y, maximum degradation was found at pH 9. At alkaline pH, high concentration of HO^- ion facilitated the formation of hydroxyl radicals, which finally diffused away and degraded the dye molecule in the bulk solution (*61*).

Simultaneous photocatalytic degradation of AB1, eosin Y and thionine in air equilibriated aqueous suspension of TiO_2 photocatalyst has been achieved by Chatterjee et al. (*59*). They also proposed the formation of reactive $^{\bullet}O_2^-/^{\bullet}HO_2$ as a major route of dye degradation. Degradation mechanism of AB1 in presence of eosin Y and thinine is quite different. Thionine accepts electron at the excited state whereas eosin Y usually releases electron. So, thionine can oxidize AB1 to $AB1^+$ even in the absence of TiO_2 and in the process it is converted into a colorless species, semithionine. However, eosin Y cannot undergo such oxidation (AB1 to $AB1^+$) upon irradiation in absence of TiO_2. Scheme 1 and Scheme 2 were proposed for the degradation of AB1-EY and AB1-Th (*59*).

In case of dye-sensitized TiO_2 photocatalyst, the dye molecules were only physically adsorbed on TiO_2 surface without the formation of any stable chemical bonds between them. Thus, the dye molecules were easily desorbed from TiO_2 surface during photocatalytic process thereby resulting in poor photocatalytic activity. Jing et al. synthesized a dye modified TiO_2 using Chrysoidine G (CG), tolylene-2,4-diisocyanate (TDI) and Degussa P25 as raw materials. TDI linked together TiO_2 and CG via –NHCOOTi and –NHCONH- bonds to produce a dye modified TiO_2 catalyst. The dye modified TiO_2 showed strong absorption in visible region and improved adsorption capacity to methylene blue (MB) As a result, the photocatalytic degradation of MB by dye-modified TiO_2 showed much higher activity than bare TiO_2 (*63*).

$$\mathrm{TiO_2 \ldots (EY)_s \xrightarrow{h\vartheta>420\,nm} TiO_2 \ldots EY^*} \quad \text{(i)}$$

$$\mathrm{TiO_2 \ldots EY^* \rightarrow TiO_2 \ldots EY^+ + e^-_{CB}} \quad \text{(ii)}$$

$$\mathrm{TiO_2 \ldots EY^+ + AB1 \rightarrow TiO_2 \ldots (EY)_s + AB1^{+\cdot}} \quad \text{(iii)}$$

$$\mathrm{e^-_{CB} + O_2 \rightarrow O_2^{-\cdot}} \quad \text{(iv)}$$

$$\mathrm{O_2^{-\cdot} + H^+ \rightarrow HO_2^{\cdot}} \quad \text{(v)}$$

$$\mathrm{O_2^{-\cdot}/HO_2^{\cdot} + AB1/AB1^{+\cdot} \rightarrow Products} \quad \text{(vi)}$$

$$\mathrm{O_2^{-\cdot}/HO_2^{\cdot} + EY/EY^{+\cdot} \rightarrow Products} \quad \text{(vii)}$$

Scheme 1

$$\mathrm{TiO_2 \ldots (Th)_s \xrightarrow{h\vartheta>420\,nm} TiO_2 \ldots Th^*} \quad \text{(i)}$$

$$\mathrm{TiO_2 \ldots Th^* + AB1 \rightarrow TiO_2 \ldots Th^{-\cdot} + AB1^{+\cdot}} \quad \text{(ii)}$$

$$\mathrm{TiO_2 \ldots Th^{-\cdot} \rightarrow TiO_2 \ldots (Th)_s + e^-_{CB}} \quad \text{(iii)}$$

$$\mathrm{e^-_{CB} + O_2 \rightarrow O_2^{-\cdot}} \quad \text{(iv)}$$

$$\mathrm{O_2^{-\cdot} + H^+ \rightarrow HO_2^{\cdot}} \quad \text{(v)}$$

$$\mathrm{O_2^{-\cdot}/HO_2^{\cdot} + AB1/AB1^{+\cdot} \rightarrow Products} \quad \text{(vi)}$$

$$\mathrm{O_2^{-\cdot}/HO_2^{\cdot} + Th \rightarrow Products} \quad \text{(vii)}$$

Scheme 2

Phenol Degradation

Photodegradation of phenol with dye-sensitized TiO_2 showed promising results under visible light irradiation (*64*, *65*). Grandos et al. utilized Zn (II) and Co (II) tetracarboxyphthalocyanine (TcPcM) to sensitize TiO_2 and TiO_2/Pt for the degradation of phenol in visible light. They expressed the degradation of phenol with TcPcM/TiO_2 in terms of photonic efficiency [$\eta = (kC_0/I_0)$]; where C_0 was the initial phenol concentration and I_0 was the incident photonic flow per unit volume. The photonic efficiencies (η) of phenol degradation were 4.3% and 3.3% on TcPcCo/TiO_2 and TcPcZn/TiO_2 respectively. Again, photodeposition of platinum on TcPcM/TiO_2 increased the photo-efficiency up to $\eta \cong 0.1$ (*64*).

Chowdhury et al. (*65*), also reported eosin Y-sensitized TiO_2/Pt catalyst for the degradation of phenol in visible light. About 93% degradation of phenol (C_0=40 mg L^{-1}) was achieved within 90 minutes under optimum reaction conditions such as pH 7.0, catalyst loading of 0.8 g L^{-1}, electron donor (triethnolamine) concentration of 0.2 M, 0.5% Pt loading on TiO_2, and visible solar light intensity of 100 mW cm^{-2}. They considered the formation of superoxide ($^{\bullet}O_2^-$) ions followed by HO$^{\bullet}$, which subsequently proceeded to the complete mineralization of phenol and other intermediates. They also found a significant effect of the Pt loading on phenol degradation. Eosin Y-sensitized TiO_2 without Pt metal can degrade phenol (~ 67 %) in visible light, but with the incorporation of Pt (0.5 %) on TiO_2, complete degradation was achieved.

Chlorophenol Degradation

Ghosh et al. (*66*) reported the degradation of 4-chlorophenol in aqueous medium with coumerine (C-343, λ_{max}= 446 nm) sensitized TiO_2 photocatalyst using 436 nm LED based photoreactor. LEDs convert high electrical to light energy generating a low heat. It also offers forward directed output facilitating delivery to a target, longer lifetime, and DC operation which might be helpful in using in remote locations. Additionally, the LED reactor efficiency for 4-chlorophenol degradation was fairly comparable with a conventional multi-lamp reactor or sun light. The coumarin dye can accumulate a larger fraction of energy from the well-matched LED. The reaction rate fitted approximately to first order kinetics and moved towards a limiting value at a catalyst concentration of 0.3 g L^{-1}. They mentioned about HO$^{\bullet}$ radical initiated reaction mechanism although the formation of chlorocatechol was not reported. The only supporting fact regarding this was HO$^{\bullet}$ radical scavenging experiment. Formation of superoxide ion ($^{\bullet}O_2^-$) was also assumed, which would involve the attack to benzene ring with loss of chlorine from chlorophenol (*66*, *67*).

Benzyl Alcohol Degradation

The photocatalytic oxidation of benzyl alcohol has been investigated by Hussein and Alkateeb, under natural weathering conditions (sunlight and oxygen) in presence of TiO_2 or sensitized TiO_2. Sensitization of TiO_2 was achieved by the impregnation of riboflavin (RF), safranine O (SO), methyl red (MR), eosin B (EB), and methyl blue (MB) on anatase (TiO_2 P25). Benzaldehyde was identified as the only reaction product through HPLC, FTIR, and spectrophotometric analysis. The photo-oxidation of benzyl alcohol was found to follow the sequence: RF > SO > naked TiO_2 > MR > EB > MB. The formation of benzaldehyde during the photocatalytic reaction was studied by using FTIR, HPLC and spectrophotometric measurements (*68*).

These degradation processes are summarized in Table 3, and 4.

Table 3. Dye-Sensitized Photodegradation of Organic Compounds

No	*Organic compounds*	*Initial substrate concentration*	*Sensitizer/ Photocatalyst*	*Light source and accessories*	*Other experimental details*	*Results/comments*	*Reference*
1.	CCl_4	10 mM	Ruthenium bipyridyl complex ($Ru^{II}(bpy)_3$)/nonionic surfactant Brij-35 .	450 W Xe arc lamp (Oriel); 10 cm IR water filter; UV cutoff filter (λ>420 nm)	Irradiation time 120 min; Air equilibrated.	i) CCl_4 degradation rate was enhanced in absence of O_2	Cho et al. (*52*)
2.	CCl_4	3 – 4 ppm	Silylated $[Ru^{II}(py\text{-}pzH)_3]^{2+}/TiO_2$. py-pzH ≡3-(2′-pyridyl)	100 W tungsten lamp (Oriel) (λ>450 nm) ; IR filter.	Irradiation time 30 min; pH 6.5-7.0; purged with air (for 15 min); I^- as sacrificial electron donor.	i) Silyl linkage is highly stable in extream pH range (pH 1.0 – 12.0).	Fung et al. (*53*)
3.	CCl_4	1 mM	Tris (4, 4′-dicarboxy-2,2′-bipyridyl) ruthenium (II) chloride	450 W Xe-arc lamp (Oriel); 10 cm IR water filter; UV cut off filter (λ>420 nm).	Irradiation time 6 h; N_2 saturated system; TiO_2=0.5g L^{-1}; pH 3; 2-propanol as electron donor.	Quantum yield of CCl_4 dechlorination 10^{-3}	Cho et al. (*54*)
4.	i) CCl_4 ii) trichloro-acetate (TCA)	i) 1 mM ii) 1 mM	Tris (4, 4′-dicarboxy-2, 2′-bipyridyl)-ruthenium (II) complex.	450 W Xe-arc lamp (Oriel); 10 cm IR water filter; UV cut off filter (λ>420 nm).	Irradiation time 1 h (CCl_4) and 2 h (TCA); O_2 was removed by N_2 sparging before reaction; pH 3	Photocatalyst was completely inactive in presence of dissolved oxygen.	Bae and Choi (*55*)

No	*Organic compounds*	*Initial substrate concentration*	*Sensitizer/ Photocatalyst*	*Light source and accessories*	*Other experimental details*	*Results/comments*	*Reference*
5.	Trichloro-ethylene (TCE)	2-500 ppm	Natural anthrocyanine dye (obtained from fruits)/buoyant TiO_2 coated microsphere.	150 W Xe lamp;	Irradiation time 15 - 1440 min	i) Mineralization product: HCl; ii) 93 % TCE degradation was reported.	Alexander and Rosentreter (*56*)
6.	Pesticides (bromacil (BR))	30-100 ppm	Methylene blue (MB), Red 22 (R).	UV: HBO 200 W; natural sunlight	TiO_2 concentration 100 mg L^{-1}; MB/BR molar ratio varied from 0.05-0.2 ; R/BR molar ratio was 0.143	Highest enhancement factor (EF) 2.27 was achieved at MB/BR molar ratio of 0.1	Muszkat et al. (*57*)
7.	Hydrazine	1 mmol	$Ru^{III}(EDTA)(H_2O)$-/ Pt/TiO_2 P25	250 W Xe lamp	Irradiation time 8 h; pH 3.1; Temperature 25°C.	i) Hydrazine is reduced to ammonia; ii) Yield of ammonia production 1.98×10^{-5} mol h^{-1}. ii) Declined yield after 5 h	Chatterjee (*58*)

Continued on next page.

Table 3. (Continued). Dye-Sensitized Photodegradation of Organic Compounds

No	*Organic compounds*	*Initial substrate concentration*	*Sensitizer/ Photocatalyst*	*Light source and accessories*	*Other experimental details*	*Results/comments*	*Reference*
8.	Phenol	100 ppm	Zn(II) and Co(II) tetracarboxphthalo-cyanine/Pt/TiO_2.	100 W halogen lamp (Osram); 1M $K_2 Cr_2O_7$ as UV filter.	Irradiation time 60 min; Temperature 20±2°C; Air supply into reaction medium.	i) 33 % phenol degradation with TcPcZn/TiO_2/Pt ii) 65 % phenol degradation with TcPcCo/TiO_2/Pt.	Granados O et al. (*64*)
9.	4-chloro-phenol (4ClPh)	40 ppm	Coumarin-343/TiO_2 P25	81 Gilway "super bright" (Peabody, MA) E472 µW (max output); 1mM $K_2 Cr_2O_7$ in 0.22 M Na_2CO_3 aqueous solution as UV filter.	Irradiation time 8 – 10 h; Temperature 20±2°C;	i) Reaction intermediate:1,2,4-benzenetriol and d ihydroxymaleic acid; ii) First order reaction kinetics followed.	Ghosh et al. (*66*)

No	*Organic compounds*	*Initial substrate concentration*	*Sensitizer/ Photocatalyst*	*Light source and accessories*	*Other experimental details*	*Results/comments*	*Reference*
10.	i) Chloro-phe-nol ii) 1,2-dichloroethane iii) 1,4-dichloroben-zene iv) trichloro-ehylene	i) 1 mmol ii) 1 mmol iii) 0.1 mmol iv) 1 mmol	Thionine, Eosin Y, Rhodamine B, methylene Blue, Nileblue A, Safranine O/ TiO_2.	150 W Xenon lamp (Oriel Instruments); UV filter solution ($NaNO_3$+$CuSO_4$+NH_4OH).	O_2 was bubbled in the photoreactor.	i) Product of CO_2, Cl^-; ii) 55-72 % degradation of pollutant was achieved.	Chatterjee et al. (*59*)
11.	Benzyl alcohol	-	Riboflavin (RF), Safranine O (SO), Methyl red (MR), Eosin B (EB), Methylene blue (MB)/TiO_2 P25	Sunlight	Irradiation time 75 min; Temperature 316-321 K; Reaction volume 150 cm^3.	i) Benzaldehyde was the photooxidation product; ii) The sequence of dye in the following superiority order: RF>SO>naked TiO_2>MR>EB>MB;	Hussein et al. (*68*)

Table 4. Dye-Sensitized Photodegradation of Organic Dyes

No.	*Organic dyes*	*Initial substrate concentration*	*Sensitizer/ Photocatalyst*	*Light source and accessories*	*Other experimental details*	*Results/Comments*	*Reference*
1.	Methylene blue	50 ppm	Chrysoidine G/TiO_2 P25.	250 W metal halide lamp (Philips); UV cut off filter.	Irradiation time 12 h.	special organic complexes were formed on the TiO_2 surface via stable π-conjugated chemical bonds between TiO_2 and dye molecules	Jiang et al. (*63*)
2.	Bromothy-mol blue (BTB)	0.25 mM	TiO_2 P25	125 W medium pressure mercury lamp; IR and UV filter.	Catalyst dose 1 g L^{-1}; Irradiation time 90 min; electron acceptor ($KBrO_3$, H_2O_2 etc).	Electron acceptors showed positive effect on BTB degradation	Haque and Muneer (*60*)
3.	Acid blue 1 (AB1)	25 ppm	Eosin Y, Thionine/TiO_2.	150 W Xenon lamp (Oriel Instruments); Copper sulfate and ammonium hydroxide as UV filter (λ> 420 nm).	Irradiation time 5 h; O_2 was bubbled in photoreactor; pH 5.0.	decoloration of eosinY (EY) or thionine (Th) due to self-sensitized degradation was also noticed	Chatterjee et al. (*59*)
4.	Chrysoidi-ne Y	0.25 mM	Chrysoidine Y/TiO_2 P25.	125 W medium pressure Hg lamp.	Irradiation time 120 min; O_2 purging and stirring; pH 3.0-9.0.	Degussa P25 was more efficient than that of ZnO	Qamar et al. (*61*)

Application of Dye-Sensitized Photocatalyst for Hydrogen Generation

Basic Requirement for Hydrogen Generation in Visible Light

When a semiconductor material is illuminated with photons having an energy hν equal to or larger than the semiconductor band gap, the result is formation of electronic charge carriers, electrons in the conduction band and holes in the valance band.

$$\text{Semiconductor} + 2h\vartheta \rightarrow 2h^+ + 2e^- \qquad (4)$$

$$H_2O + 2h^+ \rightarrow \frac{1}{2}O_2 + 2H^+\left(E_{O_2/H_2O}\right) \qquad (5)$$

$$2H^+ + 2e^- \rightarrow H_2\left(E_{H_2O/H_2}\right) \qquad (6)$$

$$H_2O + h\vartheta \rightarrow H_2(g) + \frac{1}{2}O_2(g) \qquad (7)$$

The photoelectrons and holes reduce and oxidize water to produce the stoichiometric 2:1 mixture of H_2 and O_2 by the above reaction (equations 4-7) (*69*).

The Gibb's free energy (ΔG) at STP is positive for water splitting reaction (equation (7)). Such a non-spontaneous reaction will be possible if the energy of the incident photon's energy is equivalent to the change in the Gibb's free energy (ΔG) of the reaction and that must be supplied. Under standard conditions water can be electrolyzed reversibly into hydrogen and oxygen (2:1) at a potential of 1.23 V which is derived from the equation of Gibb's free energy:

$$\Delta G^0 = -nF\,(\Delta E^0) \qquad (8)$$

Where, ΔG^0 and ΔE^0 are standard Gibbs free energy change and standard electrical potential of the reaction. The standard Gibbs free energy change (ΔG^0) is the negative value of maximum electrical work corresponding to 237.14 kJ mol^{-1} or 2.46 eV for equation (7). Since this is a two electron redox process, photocatalytic water splitting is possible if the semiconductor photocatalyst possess a band gap energy (E_g) greater than 1.23 eV (*70, 71*).

Now, the semiconductor photocatalyst will be visible light active if the band gap energy is less than 3.0 eV. In addition, the band position is also an important parameter for visible light excitation. Therefore, the photocatalytic materials for visible light water splitting should have proper band position and suitable band gap energy (1.23 eV $< E_g <$ 3.0 eV) (*70*). Oxides such as TiO_2, ZnO and SnO_2 have large band gap (3-3.8 eV) and absorb only ultra-violet part of the solar emission and so has low conversion efficiencies.

Dye sensitization is an excellent method to broaden the absorbance range of photocatalysts. Remarkable increases in visible-light absorption ability were observed after dye sensitization without compromising the photocatalyst structure. Abe et al. (*72*) reported maximum improvement in visible light absorption for N3-TiO_2/Pt (Ru complex N3) followed by M-TiO_2/Pt (merocyanine NK2045) and C-TiO_2/Pt (coumarine C343). Nanotube $Na_2Ti_2O_4(OH)_2$ (*73*), has a absorption band edge of about 390 nm, which was expanded to 670 nm by sensitization with eosin Y dye. Sensitizations with eosin Y dye were also reported for N-doped TiO_2/Pt (*74*), silane-coupled TiO_2 (*75*), mesoporous TiO_2/Pt (*76*), TS-1 zeolite (*77*), and silica gel H (*78*). In all cases a red-shift of the absorption band edge to about 600 nm were observed. Li et al. (*79*) prepared multilayer-eosin Y-TiO_2 through linkage of Fe^{3+} and achieved similar improvement in visible light absorption.

Dye-Sensitized Photocatalytic Hydrogen Generation Mechanism

Dye-sensitized photocatalytic hydrogen evolution from water is pointed out in Figure 3 (*80*). Sensitizer is excited first with light, following which electron transfer occurs from the excited sensitizer to the conduction band of the semiconductor photocatalyst. Subsequently the electron is directed to the platinum site, where hydrogen evolution occurs (*16*).

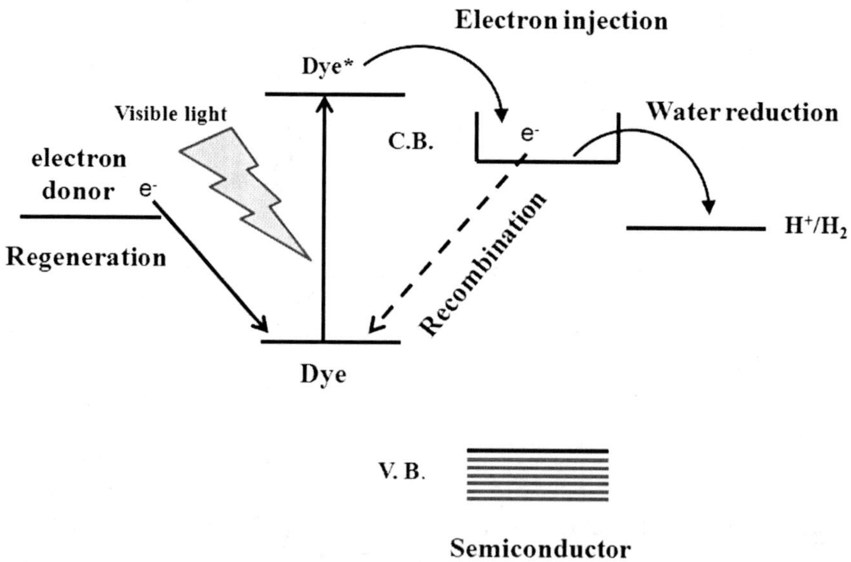

Figure 3. Dye-sensitized photocatalytic H_2 generation from water. (Reproduced with permission from reference (80). Copyright 2010 American Chemical Society).

Dye sensitization followed by photocatalytic hydrogen generation process can be expressed in few consecutive steps i) visible light absorption by dye molecule and excitation, ii) sensitization of semiconductor (TiO_2) with excited dye species, iii) hydrogen generation with the assistance of co-catalyst (Pt), and iv) regeneration of dye with the help of electron donor.

i) Light absorption:

$$\text{Dye} \xrightarrow{h\vartheta\ \text{(visible light)}} \text{Dye}^*$$

ii) Sensitization of TiO_2:

$$\text{Dye}^* + \text{TiO}_2 \rightarrow \text{Dye}^+ + \text{TiO}_2\ (e^-_{CB})$$

iii) Hydrogen generation:

$$\text{TiO}_2\ (e^-_{CB}) + \text{Pt} \rightarrow \text{TiO}_2 + \text{Pt}\ (e^-)$$

$$\text{Pt}\ (e^-) + \text{H}_2\text{O} \rightarrow \text{Pt} + \frac{1}{2}\text{H}_2 + \text{HO}^-$$

iv) Dye regeneration:

$$\text{DH} + \text{Dye}^+ \rightarrow \text{Dye} + \text{D} + \text{H}^+$$

The summarized scheme is give in Figure 4 (*81*), which shows that the overall mechanism induced by light is reduction of water and oxidation of the third component employed as an electron donor.

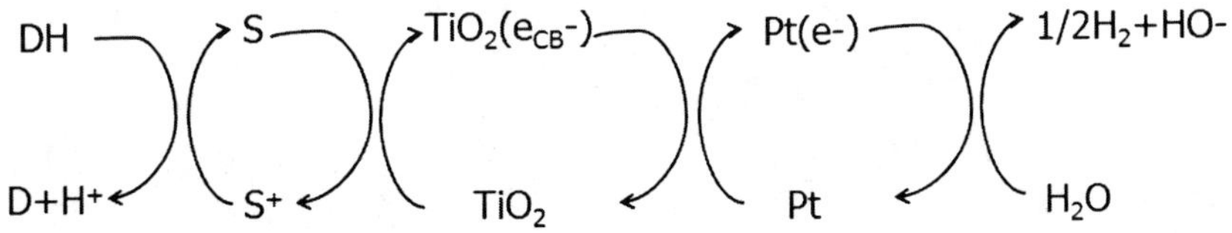

DH : electron donor; S : dye

Figure 4. Scheme for visible light-driven hydrogen generation by dye-sensitized TiO_2/Pt.

Sensitization with Different Dyes for Photocatalytic Hydrogen Generation

Different types of dye molecules have been used so far as sensitizer generating hydrogen in visible light. In early studies, $Ru(bpy)_3^{2+}$ and its derivatives were used as sensitizer for Pt/RuO_2-loaded TiO_2 particles for overall water cleavage (*82–84*). Sacrificial hydrogen generation was also reported for $Ru(bpy)_3^{2+}$ and $Ru^{II}(acid)_3$ sensitized TiO_2/Pt in visible light using EDTA as electron donor. Activities of Tris(2,2′-bipyridine-4,4′-dicarboxylic acid) ruthenium(II) ($Ru(dcbpy)_3$) and $Ru(bpym)_3^{2+}$ were better for hydrogen generation in visible light compared to $Ru(bpy)_3^{2+}$ (*85*). Enhanced dye adsorption on the photocatalyst surface was mainly projected as the explanation of such improvement. Strongly bound dye molecule can be stabilized near semiconductor surfaces, which leads to fast electron-injection of the excited dye into substrate. The photocatalytic activity and stability of $Ru(bpy)_3^{2+}$-sensitized TiO_2 depends on surface anchoring group. According to Bae et al. (*86*) phosphonate group based sensitizer exhibited higher photocatalytic activity for hydrogen production from water than carboxylate group based sensitizer. Anchoring through phosphonate group assisted faster regeneration and thus acted as better ruthenium sensitizer linkage to the TiO_2 surface in aqueous solution.

Again, close proximity of dye molecule and substrate also increases the probability of electron/oxidized dye recombination reaction. Kajiwara et al. (*87*) also believed that the thermal re-orientational motion of the loosely attached molecules was responsible for increased electron transfer. Peng et al. (*88*) did a comparative photosensitizing activity study with three kinds of dyes, having different terminal groups that can attach to TiO_2 through firm or loose linkage. Among them, N719 [$(n\text{-}Bu4N)_2\text{-}cis\text{-}Ru(dcbpy)_2(SCN)_2$], one of the best sensitizer for DSSCs has the same structure as N3 dye. $Ru(bpy)_2(him)_2\text{–}NO_3$ and $Ru_2(bpy)_4L_1\text{–}PF_6$ have no terminal group like N719 and can just loosely link with TiO_2. $Ru_2(bpy)_4L_1\text{–}PF_6$ showed more steady and higher increases in H_2 generation with extended irradiation time than the strongly linked N719. Therefore, the dynamic equilibrium between the ground state dye-TiO_2 attachment and oxidized dye-TiO_2 detachment played a crucial role in the photochemical behavior during the photocatalyst sensitization process (*89*).

Higher cost constrains and toxicity of metal based dyes (especially ruthenium based dyes) forced researchers to use less expensive organic dyes for dye sensitization. Photosensitization activity of five different dyes namely acriflavin, eosin blue, fluorescein, rhodamine B, rose bengal, was examined for hydrogen production by SnO_2/Pt (*90*). Among them eosin blue provided the maximum efficiency to sensitize SnO_2 and produced a greater amount of hydrogen, even higher than $Ru(bpy)_3^{2+}$. Photosensitization abilities of eosin Y dye for sacrificial hydrogen generation were studied extensively by several authors (*73–79*). Sreethawong et al. (*76*) synthesized mesoporous-assembled TiO_2/Pt by a single step sol-gel process with the aid of structure-directing surfactant. The activity of the photocatalyst with eosin Y sensitization was studied in diethanolamine (30 vol %) aqueous solution. Platinum loaded (0.6 wt %) TiO_2 calcined at 500°C showed highest photocatalytic performance for hydrogen generation from aqueous diethanolamine solution containing 2 mM eosin Y. Li et al. (*74*) prepared a nitrogen-doped TiO_2/Pt (N-TiO_2) by calcination of the hydrolysis product of $Ti(SO_4)_2$ with aqueous ammonia followed by platinum loading. Then, N-TiO_2/Pt was sensitized with anhydrous ethanol solution of eosin Y and used for hydrogen generation in presence of triethanolamine. Nitrogen doping created surface oxygen defects which facilitated the adsorption of eosin Y. Thus, under similar experimental conditions eosin Y-N-TiO_2/Pt showed higher photocatalytic activity compared to that of eosin Y-TiO_2/Pt. A chemical fixation of eosin Y on TiO_2/Pt was tried by Abe et al. (*75*), with silane-coupling in order to get steady hydrogen generation from aqueous triethanolamine solution. The turnover number was quite high (10^4), which represented high catalyst stability, although the initial hydrogen generation rate was low, compared to that of physically mixed system of eosin Y and TiO_2/Pt. The quantum yield at 520 nm was found to be about 10 %. Eosin Y-sensitized platinum loaded nanotube $Na_2Ti_2O_4(OH)_2$, with a quantum yield up to 14.97 %, showed good photocatalytic stability for hydrogen production over 100 h after 10 consecutive runs. Li et al. (*79*) prepared a multilayer-eosin Y-sensitized TiO_2 through Fe^{3+} coupling between TiO_2-eosin Y and different eosin Y molecules. The photocatalyst had very high light harvesting efficiency and photocatalytic activity for visible light driven hydrogen generation from aqueous triethanolamine solution. The apparent quantum yield for hydrogen evolution was 19.1 %. Eosin Y-sensitized TS-1 zeolite (*77*), silica gel H (*78*), and multiwalled carbon nanotube/Pt (*91*), were also able to generate hydrogen from aqueous triethanolamine solution in visible light with apparent quantum efficiency of 9.4 %, 10.4 % and 12.4 % respectively.

Effect of Different Parameters on the Photocatalytic Activity for Hydrogen Generation

Effect of Co-Catalyst Loading

Till date platinum has shown the highest photocatalytic activity as co-catalyst in water splitting system in visible light. There are few other noble metals from group 8-11 in periodic table (Ru, Rh, Pd, Ag and Au) which have also been reported as effective co-catalyst for such system. Higher electron flow from semiconductor to metal can be achieved with the smaller Schottky barrier height at the metal/semiconductor junction which leads to higher photocatalytic activity for hydrogen production (*92*).

Effect of Electron Donor

In case of band UV photocatalysis or band gap excitation process sacrificial agents are used to stop e^-/h^+ pair recombination. They are basically electron donors or hole scavengers. These react irreversibly with the photo-generated valance band hole and can enhance the e^-/h^+ pair separation and provide higher quantum efficiency. Aliphatic organic compounds such as ethylenediamine tetra-acetic acid (EDTA), carboxylic acids, aldehydes, and alcohols are mostly used as sacrificial agents to enhance hydrogen production (*2*). The degree of enhancement capability was found in the following order by Nada et al. (*93*): EDTA > methanol > ethanol > lactic acid. Inorganic ions such as S^{2-}/SO_3^{2-}, IO_3^-/I^- and Ce^{4+}/Ce^{3+} were also used as sacrificial agent for hydrogen production (*94–96*). In dye-sensitized system under visible light, practically no hole is present. In this case the function of electron donor may be viewed in light of its dual function since it could quench the oxidized form of the dye (dye^+) or it could extend its lifetime in the photosensitized system providing long-term hydrogen generation. Different electron donors such as EDTA, acetonitrile, methanol, isopropanol, IO_3^-/I^-, diethanolamine, triethanolamine, chloroacetic acid, and oxalic acid etc. can be used.

Effect of Solution pH

Solution pH has immense influence on photocatalytic reaction which takes place over the photocatalyst surface. In aqueous slurry of TiO_2 the catalyst surface is embraced with plenty of hydroxyl groups and thus the ionization of TiO_2 is greatly affected by solution pH. Moreover, the ionization of the electron donor and regeneration of dye are also influenced by solution pH. Chowdhury et al. (*97*) reported that dye (eosin Y) molecule showed stronger interaction with solution pH in comparison with the electron donor-pH interaction. Different optimum pH was reported for dye-sensitized photocatalytic system ranging from 3 to 11. With Ru(II) based sensitizer and methyl viologen (MV^{2+}) the optimum pH was in acidic range (*82*, *83*). In case of EDTA and triethanolamine based systems the optimum pH were 7 (*73*, *75*, *85*). On the contrary, with diethanolamine based system the optimum pH was alkaline (11.5) (*76*).

Hydrogen generation processeses with different dye-sensitized photocatalytic system are summarized in Table 5.

Table 5. Dye-Sensitized Photocatalytic Hydrogen Generation

No	*Aqueous mixture details*	*Capacity/ Concentration*	*Sensitizer/ Photocatalyst*	*Light source and accessories*	*Other experimental details*	*Results/ Comments*	*Reference*
1.	Water, methyl viologen (MV^{2+})	25 mL/dye: 10^{-4} – $5x10^{-5}$ M.	$[Ru(bpy)_3]^{2+}$; Rhodamine B /TiO_2 sol loaded with Pt and RuO_2 (Pt 1mg, RuO_2 0.2 mg)	450 W –Xe lamp, 420 nm cutoff filter	N_2 saturated; pH 3; TiO_2 1g L^{-1}.	Quantum yield (QY) 30 %	Duonghong et al. (*82*)
2.	Water, methyl viologen (MV^{2+})	25 mL/ 10^{-4} M dye; $5x10^{-3}$ M MV^{2+}	$[Ru(bpy)_3]^{2+}$/colloidal TiO_2-Pt-RuO_2 (Pt 40 mg/L)	450 W Xe lamp, IR filter, 400 nm cut off filter	N_2 saturated; pH 4.7; TiO_2 0.5g/L, 4% Nb_2O_5, 0.1% RuO_2	rate of H_2 = 2 (rate of O_2) only after saturation of TiO_2 adsorption sites	Borgarello et al. (*83*)
3.	Water- EDTA (electron donor)	3 mL water; EDTA 0.01 M	$[Ru(bpy)_3]^{2+}$/TiO_2 P25/Pt and TiO-5/Pt (0.3 wt %).	500 W Xe lamp, UV cutoff filter L42	pH 7 ; TiO_2 (TiO-5): 0.05g	rate H_2 ~5.73μmol min^{-1}	Hirano et al. (*85*)
4.	Water	40 mL/10^{-4} M dye	$[Ru(dcbpy)_2(dpq)]^{2+}$/ TiO_2 P25/Pt (0.4wt %); ZnO/Pt (0.4wt %)	250 W tungsten halogen lamp; IR filter	N_2 saturated; pyrogallol solution to remove O_2; cat dose 1g L^{-1};	rateH_2: $4.16x10^{-3}$ mL H_2/min (TiO_2/Pt), $5x10^{-4}$ mL H_2/min (ZnO/Pt)	Dhanalakshmi et al. (*16*)

Continued on next page.

Table 5. (Continued). Dye-Sensitized Photocatalytic Hydrogen Generation

No	*Aqueous mixture details*	*Capacity/ Concentration*	*Sensitizer/ Photocatalyst*	*Light source and accessories*	*Other experimental details*	*Results/ Comments*	*Reference*
5.	Water-acetonitrile, NaI (I_3^-/I^-) (electron donor)	100m L/ water 5%; NaI 0.1 M	Merocyanine (M), Coumarin (C), Ru complex (N3) dye/TiO_2/Pt (0.5 wt %)	300 W Xe lamp, 410 nm cut off filter	Cat dose 0.5 g L^{-1}	Q.Y (%): N3-Pt/TiO_2 :4.5 ; C-Pt/TiO_2 :1.8 ; M-Pt/TiO2 2.5.	Abe et al. (*72*)
6.	Water-methanol (MeOH) (electron donor); Water-triethanolamine (TEOA) (electron donor)	100 mL / 20 ml electron donor	N719, $Ru(bpy)_2(him)_2$-NO3, $Ru_2(bpy)_4L_1$-PF_6/ TiO_2/Pt (1wt %)	500 W-Xe arc lamp, 420 cutoff filter, outer radiation type photo reactor	Degassed before reaction; Cat dose 0.4g L^{-1}	Highest H_2 generation rate was observed with $Ru_2(bpy)_4L_1$-PF_6; rateH_2: 0.58 μmol H_2/min (with TEOA) and 0.04μmol H_2/min (with methanol)	Peng et al. (*88*)
7.	Water- MeOH (electron donor)	100 mL /20 ml MeOH	N719, $Ru(bpy)_2(him)_2$-NO_3, $Ru_2(bpy)_4L_1$-PF_6/ mesoporous TiO_2/Pt (1 wt%)	250 W Xe lamp 420 nm cutoff filter, outer radiation type photo reactor	Degassed before reaction; Cat dose 0.4g L^{-1}	rateH_2 :0.24 μmol min^{-1} (300° C calcination); 0.16 μmol min^{-1} (500 °C calcination)	Peng et al. (*89*)

No	*Aqueous mixture details*	*Capacity/ Concentration*	*Sensitizer/ Photocatalyst*	*Light source and accessories*	*Other experimental details*	*Results/ Comments*	*Reference*
8.	Water-TEOA (electron donor)	80 mL /TEOA 15 vol%,	Eosin Y/nanotube $Na_2Ti_2O_4(OH)_2$/Pt (0.5wt %)	300 W tungsten halogen lamp, 420 nm cut off filter	N_2 saturated; dye:cat mass=1:1; pH 7; Cat dose: 0.5g L^{-1};	rateH_2:1.26 μmol/min; QY:14.97 %	Li et al. (*73*)
9.	Water-TEOA (electron donor)	80 mL/ TEOA 0.79 M	Eosin Y/N-TiO_2/Pt (0.5 wt%)	400 W high pressure Hg lamp; 420 nm cut off filter	N_2 saturated; pH 7; Cat dose: 1.25 g L^{-1}	rateH_2~1.33 μmol min^{-1};	Li et al. (*74*)
10.	Water-TEOA (electron donor)	250 mL/ TEOA	Eosin Y/TiO_2/Pt (0.1 wt%) (fixed with silane coupling agent)	300 W Xe lamp; 460 nm cut off filter;	N_2 saturated; pH 7; Cat dose: 1.2g L^{-1};	rateH_2~11.6 μmol min^{-1}; QY:10 %	Abe et al. (*75*)
11.	Water-diethanolamine (electron donor)	150 mL/dierthanolamine 30vol %	Eosin Y/ mesoporous-assembled TiO_2/Pt (0.6 wt%)	300 W Xe arc lamp; 400 nm cut off filter;	N_2 saturated; pH 11.5; Cat dose: 3.33g L^{-1};	rate H_2 ~ 0.033 ml min^{-1}	Sreethawong et al. (*76*)
12.	Water-TEOA (electron donor)	60 mL/ TEOA 15wt %	Eosin Y/TS-1 zeolite/Pt (1 wt%)	300 W tungsten halogen lamp; 420 nm cut off filter;	N_2 saturated; pH 7; dye:cat mass=1:8; Cat dose: 0.6g L^{-1};	rate H_2: 0.56 μmol /min; QY:9.4 %	Zhang et al. (*77*)
13.	Water-TEOA (electron donor)	60 mL/ TEOA 15wt %	Eosin Y/silica gel/Pt	300 W tungsten halogen lamp; 420 nm cut off filter;	N_2 saturated; pH 7; dye:cat mass=1:3; Cat dose: 1g L^{-1};	rate H_2: 0.72 μmol min^{-1}; QY:10.4 %	Zhang et al. (*78*)
14.	Water-TEOA (electron donor)	80 mL/ TEOA 0.79 M	Eosin Y/ TiO_2/Pt (1 wt%) (linkage via Fe^{3+} coupling)	400 W metal halide lamp; 420 nm filter;	N_2 saturated; pH 7; Cat dose:1.25g L^{-1}	rate H_2: 4.58 μmol /min; QY:19.1 %	Li et al. (*79*)

Conclusions

In this review the basic principle of dye-sensitization and effect of different parameters on the process were discussed. Surface anchoring group, energy level and ground state redox potential of dye molecule were recognized as the most important parameters for successful electron injection to the conduction band of semiconductor. Ruthenium based sensitizers were used extensively in dye sensitization field but because of their higher cost and toxicity, future researchers are gradually focusing on non-toxic organic dyes as well as natural dyes from fruits, flowers and vegetables. The research is moving towards green chemistry applications.

The presence of novel metals on semiconductor surface was crucial for hydrogen generation and also had positive effect on organic degradation. Different methods of co-catalyst loading have been discussed in details. Among the five methods, atomic layer deposition (ALD) was described as the best method in terms of precise control over both composition and thickness of loaded metal. This however is an expensive method, and further research is needed to find a simpler method such as photodeposition that can utilize solar radiation instead of UV lamps to minimize the operation cost.

Finally, the applications of different dye-sensitized photocatalysts under visible light were presented for treatment of organic compounds and hydrogen generation. Ru (II) based dye-sensitized photocatalysts were able to degrade carbon tetrachloride (CCl_4) and hydrazine in high acidic pH. They also performed water splitting in visible light with high quantum yield. Organic and natural dye-sensitized photocatalysts were efficient for the degradation of pesticides, trichloroethylene, phenol and chlorophenols. Excellent hydrogen generation activity was observed for metal free dyes such as xanthane dyes, merocyanine dyes and coumarin dyes in visible light.

Although extensive work in this field has been carried out, several topics are yet to be explored. Natural dyes need to be studied properly for dye sensitization purpose. We need to consider the utilization of complete solar spectrum rather than only visible light in those experiments. Effect of light intensity has not been studied for dye-sensitized photocatalytic hydrogen generation. If the effect of light intensity on hydrogen generation is not properly selected, a large portion of photons energy will be dissipated in the form of heat.

References

1. Fujishima, A.; Honda, K. *Nature (London)* **1972**, *238*, 37–38.
2. Ni, M.; Leung, M. K. H.; Leung, D. Y. C.; Sumathy, K. *Renewable Sustainable Energy Rev.* **2007**, *11*, 401–425.
3. Gratzel, M.; Kalyanasundaram, K. Metal Complexes as Photosensitizers in Photoelectrochemical Cells. In *Photosensitization and Photocatalysis Using Inorganic and Organometallic Compounds*; Kalyanasundaram, K., Gratzel, M., Eds.; Kluwer Academic Publishers: 1993; Vol. 14, pp 247–271.

4. Nazeeruddin, K. M.; Gratzel, M. Semiconductor Electrodes and Photoelectrochemistry. In *Encyclopedia of Electrochemistry*; Licht, S., Ed.; Wiley-VCH: Germany, 2002; Vol. 6, pp 407–431.
5. Gratzel, M. *J. Photochem. Photobiol., C* **2003**, *4*, 145–153.
6. Tennakone, K.; Kumarasinghe, A. R.; Sirimanne, P. M. *Semicond. Sci. Technol.* **1993**, *8*, 1557–1560.
7. Putzeiko, E. K.; Terenin, A. N. *Zh. Fiz. Khim.* **1949**, *23*, 676–688.
8. Galoppini, E. *Coordin. Chem. Rev.* **2004**, *248* (13), 1283–1297.
9. Matsumura, M.; Nomura, Y.; Tsubomura, H. B. *Chem. Soc. Jpn.* **1977**, *50*, 2533–2537.
10. Houlding, V. H.; Gratzel, M. *J. Am. Chem. Soc.* **1983**, *105*, 5695–5696.
11. Duncan, W. R.; Prezhdo, O. V. *Annu. Rev. Phys. Chem.* **2007**, *58*, 143–184.
12. Polo, A. S.; Itokazu, M. K.; Murakami Iha, N. Y. *Coord.Chem. Rev.* **2004**, *248*, 1343–1361.
13. Wongcharee, K.; Meeyoo, V.; Chavadej, S. *Sol. Energy Mater. Sol. Cells* **2007**, *91*, 566–571.
14. Polo, A. S.; Murakami Iha, N. Y. *Sol. Energy Mater. Sol. Cells* **2006**, *90*, 1936–1944.
15. Garcia, C. G.; Polo, A. S.; Murakami Iha, N. Y. *J. Photochem. Photobiol., A* **2003**, *160*, 87–91.
16. Dhanalakshmi, K. B.; Latha, S.; Anandan, S.; Maruthamuthu, P. *Int. J. Hydrogen Energy* **2001**, *26*, 669–674.
17. Nian, J. N.; Hu, C. C.; Teng, H. *Int. J. Hydrogen Energy* **2008**, *33*, 2897–2903.
18. Nada, A. A.; Hamed, H. A.; Barakat, M. H.; Mohamed, N. R.; Veziroglu, T. N. *Int. J. Hydrogen Energy* **2008**, *33*, 3264–3269.
19. Mane, R. S.; Lee, W. J.; Pathan, H. M.; Han, S. H. *J. Phys. Chem. B* **2005**, *109*, 24254–24259.
20. Lincot, D. *Thin Solid Films* **2005**, *487*, 40–48.
21. Karuppuchamy, S.; Nonomura, K.; Yoshida, T.; Sugiura, T.; Minoura, H. *Solid State Ionics* **2002**, *151*, 19–27.
22. Yoshida, T.; Pauporte, T.; Lincot, D.; Oekermann, T.; Minoura, H. *J. Electrochem. Soc.* **2003**, *150*, C608–C615.
23. Wang, C. C.; Ying, J. Y. *Chem. Mater.* **1999**, *11*, 3113–3120.
24. Mao, L.; Li, Q.; Dang, H.; Zhang, Z. *Mater. Res. Bull.* **2005**, *40*, 201–208.
25. Venkatachalam, N.; Palanichamy, M.; Murugesan, V. *Mater. Chem. Phys.* **2007**, *104*, 454–459.
26. Ristic, M.; Music, S.; Ivanda, M.; Popovic, S. *J. Alloy Compd.* **2005**, *397*, L1–L4.
27. Rani, S.; Suri, P.; Shishodia, P. K.; Mehra, R. M. *Sol. Energy Mater. Sol. Cells* **2008**, *92*, 1639–1645.
28. Choi, W.; Termin, A.; Hoffmann, M. R. *J. Phys. Chem.* **1994**, *98*, 13669–13679.
29. Karakitsou, K. E.; Verykios, X. E. *J. Phys. Chem.* **1993**, *97*, 1184–1189.
30. Soria, J.; Conesa, J. C.; Augugliaro, V.; Palmisano, L.; Schiavello, M.; Sclafani, A. *J. Phys. Chem.* **1991**, *95*, 274–282.
31. Choi, J.; Park, H.; Hoffmann, M. R. *J. Phys. Chem. C* **2010**, *114*, 783–792.

32. Kennedy, J. C.; Datye, A. K. *J. Catal.* **1998**, *179*, 375–389.
33. Vorontsov, A. V.; Stoyanova, I. V.; Kozlov, D. V.; Simagina, V. I.; Savinov, E. N. *J. Catal.* **2000**, *189*, 360–369.
34. Falconer, J. L.; Magrini-Bair, K. A. *J. Catal.* **1998**, *179*, 171–178.
35. He, C.; Xiong, Y.; Zhu, X.; Li, X. *Appl. Catal., A* **2004**, *275*, 55–60.
36. Kraeutler, B.; Bard, A. J. *J. Am. Chem. Soc.* **1978**, *100*, 4317–4318.
37. Kryukova, G. N.; Zenkovets, G. A.; Shutilov, A. A.; Wilde, M.; Gunther, K.; Fassler, D.; Richter, K. *Appl. Catal., B* **2007**, *71*, 169–176.
38. Sivalingam, G.; Nagaveni, K.; Hegde, M. S.; Madras, G. *Appl. Catal., B* **2003**, *45*, 23–38.
39. Mills, A.; Lee, S. K. *Platinum Met. Rev.* **2003**, *47*, 2–12.
40. Zhou, Y.; King, D. M.; Liang, X.; Li, J.; Weimer, A. W. *Appl. Catal., B* **2010**, *101*, 54–60.
41. Nakamatsu, H.; Kawai, T.; Koreeda, A.; Kawai, S. *J. Chem. Soc., Faraday Trans.* **1986**, *82*, 527–531.
42. Li, F. B.; Li, X. Z. *Chemosphere* **2002**, *48*, 1103–1111.
43. Yang, J. C.; Kim, Y. C.; Shul, Y. G.; Shin, C. H.; Lee, T. K. *Appl. Surf. Sci.* **1997**, *121*, 525–529.
44. Spieker, W. A.; Regalbuto, J. R. *Chem. Eng. Sci.* **2001**, *56*, 3491–3504.
45. Bavykin, D. V.; Lapkin, A. A.; Plucinski, P. K.; Torrente-Murciano, L.; Friedrich, J. M.; Walsh, F. C. *Top. Catal.* **2006**, *39*, 151–160.
46. Garcia, J. R. V.; Goto, T. *Mater. Trans.* **2003**, *44*, 1717–1728.
47. Li, W.; Liang, C.; Zhou, W.; Qiu, J.; Zhou, Z.; Sun, G.; Xin, Q. *J. Phys. Chem. B* **2003**, *107*, 6292–6299.
48. Mei, Y.; Sharma, G.; Lu, Y.; Ballauff, M.; Drechsler, M.; Irrgang, T.; Kempe, R. *Langmuir* **2005**, *21*, 12229–12234.
49. Floro, J. A.; Hearne, S. J.; Hunter, J. A.; Kotula, P.; Chason, E.; Seel, S. C.; Thompson, C. V. *J. Appl. Phys.* **2001**, *89*, 4886.
50. Liang, X.; George, S. M.; Weimer, A. W.; Li, N. H.; Blackson, J. H.; Harris, J. D.; Li, P. *Chem. Mater.* **2007**, *19*, 5388–5394.
51. Hakim, L. F.; King, D. M.; Zhou, Y.; Gump, C. J.; George, S. M.; Weimer, A. W. *Adv. Funct. Mater.* **2007**, *17*, 3175–3181.
52. Cho, Y.; Park, Y.; Choi, W. *J. Ind. Eng. Chem.* **2008**, *14*, 315–321.
53. Fung, A. K. M.; Chiu, B. K. W.; Lam, M. H. W. *Water Res.* **2003**, *37*, 1939–1947.
54. Cho, Y.; Choi, W.; Lee, C. H.; Hyeon, T.; Lee, H. I. *Environ. Sci. Technol.* **2001**, *35*, 966–970.
55. Bae, E.; Choi, W. *Environ. Sci. Technol.* **2003**, *37*, 147–152.
56. Alexander, M. V.; Rosentreter, J. J. *Microchem. J.* **2008**, *88*, 38–44.
57. Muszkat, L.; Feigelson, L.; Bir, L.; Muszkat, K. A. *Pest Manage. Sci.* **2002**, *58*, 1143–1148.
58. Chatterjee, D. *J. Mol. Catal. A: Chem.* **2000**, *154*, 1–3.
59. Chatterjee, D.; Dasgupta, S.; Dhodapkar, R. S.; Rao, N. N. *J. Mol. Catal. A: Chem.* **2006**, *260*, 264–268.
60. Haque, M. M.; Muneer, M. *Dyes Pigm.* **2007**, *75*, 443–448.
61. Qamar, M.; Saquib, M.; Muneer, M. *Desalination* **2005**, *171*, 185–193.

62. Kojima, M.; Takahashi, K.; Nakamura, K. *Photochem. Photobiol.* **2001**, *74*, 369–377.
63. Jiang, D.; Xu, Y.; Wu, D.; Sun, Y. *J. Solid State Chem.* **2008**, *181*, 593–602.
64. Granados, O G.; Paez, M C. A.; Martinez, O F.; Paez-Mozo, E. A. *Catal. Today* **2005**, *107-108*, 589–594.
65. Chowdhury, P.; Moreira, J.; Gomaa, H.; Ray, A. K. *Ind. Eng. Chem. Res.* **2012**, *51*, 4523–4532.
66. Ghosh, J. P.; Langford, C. H.; Achari, G. *J. Phys. Chem. A* **2008**, *112*, 10310–10314.
67. Li, X.; Cubbage, J. W.; Tetzlaff, T. A.; Jenks, W. S. *J. Org. Chem.* **1999**, *64*, 8509–8524.
68. Hussein, F. H.; Alkhateeb, A. N. *Desalination* **2007**, *209*, 350–355.
69. Salvador, P. *New J. Chem.* **1988**, *12*, 35–43.
70. Lee, J. S. *Catal. Surv. Asia* **2005**, *9*, 217–227.
71. Grimes, C. A.; Varghese, O. K.; Ranjan, S. *Light, Water, Hydrogen – The Solar Generation of Hydrogen by Water Photoelectrolysis*; Springer Science: New York, 2008; pp 1–10.
72. Abe, R.; Sayama, K.; Sugihara, H. *J. Sol. Energy Eng.* **2005**, *127*, 413–416.
73. Li, Q.; Lu, G. *J. Mol. Catal. A: Chem.* **2007**, *266*, 75–79.
74. Li, Y.; Xie, C.; Peng, S.; Lu, G.; Li, S. *J. Mol. Catal. A: Chem.* **2008**, *282*, 117–123.
75. Abe, R.; Hara, K.; Sayama, K.; Domen, K.; Arakawa, H. *J. Photochem. Photobiol., A* **2000**, *137*, 63–69.
76. Sreethawong, T.; Junbua, C.; Chavadej, S. *J. Power Sources* **2009**, *190*, 513–524.
77. Zhang, X.; Jin, Z.; Li, Y.; Li, S.; Lu, G. *Appl. Surf. Sci.* **2008**, *254*, 4452–4456.
78. Zhang, X.; Jin, Z.; Li, Y.; Li, S.; Lu, G. *J. Power Sources* **2007**, *166*, 74–79.
79. Li, Y.; Guo, M.; Peng, S.; Lu, G.; Li, S. *Int. J. Hydrogen Energy* **2009**, *34*, 5629–5636.
80. Chen, X.; Shen, S.; Guo, L.; Mao, S. S. *Chem. Rev.* **2010**, *110*, 6503–6570.
81. Kalyanasundaram, K.; Kiwi, J.; Gratzel, M. *Helv. Chim. Acta* **1978**, *61*, 2720–2730.
82. Duonghong, D.; Borgarello, E.; Graetzel, M. *J. Am. Chem. Soc.* **1981**, *103*, 4685–4690.
83. Borgarello, E.; Kiwi, J.; Pelizzetti, E.; Visca, M.; Graetzel, M. *J. Am. Chem. Soc.* **1981**, *103*, 6324–6329.
84. Borgarello, E.; Kiwi, J.; Pelizzetti, E.; Visca, M.; Gratzel *Nature* **1981**, *289*, 158–160.
85. Hirano, K.; Suzuki, E.; Ishikawa, A.; Moroi, T.; Shiroishi, H.; Kaneko, M. *J. Photochem. Photobiol., A* **2000**, *136*, 157–161.
86. Bae, E.; Choi, W.; Park, J.; Shin, H. S.; Kim, S. B.; Lee, J. S. *J. Phys. Chem. B* **2004**, *108*, 14093–14101.
87. Kajiwara, T.; Hashimoto, K.; Kawai, T.; Sakata, T. *J. Phys. Chem.* **1982**, *86*, 4516–4522.
88. Peng, T.; Dai, K.; Yi, H.; Ke, D.; Cai, P.; Zan, L. *Chem. Phys. Lett.* **2008**, *460*, 216–219.

89. Peng, T.; Ke, D.; Cai, P.; Dai, K.; Ma, L.; Zan, L. *J. Power Sources* **2008**, *180*, 498–505.
90. Gurunathan, K.; Maruthamuthu, P.; Sastri, M. V. C. *Int. J. Hydrogen Energy* **1997**, *22*, 57–62.
91. Li, Q.; Chen, L.; Lu, G. *J. Phys. Chem. C* **2007**, *111*, 11494–11499.
92. Gurunathan, K. *Int. J. Hydrogen Energy* **2004**, *29*, 933–940.
93. Nada, A. A.; Barakat, M. H.; Hamed, H. A.; Mohamed, N. R.; Veziroglu, T. N. *Int. J. Hydrogen Energy* **2005**, *30*, 687–691.
94. Koca, A.; Sahin, M. *Int. J. Hydrogen Energy* **2002**, *27*, 363–367.
95. Abe, R.; Sayama, K.; Domen, K.; Arakawa, H. *Chem. Phys. Lett.* **2001**, *344*, 339–344.
96. Bamwenda, G. R.; Arakawa, H. *Sol. Energy Mater. Sol. Cells* **2001**, *70*, 14.
97. Chowdhury, P.; Gomaa, H.; Ray, A. K. *Int. J. Hydrogen Energy* **2011**, *36*, 13442–13451.

Chapter 14

Hierarchical Hybrid K-OMS-2/TiO_2 Nanofibrous Membrane for Water Treatment

Tong Zhang and Darren Delai Sun*

School of Civil and Environmental Engineering,
Nanyang Technological University, Singapore 639798
***E-mail: ddsun@ntu.edu.sg. Tel: +65 6790 6273**

This chapter presents a novel free-standing nanofibrous microfiltration membrane made by hierarchical hybrid K-OMS-2/TiO_2 nanowires for the removal of organic pollutants in water. The K-OMS-2/TiO_2 nanowire was around 10 μm in length with the high density secondary TiO_2 hair-like structures (20-40 nm) anchored on surfaces of primary underlying K-OMS-2 nanowires. At 30 $L/m^2 \cdot h$ (LMH) membrane flux, the color and total organic carbon (TOC) removal rates were 90 % and 50%, respectively. Clearly, the organics (AO 7) will be filtered directly by membrane pores with external pressure, and the smaller ones would be adsorbed onto the nanowires via an adsorption process that is greatly enhanced by the hydrophobic nature of the supporting K-OMS-2 nanowires, as well as large surface area of the hierarchical nanowires. The organic pollutants on the photocatalytic membrane were simultaneously degraded by the PCO process under a UV light, alleviating membrane pore blocking by small organic molecules, thus maintaining a constant permeate flux.

Introduction

Membrane technologies have been widely applied in water treatment, dairy, food, pharmaceutical, bioengineering, chemical, nuclear-energy and electronic industries due to their excellent performance on the removal of pollutants, relative low energy cost, well arranged process conductions, and no addition of chemicals. Membranes for water purification can offer high quality clean water with small footprint and high flexibility for scaling up. Most membranes used in industrially are polymer based membrane (*1–5*). However, the membrane fouling caused by deposition of organics and inorganics on membrane surfaces is the major obstacle. In addition, these organic membranes are typically unsuitable for very high-temperature applications (*2*, *4*). Thus, there is an urgent need in searching for a multifunctional membrane which is able to purify water and concurrently remove fouling from the membrane surfaces.

Nanofabrication technology has great potential for molecular separation applications by offering more structurally controlled materials for such needs. For the past 15 years, our research group has put in a large amount of effort to develop a robust, flexible and free-standing multifunctional TiO_2 membrane using nanofabrication technology to overcome the fouling problem that arose from conventional polymeric membranes (*6*, *7*). The pioneer work in developing a multifunctional free-standing, flexible membrane to overcome the polymeric membrane problem was carried out by Zhang and co-workers (*6*, *7*). This two-dimensional (2D) nanostructured TiO_2, in the form of nanowires membrane was synthesized by hydrothermal method (*7*). The 2D nanomaterial brought in new properties such as flexibility and concurrent filtration while retaining the properties of its 1D nanomaterial.

Inorganic nanofibrous/nanowires membranes have received increasing attention in the fields of catalysis, adsorption, sensors and filtration (*8–10*). Owing to their great mechanical resistance and reusability, inorganic membranes with different properties and functions have been successfully fabricated for water purification (*11–13*). As the main function of a membrane is to discriminate species, it comes into contact within one phase and transports them to the other. Selectivity and permeability becomes the most important properties of a membrane. Selectivity is expressed as a separation factor, which is governed by the surface properties of a membrane, while permeability describes the rate at which species are transported across a membrane (*14*). The two properties are determined by the structural and morphological properties of the membrane, such as the pore size, pore structure, wettability and porosity. Thus, applications of inorganic nanofibrous membranes in separation technology will depend upon the ability to prepare membranes with desired pore size and pore structure. However, controlling of pore size and pore structure suffers from complicated technical requirements as well as high operational costs and thus severely restricting the applications of membranes. To address this problem, other techniques can be considered in the filtration process to maintain high performences of the membrane, such as adsorption and photocatatic oxidation. Membranes with higher organic adsorption capacity would enhance rejection of organic molecules; while the introduction of photocatalysts as membrane components would endow a membrane with self-cleaning property. It is desirable to synthesize hybrid nanofibrous membranes with efficient adsorption and photocatalytic activity, which can further enhance the selectivity and permeability of the membranes.

In this chapter, we introduce a 2D hierarchical hybrid K-OMS-2/TiO_2 nanowires membrane. The advantages of the 2D hierarchical hybrid K-OMS-2/TiO_2 nanowires membrane are: (1) full surface exposure to UV or solar for self-regeneration through TiO_2 photocatalytic oxidation (PCO) reaction, which effectively eliminates the membrane fouling problem; (2) concurrent membrane filtration and PCO oxidation; (3) high surface area and hybrid structure, which allows higher adsorption rate and better PCO oxidation of various trace organics for improving water quality; (4) higher acid/basic and temperature controlled resistance; (5) flexible property which enables the membrane to be formed into various membrane modules for larger commercial application. A free-standing and flexible manganese oxide based nanofibrous TiO_2 hair-like membrane will be investigated by evaluating its permeability and photocatalytic activity.

Materials and Methods

In order to obtain the basic supporting material of the membrane, K-OMS-2 nanowires were synthesized via a hydrothermal method (*15*). In a typical procedure, 12 mmol of K_2SO_4, 12 mmol of $K_2S_2O_8$, and 8 mmol of $MnSO_4 \cdot H_2O$ were dissolved in 70 ml of deionized water. The solution was then transferred to a 125 ml Teflon-lined stainless-steel autoclave. The autoclave was sealed and heated in an oven at 250 °C for 4 days. The resulting black precipitate was suspended in 1000 ml deionized water, and stirred vigorously for 12 h. After thorough washing with deionized water to remove remaining ions present in the product, the sample was dried at 105 °C for 24 h.

To synthesize the K-OMS-2/TiO_2 heterojunctions, 50 mg of the synthesized K-OMS-2 nanowires, 400 mg titanous sulfate, and 0.4 ml concentrated H_2SO_4 (98%) were dissolved in 70 ml deionized water. After homogenization by an ultrasonic homogenizer for 10 mins, the solution was transferred to the 125 ml Teflon-lined autoclave again and heated at 105 °C for 20 h. Then, the product was washed several times before being separated using a centrifuge, and the resulting product was dried in vacuum for 48 h.

To synthesize the K-OMS-2/TiO_2 nanofibrous membrane, suspension of the synthesized hierarchical K-OMS-2/TiO_2 nanowires was vigorously stirred for 10 min. Then, the suspension was filtered on a vacuum-filtration setup with a glass membrane (ADVANTEC, GC-50, 0.45 μm), and the hierarchical nanowires would form a compact cake layer on the membrane. The glass membrane was then placed in an oven for drying at 105 °C. After removing the glass membrane, a free-standing membrane was formed. The membrane was further pressurized under 5 bar at 120 °C on a customized hot press for 2 mins. Finally, the membrane was calcinated at 550 °C for 1 h. K-OMS-2 nanowire membrane was also fabricated via a similar process for comparison.

Results and Discussion

Since organic pollutants have higher affinity towards the hydrophobic K-OMS-2 nanowires, this material can be used as a good adsorbent for the removal of organic pollutants (*16*). Furthermore, the nanowires can also act as a template, and other nanostructures (e.g. TiO_2) can be deposited onto the surface to form hierarchical heterojunctions. FESEM and TEM images of K-OMS-2 nanowires are shown in Figure 1a and 1b respectively, indicating that the K-OMS-2 nanowires have diameters of about 40-100 nm and are over several hundreds of microns long. Figure 1c and 1d shows that high density secondary TiO_2 nanostructures were uniformly anchored on surfaces of the primary underlying K-OMS-2 nanowires, making a hybrid inorganic heterojunctions. It can be seen from Figure 1d that the TiO_2 nanostructures are of lengths ranging from 20-40 nm with much smaller diameter as compared to the K-OMS-2 nanowire. The BET surface area of the synthesized K-OMS-2 nanowires and K-OMS-2/TiO_2 heterojunctions are recorded at 25.08 m^2/g and 111.58 m^2/g, respectively. It is noteworthy that the loading of TiO_2 nanostuctures on the K-OMS-2 nanowires has greatly increased the specific area of the material. This phenomenon can be attributed to the presence of K-OMS-2 scaffold, which acts as a dispersing template to downsize the TiO_2 nanostructures during the synthesis process (*17*).

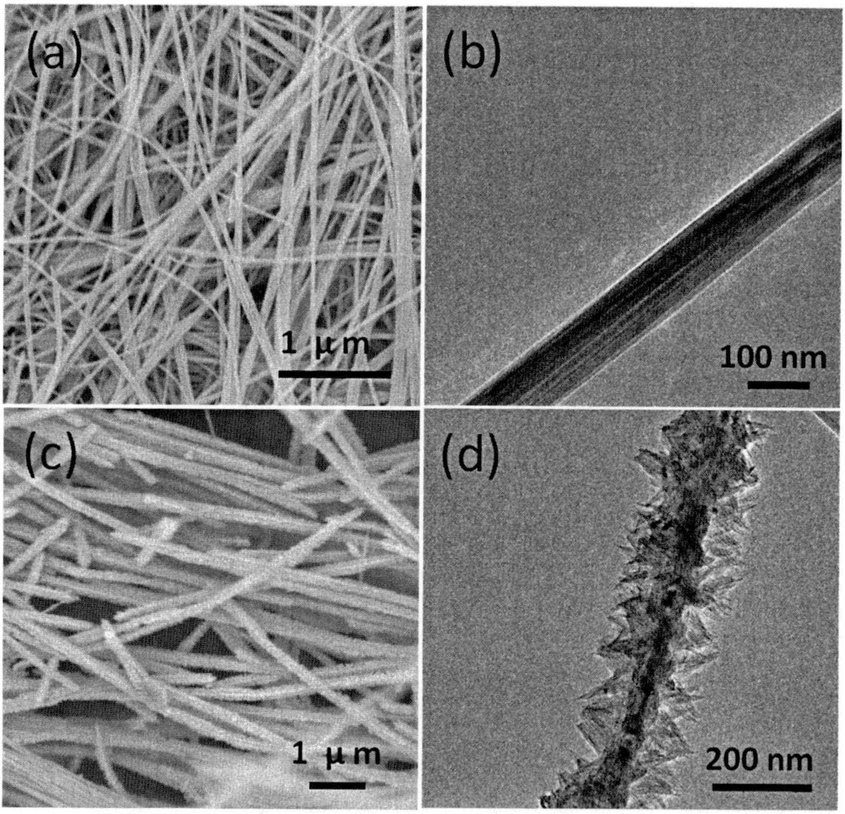

Figure 1. (a) FESEM image of K-OMS-2 nanowires. (b) TEM image of K-OMS-2 nanowire. (c)FESEM image of K-OMS-2/TiO2 heterojunctions. (d)TEM image of K-OMS-2/TiO_2 heterojunctions.

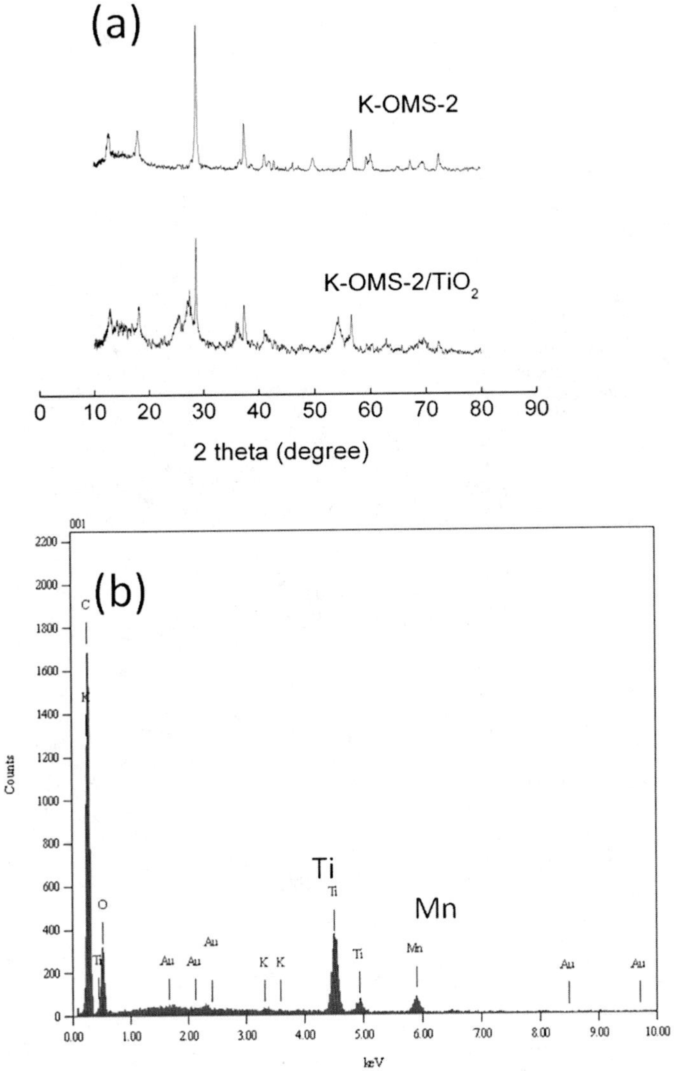

Figure 2. (a)XRD patterns of the synthesized K-OMS-2 nanowires and K-OMS-2/TiO₂ heterojunctions. (b) EDX spectrum of the synthesized K-OMS-2/TiO₂ heterojunctions. (Reproduced from reference (17). Copyright 2012 Royal Society of Chemistry).

XRD analysis displays the crystal phase of the synthesized materials. All diffraction peaks of the upper curve in Figure 2a can be perfectly indexed to the K-OMS-2 crystalline phase (JCPDS 44-1386). The XRD pattern of synthesized K-OMS-2/TiO_2 nanostructures shows that TiO_2 structures are present in the synthesized K-OMS-2/TiO_2 heterojunction, indicating the successful coating of TiO_2 nanostructure. The diffraction peaks at 27°, 36° and 55° indicate that the rutile phase of TiO_2 (JCPDS 21-1276) is present in the synthesized material, and the diffraction peaks at 25° and 48° evidenced the co-existence of anatase TiO_2 (JCPDS 21-1272). Moreover, from EDX result in Figure 2b, we can confirm that the external surface of the synthesized K-OMS-2/TiO_2 heterojunction contains K, Mn, Ti, and O.

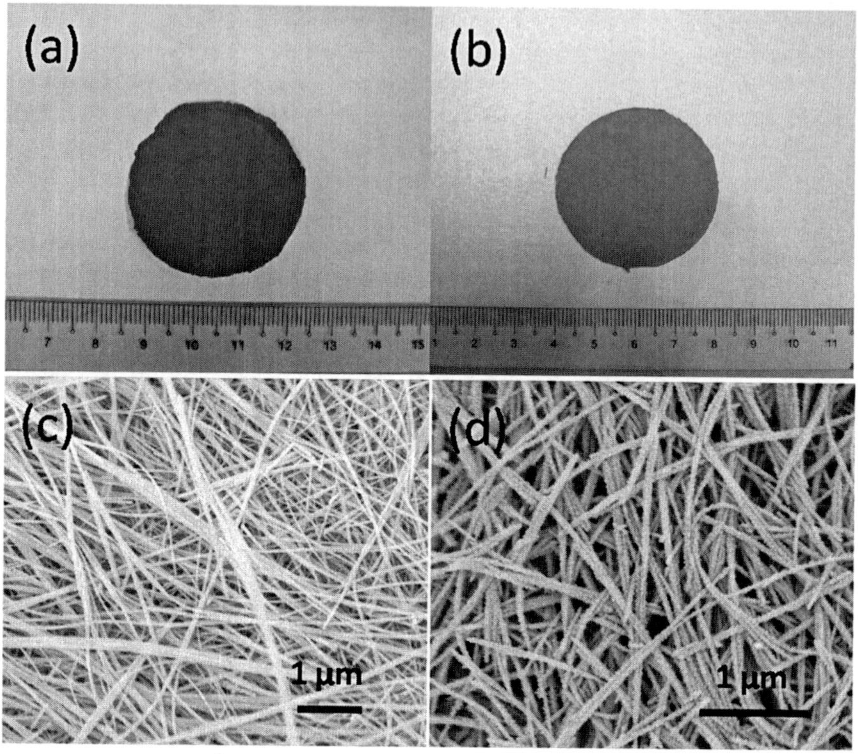

Figure 3. (a) Digital photo of the K-OMS-2 membrane. (b) Digital photo of the K-OMS-2/TiO_2 membrane. (c) Top view FESEM image of the K-OMS-2 membrane. (d) Top view FESEM image of the K-OMS-2/TiO_2 membrane. (Figure 4c is reproduced from reference (18); Figure 4d are reproduced from reference (17). Copyright 2012 Royal Society of Chemistry).

The digital photos of the synthesized free-standing membranes are shown in Figure 3a and 3b. The thickness of the membrane can be adjusted by adjusting the dosage of the materials, and the diameter of the membrane can be controlled by our filtration apparatus. In this work, the thickness and diameter of the membrane are about 300 µm and 40 mm, respectively. Figure 3c and 3d shows the top view FESEM images of K-OMS-2 membrane and K-OMS-2/TiO_2 membrane, respectively. Both of them have relatively flat topology with no observed cracks or pinholes. From Figure 3d, it can be found that the open porous network was formed by overlapping and interweaving of the ultra long hierarchical K-OMS-2/TiO_2 heterojunctions, which enhance the permeability of the membrane in a filtration process. Furthermore, the pore size of the synthesized K-OMS-2/TiO_2 membrane is around 0.1 μm, which classify the membrane under the microfiltration category.

The synthesized K-OMS-2/TiO_2 membrane was investigated by XPS to identify the surface elemental compositions and binding energies, and the result indicates the sample contains K, Mn, O and Ti. The high-resolution XPS spectrum of Mn 2p taken on the surface of K-OMS-2/TiO_2 heterojunctions is shown in Figure 4a and Gaussian curve fitting of Mn $2p_{3/2}$ was performed to describe the oxidation states of Mn more specifically. The peaks are centred at 641.7 eV and 643.0 eV, indicating the presence of Mn^{3+} and Mn^{4+} respectively (*19*). It is noteworthy that the mixed-valences of Mn is important in electron transfer and lead to conductivity of resultant material, and thus enhance the photocatalytic activity of the synthesized material (*20*). Figure 4b shows that the peaks for Ti $2p_{3/2}$ and Ti $2p_{1/2}$ are centered at 458.8 eV and 464.4 eV, representing the presence of a Ti^{4+} oxidation state in the sample (*21*), which further confirmed the presence of TiO_2 on the surface of K-OMS-2 nanowires. Importantly, the presence of TiO_2 can endow the synthesized membrane with self-cleaning property under UV light or solar light irradiation.

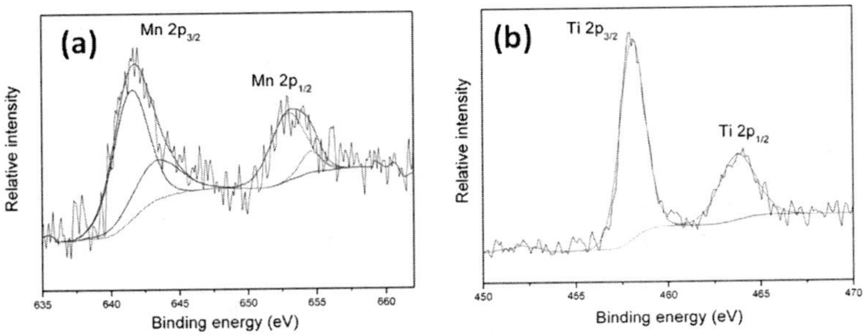

Figure 4. High-resolution XPS spectra of Mn 2p (a) and Ti 2p region (b) taken on the K-OMS-2/TiO_2 membrane.

To investigate the performance of the synthesized membrane, the adsorption capacity of the synthesized K-OMS-2/TiO_2 heterojunctions were firstly evaluated using acid orange 7 (AO 7) as a pollutant. The K-OMS-2 nanowire and TiO_2 nanostructure were also used for comparison. As shown in Figure 5, the synthesized K-OMS-2/TiO_2 exhibited much higher adsorption capacity than TiO_2 which was prepared using an identical procedure as that of K-OMS-2/TiO_2. The BET surface area of the synthesized K-OMS-2, TiO_2 and K-OMS-2/TiO_2 are 25.08 m^2/g, 72.74 m^2/g and 111.58 m^2/g respectively. The increased BET surface area of K-OMS-2/TiO_2 may partly explain the enhanced adsorption capacity. Figure 5 also shows that K-OMS-2 nanowires possess excellent adsorption capacity for the removal of AO 7, which can be attributed to the hydrophobic nature of the nanowires. Previous studies indicated that the surrounding environment of the constructed TiO_2 nanostructures can be affected by the hydrophobic supporting materials (*22*). Thus, the hydrophobic nature of K-OMS-2 nanowires may be another important factor for the enhanced adsorption capacity.

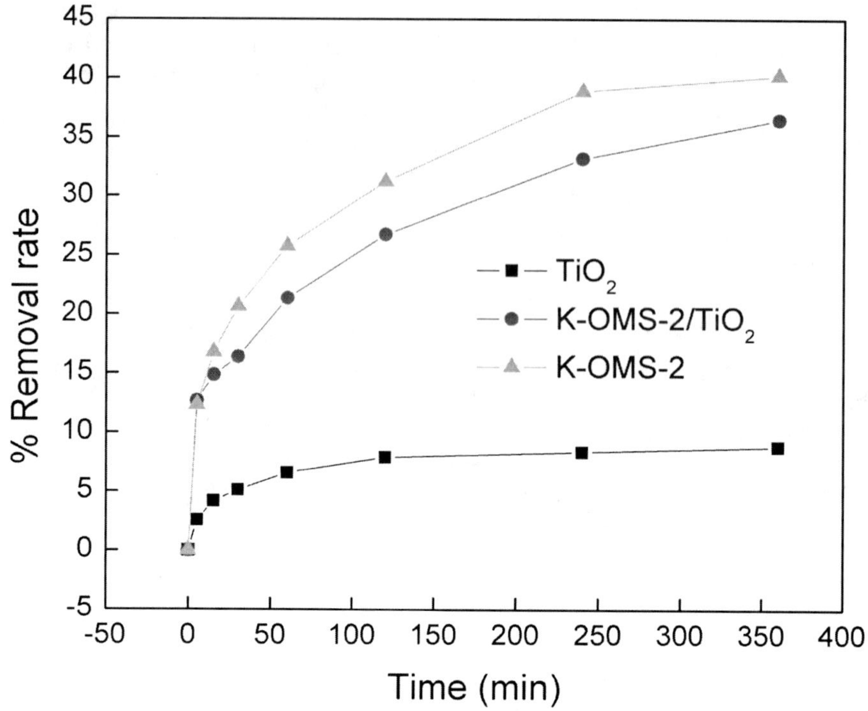

Figure 5. Adsorption of AO 7 onto the prepared K-OMS-2, TiO_2 and K-OMS-2/TiO_2 at pH 5. (Reproduced from reference (17). Copyright 2012 Royal Society of Chemistry).

The performance of the synthesized K-OMS-2/TiO_2 nanofibrous membrane was evaluated using a dead-end filtration equipment (Figure 6). When a flux of 30 L/m^2·h was applied in the system, only 8.7 % of AO 7 was removed using the synthesized nanofibrous membrane alone, which can be attributed to the adsorption equilibrium of the membrane was gradually reached during the filtration process. When UV irradiation was concurrently applied to the membrane, an AO 7 removal rate of 96.3 % was achieved, owing to combined adsorption and PCO effects. However, only 54.1 % of total organic carbon (TOC) was removed in this process due to the incompletely photocatalytic mineralization of the AO 7 molecules. In addition, flux is another factor which can affect the removal rate of AO 7 and TOC. According to our study, with the increase of flux, the removal rate of AO 7 and TOC decreased, and 60 L/m^2·h was the critical point in establishing an optimal balance between photocatalytic degradation of AO 7 and flux in our experimental conditions.

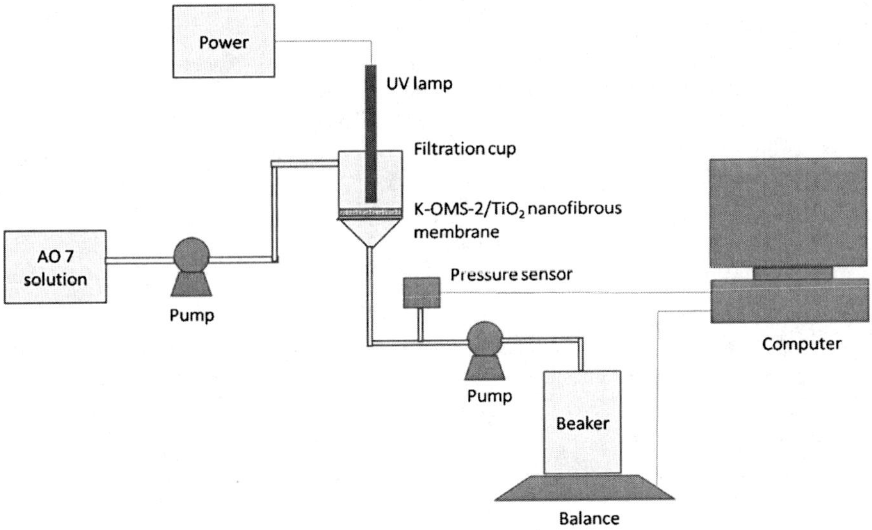

Figure 6. The setup of concurrent filtration, adsorption and photocatalytic oxidation. (Reproduced from reference (17). Copyright 2012 Royal Society of Chemistry).

Conclusion

In this study, we synthesized a novel free-standing nanofibrous microfiltration membrane via a vacuum-filtration process. Importantly, the synthesized membrane can reject organic pollutants via a combination effect of interception and adsorption. A concurrent filtration, adsorption and photocatalytic oxidation experiment showed that the nanofibrous membrane possessed excellent performance in the removal of organic pollutants. The synthesized membrane may have potential in extensive applications of membrane filtration and water purification technologies.

References

1. Gabelman, A.; Hwang, S. T. *J. Membr. Sci.* **1999**, *159*, 1–2.
2. Heitner-Wirguin, C. *J. Membr. Sci.* **1996**, *120*, 1–33.
3. Kwak, S. Y.; Kim, S. H.; Kim, S. S. *Environ. Sci. Technol.* **2001**, *35*, 2388–2394.
4. Nghiem, L. D.; Schafer, A. L.; Elimelech, M. *Environ. Sci. Technol.* **2004**, *38*, 1888–1896.
5. Lin, H.; Van Wagner, E.; Freeman, B. D.; Toy, L. G.; Gupta, R. P. *Science* **2006**, *311*, 639–642.
6. Zhang, X.; Du, A. J.; Lee, P.; Sun, D. D.; Leckie, J. O. *J. Membr. Sci.* **2008**, *313*, 44–51.
7. Zhang, X.; Zhang, T.; Ng, J.; Sun, D. D. *Adv. Funct. Mater.* **2009**, *19*, 3731–3736.

8. Choi, S. W.; Jo, S. M.; Lee, W. S.; Kim, Y. R. *Adv. Mater.* **2003**, *15*, 2027–2032.
9. Sholl, D. S.; Johnson, J. K. *Science* **2006**, *312*, 1003–1004.
10. Tan, K.; Obendorf, S. K. *J. Membr. Sci.* **2007**, *305*, 287–298.
11. Gopal, R.; Kaur, S.; Ma, Z.; Chan, C.; Ramakrishna, S.; Matsuura, T. *J. Membr. Sci.* **2006**, *281*, 581–586.
12. Yuan, J.; Liu, X.; Akbulut, O.; Hu, J.; Suib, S. L.; Kong, J.; Stellacci, F. *Nat. Nanotechnol.* **2008**, *3*, 332–336.
13. Liang, H. W.; Cao, X.; Zhang, W. J.; Lin, H. T.; Zhou, F.; Chen, L. F.; Yu, S. H. *Adv. Funct. Mater.* **2011**, *21*, 3851–3858.
14. Gopal, R.; Kaur, S.; Ma, Z.; Chan, C.; Ramakrishna, S.; Matsuura, T. *J. Membr. Sci.* **2006**, *281*, 581–586.
15. Yuan, J.; Laubernds, K.; Villegas, J.; Gomez, S.; Suib, S. L. *Adv. Mater.* **2004**, *16*, 1729–1732.
16. Luo, J.; Zhang, Q.; Garcia-Martinez, J.; Suib, S. L. *J. Am. Chem. Soc.* **2008**, *130*, 3198–3207.
17. Zhang, T.; Wang, Y.; Ng, J.; Sun, D. D. *RSC Adv.* **2012**, *2*, 3638–3641.
18. Zhang, T.; Liu, J.; Sun, D. D. *RSC Adv.* **2012**, *2*, 5134–5137.
19. Zou, Z. Q.; Meng, M.; Zha, Y. Q. *J. Phys. Chem. C* **2010**, *114*, 468–477.
20. Suib, S. L. *J. Mater. Chem.* **2008**, *18*, 1623–1631.
21. Thind, S. S.; Wu, G.; Chen, A. *Appl. Catal., B* **2012**, *111-112*, 38–45.
22. Kuwahara, Y.; Kamegawa, T.; Mori, K.; Yamashita, H. *Curr. Org. Chem.* **2010**, *14*, 616–629.

Chapter 15

Visible-Light-Induced Activity of AgI-BiOI Composites for Removal of Organic Contaminants from Water and Wastewater

Chun He,[*,1,3] Qiong Zhang,[1] Jingling Yang,[1] Zuocheng Xu,[1] Dong Shu,[*,2] Chun Shan,[4] Linfei Zhu,[1] Weicheng Liao,[1] and Ya Xiong[1,3]

[1]School of Environmental Science and Engineering, Sun Yat-sen University, Guangzhou 510275, China
[2]Base of Production, Education & Research on Energy Storage and Power Battery of Guangdong Higher Education Institutes, School of Chemistry and Environment, South China Normal University, Guangzhou 510006, China
[3]Guangdong Provincial Key Laboratory of Environmental Pollution Control and Remediation Technology, Guangzhou 510275, China
[4]Guangdong Polytechnic Normal University, Guangzhou 510665, China
[*]Tel.: +86 20 39332690; Fax: +86 20 39332690. E-mail: hechun@mail.sysu.edu.cn (C.H.) or dshu@scnu.edu.cn (D.S).

The aim of this paper is to: (i) prepare AgI-BiOI composites by solvothermal methods, (ii) demonstrate the enhanced photocatalytic activity of AgI-BiOI composites compared to BiOI in the removal of AOII, (iii) provide the evidences for the efficient electron transfer between the interface of AgI and BiOI in studied system by measuring the electrochemical impedance spectroscopy (EIS), and (iv) determine the suitability of AgI-BiOI composites in real applications in term of its degradation efficiency and durability in the repeated uses. The experimental results indicated that the prepared AgI-BiOI powder was high purity and composed of nanoplates. The DRS results indicated that AgI-BiOI composites have the significant adsorption and the red-shift in the visible light region compared to BiOI. In addition, the photocatalytic degradation efficiency of AOII using AgI-BiOI composites was more than 85% within 120 min, which was higher than that of individual AgI and

BiOI. The enhanced photocatalytic activity was due to the decreased impedance value of AgI-BiOI, thus, leading to the enhanced photocatalytic activity of AgI-BiOI composites.

Introduction

Pollution of water dyes is a serious problem in the developing countries. It was reported that the dye wastewater emission every day in china was about 300 ~ 400 million tons (*1*). Moreover, the removal of dyes from wastewater is a challenge to the related industries, because the synthetic dyes used are stable compounds, difficult to destroy by common treatments. Thus, the physical, chemical, and biological methods are presently available for treatment of wastewater discharged from various industries (*2*). The main currently used techniques for decontamination of waters proceeding from the textile industry involve adsorption processes with activated carbon. However, these processes may be costly, inefficient and often produce a high amount of secondary wastes. Ozone and hypochlorite oxidations are efficient decolorizing methods, but they are not desirable because of the high cost of equipment, operating costs and the secondary pollution arising from the residual chlorine (*3*).

Since the discovery of photocatalytic splitting of water on the TiO_2 electrode by Honda and Fujishima in 1972 (*4*), a large number of investigations have focused on the semiconductor photocatalysts for wide application in environmental purification and solar energy conversion (*5*). Among various semiconductor photocatalysts, TiO_2 was intensively investigated because of its strong oxidative power, biological and chemical inertness, stability against photo- and chemical-corrosion, low cost, and non-toxicity (*6*). However, TiO_2 is effective only under ultraviolet irradiation ($\lambda < 380$ nm) due to its large band gap, thus severely restricting the use of sunlight for photocatalysis (*7–9*). Therefore, it is of great interest to develop visible-light-induced photocatalysts to extend the adsorption wavelength range into the visible light region.

Recently, the Bi-based structured compounds, including Aurivillius family (*10–12*), Sillén family (*13–15*) and Aurivillius–Sillén intergrows (*16*), have been extensively investigated as visible-light-driven photocatalysts due to their unique layered structure and high activity. BiOI, belonging to the Sillén family expressed by $[M_2O_2][I_m]$ (m = 2), has drawn increasing attention for the application as visible-light-induced photocatalyst owing to its small band gap (1.77–1.92 eV) and strong absorption in visible-light region. Up to now, different morphologies of BiOI, such as 3-D hierarchical structures (*17–19*), nanosheets (*20*, *21*) and nanolamellas (*22*) have been fabricated successfully by a variety of methods. Also a few model pollutants,such as methyl orange (MO), methylene blue (MB), rhodamine (RhB), sodium pentachlorophenol (PCP-Na) and phenol etc., have been used to evaluate their photocatalytic activity. For example, Xia et al. demonstrated the photocatalytic activity of BiOI prepared by solvothermal processes for the photodegradation of methyl orange under visible light (*23*). Zhang et al. prepared BiOI by one-pot solvothemal process and BiOI showed the photocatalytic activity on the degradation of methyl orange under visible light irradiation (*24*). However, the photoinduced electrons and holes can be recombined to decrease the photocatalytic activity (*25*, *26*). In order to enhance the photocatalytic activity, some approaches were to deposit BiOI with noble metal such as platinum and silver (*25*, *27*, *28*) or the fabrication of the composite semiconductor to facilitate the separation of photoinduced carrier. The composite of BiOI/TiO_2 and BiOI/Bi_2O_3 have been proven to be efficient photocatalysts (*29*, *30*).

In this study, AgI-BiOI composites were prepared by solvothermal process. The photocatalytic activity of AgI-BiOI was evaluated by photocatalytic degradation of acid orange II (AOII) under visible light irradiation ($\lambda > 400$ nm). The enhanced photocatalytic activity mechanism was further studied by the electrochemical impedance spectroscopy (EIS).

Experimental

Chemicals

The raw chemicals of $Bi(NO_3)_3 \cdot 5H_2O$, KI and $AgNO_3$, with analytic grade, were purchased from Fuchen Chemical Reagent Co.. Analytic grade ethylene glycol (EG) was purchased from Fuchen Chemical Reagent Co.. Acid orange II was purchased from Rongqing Chemical Reagent Co. (Guangzhou, China). Other chemicals were analytical grade and were purchased as reagent grade. All chemicals were used without further purification. Doubly-distilled deionized water was used throughout this study.

Synthesis of BiOI and AgI-BiOI Composites

BiOI was prepared by EG-assisted solvothermal method based on our previous report (*28*). In a typical prepared procedure, the salt of iodine (I) and bismuth (III) were put into an EG solution with [I]:[Bi] molar ratio of 1:1. The mixture was magnetically stirred at 273 K for 30 min. Subsequently, the mixture was sonicated for 30 min to form a homogeneous phase and then was transferred into a 100 mL Teflon-lined stainless autoclave until 80% of the volume was filled. The autoclave was naturally cooled to room temperature. Finally, the precipitate was centrifuged, filtered through a 0.22 μm Millipore membrane, then washed with doubly-distilled water and ethanol for three times until there were no other possible impurities in the product. The resulting yellow product was dried in a vacuum oven at 70 °C for 12 h.

AgI-BiOI composites were prepared by depositing AgI on BiOI particles. The preparation procedure is described as follows: The BiOI powder was added into 200 ml 0.1 mol L^{-1} KI solution. The mixture was sonicated for 10 min and then stirred magnetically for 30 min to form the fine BiOI suspension, then 0.1 mol l^{-1} $AgNO_3$ solution was slowly dropped to the suspension. Finally, the precipitate was centrifuged, filtered through a 0.22 μm Millipore membrane, then washed with doubly-distilled water and ethanol for three times until there were no other possible impurities in the product. The resulting product was dried in a vacuum oven at 70 °C for 12 h.

Photocatalysis

All the photocatalytic experiments were carried out in a cylindrical Pyrex reactor with an effective volume of 350 ml as shown in Figure 1. A 400 W gallium iodine lamp equipped with a circulating-water glass jacket filled with 2.0 M $NaNO_2$ solution, was put sideward the reactor to provide the visible light irradiation ($\lambda > 400$ nm). For all typical catalytic experiments, a 200 ml solution of 100 mg L^{-1} AOII and 20 mg catalysts were prepared. Prior to irradiation, suspensions were ultrasonicated for 3 min and magnetically stirred for 1 h in the dark at room temperature to establish the adsorption/desorption equilibrium. At the given time intervals, about 5 ml of suspension was taken for further analysis after centrifugation. AOII concentration were analyzed at 484 nm on a UV-vis spectrophotometer (UNICO UV-2000).

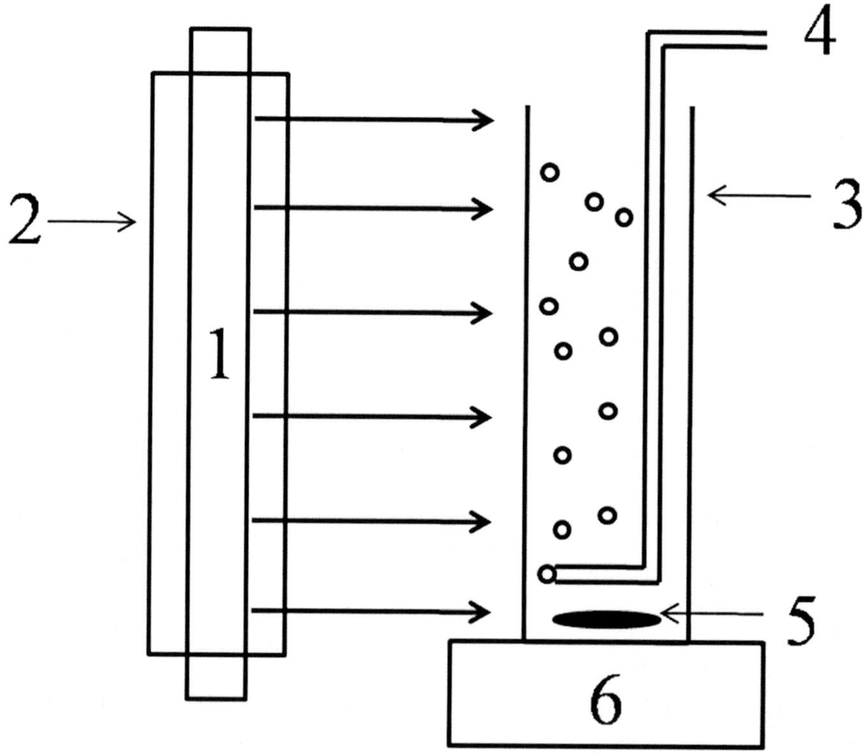

Figure 1. Schematic diagram of photocatalytic system. 1. 400 W commercial gallium iodine lamp, 2. double-walled glass cooling tube, 3. batch Pyrex glass reactor; 4. inlet air, 5. stirrer, and 6. magnetic stirring apparatus.

Results and Discussion

Phase Structures of AgI-BiOI Composites

XRD was firstly employed to study the phase structures. Figure 2 showed the XRD patterns of the as-prepared powders. It was observed in Figure 2 that the as-prepared BiOI with the high purity was successfully prepared by a hydrothermal method, and no crystalline impurity can be detected. The XRD pattern of BiOI with the diffraction peaks around 2θ of 24.3°, 29.7°, 31.8°, 45.8°, was consistent with the standard tetragonal BiOI phase (JCPDS No. 10-0445). Two peaks of AgI at 23.69° and 39.15° appeared in accompany with XRD patterns of BiOI, which can be indexed to be β-AgI phase (JCPDS No.09-0374). XRD results demonstrated that AgI-BiOI composites with a coexistence of AgI and BiOI phase were prepared. According to the Scherrer's formula, the average grain sizes of BiOI and AgI-BiOI composites were calculated to be 48.0 and 50.2 nm, respectively. This result indicated that the doping AgI to BiOI has no significant effect on the size of AgI-BiOI composites.

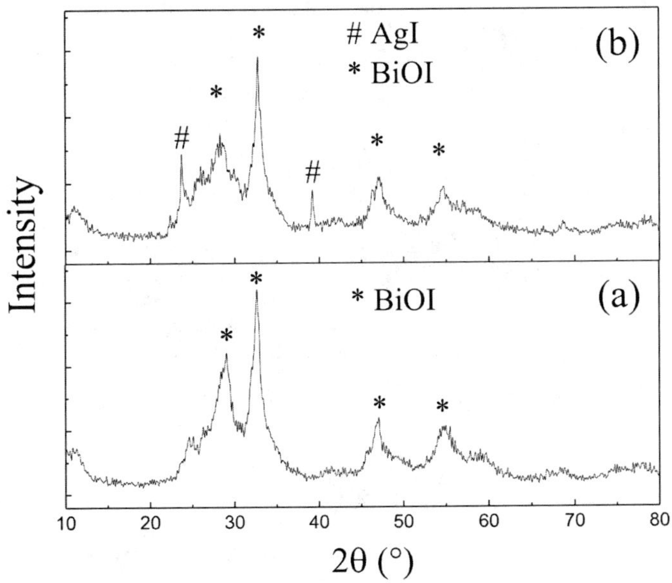

Figure 2. XRD patterns of BiOI (a) and AgI-BiOI (b).

Morphology of AgI-BiOI Composites

Figure 3 showed SEM images of BiOI and AgI-BiOI composites. It can be seen clearly that both BiOI and AgI-BiOI composites were composed of microsphere with a polyporous surface, and the diameters of these spheres were in the range of 1-4 μm. The structure of AgI-BiOI was similar to that of BiOI, indicating that the coexistence of AgI has no significantly effect on the morphologies of BiOI. In addition, it should be noted that the surface of prepared AgI-BiOI composites (Inset in Figure 3b) is less porous than that of BiOI (Figure 3a), indicating that AgI was deposited on the gap of BiOI. However, these SEM results did not provide clear information about the existing of AgI, TEM studied are necessary to further determine the existing of AgI deposits.

Figure 3. SEM images of BiOI (a) and AgI-BiOI (b).

BiOI and AgI-BiOI were examined by TEM and their images were presented in Figure 4. It can be seen that BiOI was consisted of solid structured spheres, and irregular nanoplates can be observed on the surface of BiOI microspheres. In addition, some small nanoparticles were observed in the nanoplates of AgI-BiOI, which did not appear in TEM image of pure BiOI. Thus, the small AgI nanoparticle was deposited on the surface of BiOI. It is expected that the nanostructure of AgI-BiOI would be beneficial to the charge transfer between AgI and BiOI once AgI-BiOI is excited under visible light, leading to the enhanced activity of AgI-BiOI compared to the individual AgI and BiOI.

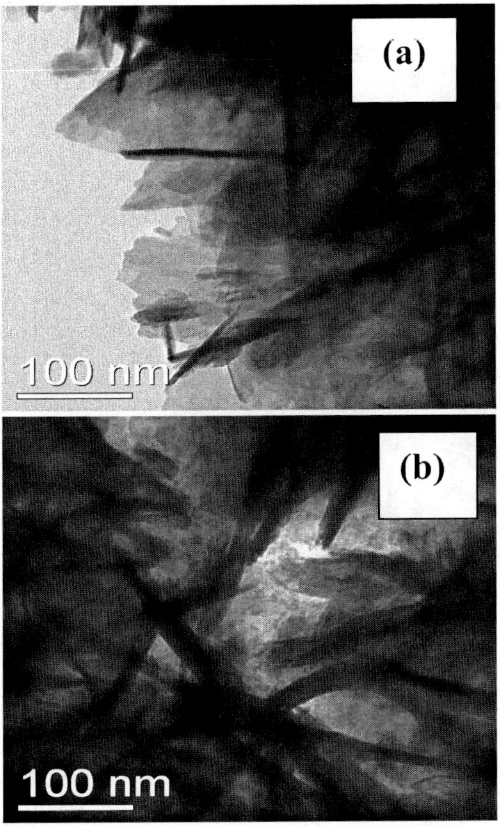

Figure 4. TEM images of BiOI (a) and AgI-BiOI (b).

DRS of AgI-BiOI Composites

Figure 5 showed UV-vis diffuse reflectance spectra (DRS) of BiOI and AgI-BiOI. It can be seen that AgI-BiOI composites had better optical absorption generally in the while visible light range. Especially, a new significant absorption peak occurred in the region 500-680 nm compared to TiO_2 and BiOI, which was attributed to the AgI nanoparticles deposited on the surface of BiOI. Thus, it is believed that the better optical absorption of AgI-BiOI composites in the visible light range is an essential condition to obtain the high photocatalytic performance under visible light.

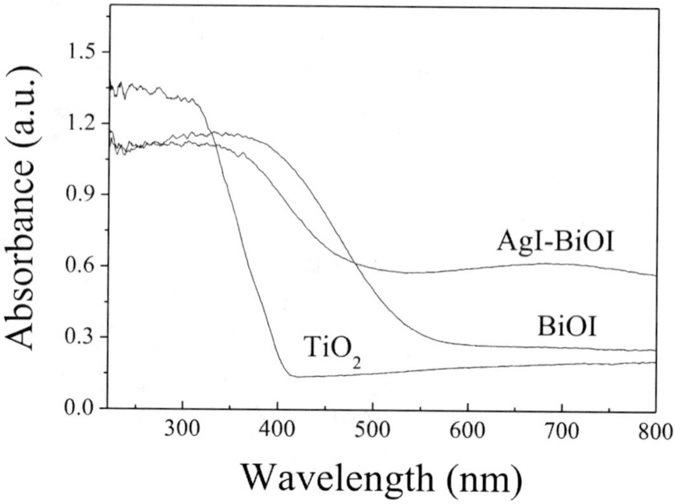

Figure 5. UV-vis diffuse reflectance spectra of TiO_2, BiOI and AgI-BiOI.

Photocatalytic Activity of BiOI and AgI-BiOI

The photocatalytic activity of BiOI and AgI-BiOI were evaluated by the photodegradation of acid orange II (AOII) in aqueous solution under visible irradiation. The blank experiment showed that AOII can not be degraded neither with the presence of AgI-BiOI in dark nor with the presence of TiO_2 under visible light (Figure 6a). However, AOII was removed more than 85% within 120 min using AgI-BiOI under visible light ($\lambda > 400$ nm). This result indicated that the presence of both visible-light irradiation and photocatalyst is indispensable for the photocatalytic degradation of AOII. In addition, it should be noted that the degradation efficiency of AOII, using individual AgI and BiOI, was only 10% and 36%, respectively. In contrast, the photocatalytic degradation efficiency of AOII was more than 85% within 120 min, higher than that of AgI and BiOI. The enhanced photocatalytic activity of AgI-BiOI was ascribed to the effective charge transfer between AgI and BiOI. Therefore, AgI-BiOI composite can be excited under visible light ($\lambda > 400$ nm) and demonstrated an effective performance for BPA degradation under visible light.

The kinetic results of photocatalytic degradation of AOII using AgI-BiOI can be analyzed using the Langmuir-Hinsehelwood (L-H) model (*31*). For a low adsorption equilibrium concentration of AOII in this investigation (much less than 10^{-3} M), L-H equation can be expressed as the pseudo first-order kinetics (*32*):

$$r = -\frac{dC}{dt} = k'C \qquad (1)$$

where r is the reaction rate (mg L^{-1} min^{-1}), k' is apparent reaction rate constants, and C is the substrate concentration in the solution (mg L^{-1}).

Figure 6b showed the first-order reaction kinetics curves for photocatalytic degradation of AOII were calculated from the experimental data in Figure 6a using a linear regression. The values of k' for AgI-BiOI composites (1.6×10^{-2} min^{-1}) were of 18.3 times and 3.7 times higher than that of individual AgI (8.9×10^{-4} min^{-1}) and BiOI (4.3×10^{-3} min^{-1}), respectively. The enhanced photocatalytic activity using AgI-BiOI composites could be attributed to the excellent optical absorbance in the visible light region and the effective separation of the photogenenratd electron and hole between the interface of AgI and BiOI.

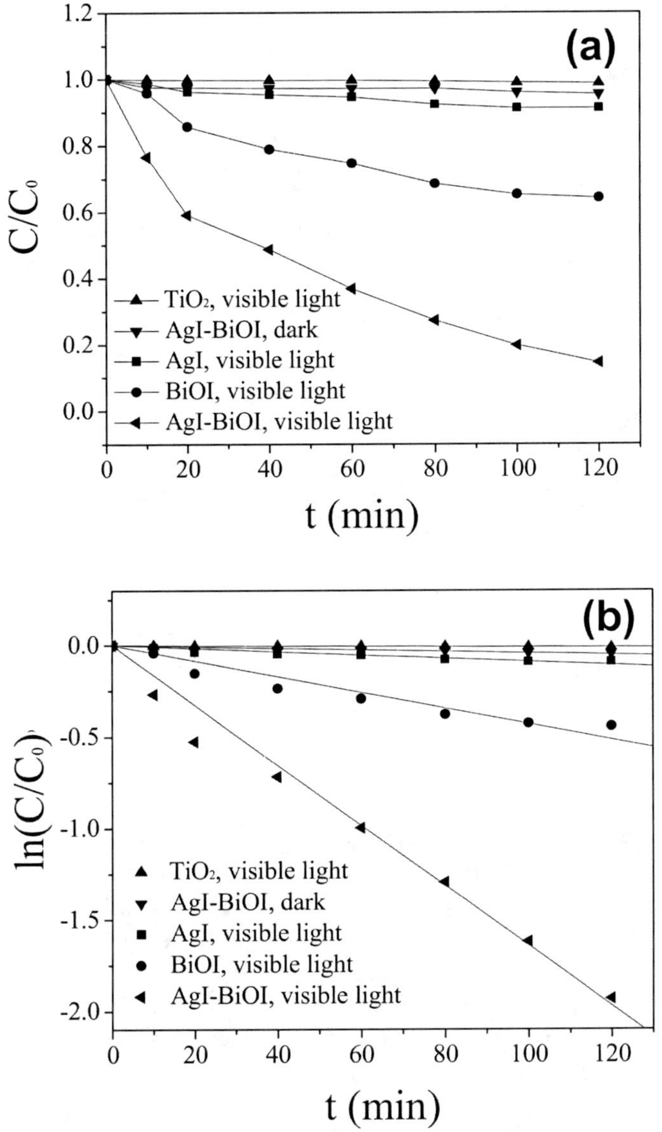

Figure 6. Removal efficiency (a) and first-order reaction kinetics curves (b) of AOII under different conditions.

The degradation of AOII catalyzed using AgI-BiOI was determined by measuring UV-vis spectral changes of AOII as a function of the reaction time (Figure 7). It can be seen that there appeared 3 absorption peaks at 484, 310 and 230 nm and one shoulder peak at 430 nm. The peaks at 310 and 230 nm were assigned to the aromatic rings. The peak at 484 nm and the shoulder peak at 430 nm were assigned to the conjugated structure formed by azo bond (*33*). As the reaction proceeds, the characteristic absorption peaks at 484 nm decreased dramatically after 120 min, indicating that the chromophore and conjugated π^* bonds were completely destructed.

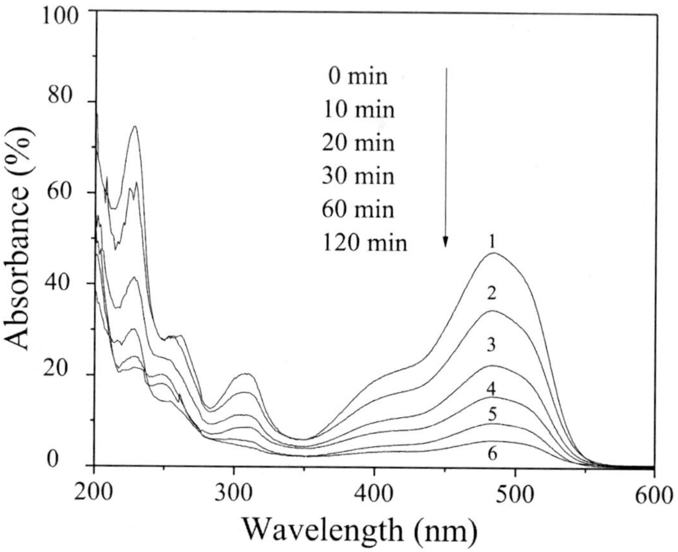

Figure 7. UV-vis spectra change of AOII in photocatalytic degradation processes using AgI-BiOI.

EIS Properties of BiOI and AgI-BiOI

To further study the charge transfer in AgI-BiOI system, the electrochemical impedance spectroscopy (EIS) spectra of BiOI and AgI-BiOI electrodes were investigated as shown in Figure 8. EIS spectrum presented as Nyquist plots has been widely applied in the study of interfacial change transfer processes occurring in the photoreaction (*34*, *35*). Generally, the smaller the arc radius on EIS Nyquist plot, the lower the charge transfer resistance is. It can be seen that the arc radius on EIS Nyquist plot of AgI-BiOI was smaller than that of BiOI in dark or under visible irradiation. This EIS results indicated that AgI-BiOI had a lower impedance than that of BiOI, which could promote a faster interfacial charge transfer process and suppress the recombination of electron-hole pairs. Therefore, AgI-BiOI would have better photocatalytic activities as a result of the effective separation of photogenerated electron-hole pairs between the interface of AgI and BiOI .

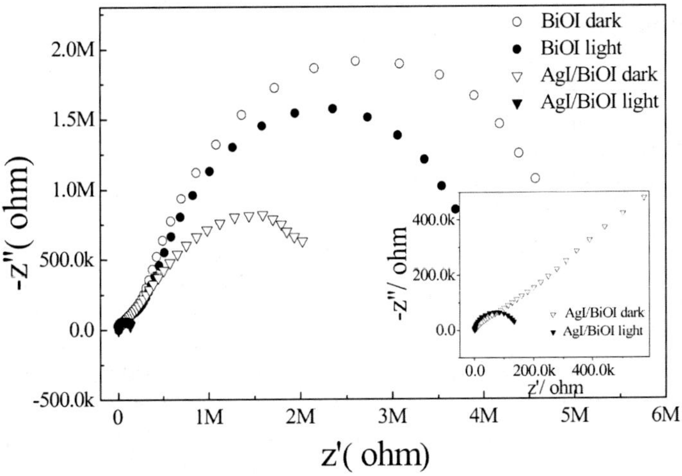

Figure 8. EIS plots of BiOI and AgI-BiOI in dark and under visible light.

Stability of AgI-BiOI Composites

The stability and reusability of the photocatalyst is a critical factor for its practical application. Thus, AgI-BiOI was tested in the photocatalytic system for its ability to degrade AOII in aqueous solution for a series of 5 identical tests. The performance was evaluated and was shown in Figure 9. It can be seen that the AgI-BiOI retained its degradation effectiveness at 83.6 ± 1.63% within the limits of experimental error. This results indicated that AgI-BiOI composites possessed chemical and catalytic stability.

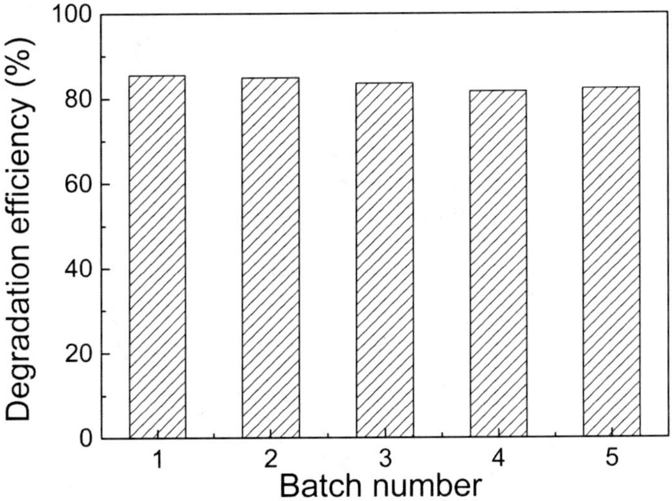

Figure 9. Degradation efficiency of AOII in five batch runs using AgI-BiOI composites under visible light.

Mechanism of Enhanced Photocatalytic Degradation

The mechanism to account for the photocatalytic degradation of AOII involves the initial photoexcitation of the catalyst, and the generation of electron-hole pairs under photoexcitation (*6*, *28*). The enhanced photocatalytic activity of AgI-BiOI should be related to the efficient charge transfer between AgI and BiOI. Pioneer studies showed that AgI and BiOI have the matching band potentials, that is, the CB potential of AgI (E_{cb}= 0.55 eV) is negative than that of BiOI (E_{cb}= 0.56 eV), while the VB potential of BiOI (E_{vb}= 2.42 eV) is positive than that of AgI (E_{vb}= 2.25 eV) (*29*). Thus, the photoinduced electrons on the surface of AgI can transfer to the surface of BiOI, and the photoinduced holes of BiOI can transfer to AgI, once AgI-BiOI is excited under visible light. The possible scheme of photoinduced charge transfer in AgI-BiOI system was given in Figure 10. Thus, this charge transfer between AgI and BiOI suppressed the combination of the photoinduced electrons and holes, which is beneficial to improve the photocatalytic efficiency of AOII. Subsequently, the photogenerated electrons can participate in the oxidation process through the capture the dissolved O_2 in solution to generate $\cdot O_2^-$, and further produce ·OH, while the photogenerated holes oxidized water molecules to form ·OH (*36*). Therefore, the enhanced photocatalytic efficiency of AOII in AgI-BiOI system under visible light could be ascribed to the effectively generated active ·OH.

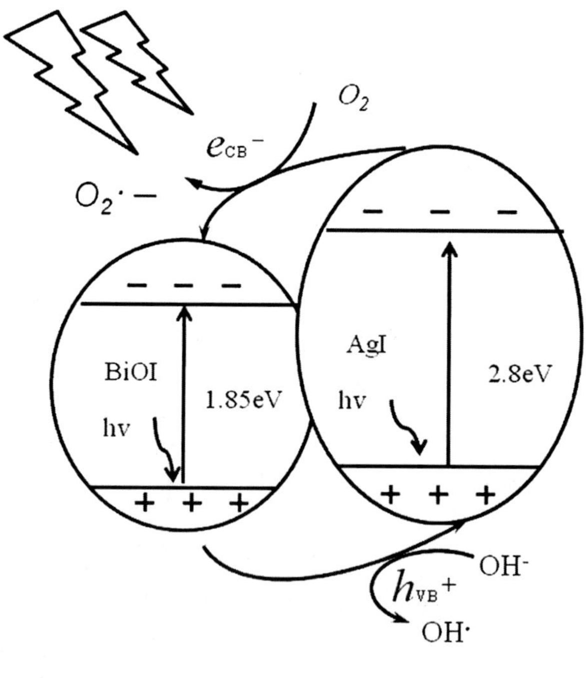

Figure 10. Possible scheme of photoinduced charge transfer in AgI-BiOI system.

Conclusions

In this paper, the nanostructured AgI-BiOI composites were synthesized in the presence of ethylene glycol by a solvothermal process. Their photocatalytic activities were evaluated by the degradation of AOII in aqueous solution. The experimental results indicated that the as-prepared AgI-BiOI powder was high purity and was composed of nanoplates. The DRS results indicated that AgI-BiOI composites have the significant adsorption and the red-shift in the visible light region compared to BiOI. The photocatalytic degradation efficiency of AOII using AgI-BiOI composites was more than 85% within 120 min, which was higher than that of individual AgI and BiOI. The enhanced photocatalytic activity of AgI-BiOI was ascribed to the efficient charge transfer between AgI and BiOI. Moreover, AgI-BiOI composites retained their stable degradation efficiencies for a series of repetitive batch runs, indicating the true photocatalytic process.

Acknowledgments

The authors wish to thank the National Natural Science Foundation of China (No. 20877025, 21273085), National Natural Science Foundation of Guangdong Province (No. S2011010001836, S2011010003416), Fundamental Research Funds for the Central Universities (No. 09lgpy20), Science and Technology Research Project from Guangzhou City (No. 12A42091603, 2010Z2-C1009) and Ministry of Science and Technology of China (No. 10C26214414753), the Research Fund Program of Guangdong Provincial Key Laboratory of Environmental Pollution Control and Remediation Technology (No. 2011K0003), Project from Education Bureau of Guangdong Province (2010-275), and the Guangdong Province Science and Technology Bureau (Grant No. 2010B090400552) for financially supporting this work.

References

1. Zhang, G. K.; Ding, X. M.; Hu, Y. J.; Huang, B. B.; Zhang, X. Y.; Qin, X. Y.; Zhou, J.; Xie, J. W. *J. Phys. Chem. C* **2008**, *112*, 17994–17997.
2. Swaminathan, K.; Sandhya, S.; Sophia, A. C.; Pachhade, K.; Subrahmanyam, Y. V. *Chemosphere* **2003**, *50*, 619–625.
3. Malik, P. K.; Saha, S. K. *Sep. Purif. Technol.* **2003**, *31*, 241–250.
4. Fujishima, A.; Honda, K. *Nature* **1972**, *238*, 37.
5. Hoffmann, M. R.; Martin, S. T.; Choi, W.; Bahnemann, D. W. *Chem. Rev.* **1995**, *95*, 69–96.
6. Yu, J. C.; Ho, W. K.; Yu, J. G.; Yip, H. Y.; Wong, P. K.; Zhao, J. C. *Environ. Sci. Technol.* **2005**, *39*, 1175–1179.
7. Anpo, M.; Takeuchi, M. *J. Catal.* **2003**, *216*, 505–516.
8. Chatterjee, D.; Dasgupta, S. *J. Photochem. Photobiol., C* **2005**, *6*, 186–205.
9. Hu, C.; Guo, J.; Qu, J. H.; Hu, X. X. *Langmuir* **2007**, *23*, 4982–4987.
10. Shimodaira, Y.; Kato, H.; Kobayashi, H.; Kudo, A. *J. Phys. Chem. B* **2006**, *110*, 17790–17797.

11. Kim, H. G.; Hwang, D. W.; Lee, J. S. *J. Am. Chem. Soc* **2004**, *126*, 8912–8913.
12. Ma, D. K.; Huang, S. M.; Chen, W. X.; Hu, S. W.; Shi, F. F.; Fan, K. L. *J. Phys. Chem. B* **2009**, *113*, 4369–4374.
13. Lin, X. P.; Huang, T.; Huang, F. Q.; Wang, W. D.; Shi, J. L. *J. Phys. Chem. B* **2006**, *110*, 24629–24634.
14. Ai, Z. H.; Ho, W. K.; Lee, S. C.; Zhang, L. Z. *Environ. Sci. Technol.* **2009**, *43*, 4143–4150.
15. Zhang, K. L.; Liu, C. M.; Huang, F. Q.; Zheng, C.; Wang, W. D. *Appl. Catal., B* **2006**, *68*, 125–129.
16. Lin, X. P; Huang, T.; Huang, F. Q.; Wang, W. D.; Shi, J. L. *J. Mater. Chem.* **2007**, *17*, 2145–2150.
17. Zhang, X.; Ai, Z. H.; Jia, F. L.; Zhang, L. Z. *J. Phys. Chem. C* **2008**, *112*, 747–753.
18. Lei, Y. Q.; Wang, G. H.; Song, S. Y.; Fan, W. Q; Pang, M.; Tang, J. K.; Zhang, H. J. *Dalton Trans.* **2010**, *39*, 3273–3278.
19. Xiao, X.; Zhang, W. D. *J. Mater. Chem.* **2010**, *20* (28), 5866–5870.
20. An, H. Z.; Du, Y.; Wang, T. M.; Wang, C.; Hao, W. C.; Zhang, J. Y. *Rare Metals* **2008**, *27*, 243–250.
21. Chang, X. F.; Huang, J.; Tan, Q. Y.; Wang, M.; Ji, G. B.; Deng, S. B.; Yu, G. *Catal. Commun.* **2009**, *10*, 1957–1961.
22. Fang, F.; Chen, L.; Wu, L. M. *Chin. J. Struc. Chem.* **2009**, *28*, 1399–1406.
23. Xia, J. X.; Yin, S.; Li, H. M; Xu, H.; Yan, Y. S.; Zhang, Q. *Langmuir* **2010**, *27*, 1200–1206.
24. Zhang, X.; Zhang, L. Z.; Xie, T. F.; Wang, D. J. *J. Phys. Chem. C* **2009**, *113*, 7371–7378.
25. Yu, C. L.; Yu, J. C.; Fan, C. F.; Wen, H. R.; Hu, S. J. *Mater. Sci. Eng., B* **2010**, *166*, 213–219.
26. Cheng, H. F.; Huang, B. B.; Dai, Y.; Qin, X. Y.; Zhang, X. Y. *Langmuir* **2010**, *26*, 6618–6624.
27. Liu, H.; Cao, W. R.; Su, Y.; Wang, Y.; Wang, X. H. *Appl. Catal., B* **2012**, *111*, 271–279.
28. Zhu, L. F.; He, C.; Huang, Y. L.; Chen, Z. H.; Xia, D. H.; Su, M. H.; Xiong, Y.; Li, S. Y.; Shu, D. *Sep. Purif. Technol.* **2012**, *91*, 59–66.
29. Li, Y. Y; Wang, J. S.; Yao, H. C; Dang, L. Y.; Li, Z. J. *Catal. Commun.* **2011**, *12*, 660–664.
30. Li, Y. Y; Wang, J. S.; Liu, B.; Dang, L. Y.; Yao, H. C.; Li, Z. J. *Chem. Phys. Lett.* **2011**, *508*, 102–106.
31. Turchi, C. S.; Ollis, D. F. *J. Catal.* **1990**, *122*, 178–192.
32. Behnajady, M. A.; Modirshahla, N.; Hamzavi, R. *J. Hazard. Mater.* **2006**, *133*, 226–232.
33. Zhang, S. J.; Yu, H. Q.; Zhao, Y. *Water Res.* **2005**, *39*, 839–846.
34. Zhang, H; Lv, X. Y.; Li, Y. M.; Wang, Y.; Li, J. H. *ACS Nano* **2009**, *4*, 380–386.
35. Zhu, S. B.; Xu, T. G.; Fu, H. B.; Zhao, J. C.; Zhu, Y. F. *Environ. Sci. Technol.* **2007**, *41*, 6234–6239.
36. Zhang, L. S.; Wong, K. H.; Yip, H. Y.; Hu, C.; Yu, J. C.; Chan, C. Y.; Wong, P. K. *Environ. Sci. Technol.* **2010**, *44*, 1392–1398.

Chapter 16

Magnetically Recyclable Gold−Magnetite Nanocatalysts for Reduction of Nitrophenols

Fang-hsin Lin and Ruey-an Doong*

Department of Biomedical Engineering and Environmental Sciences, National Tsing Hua University, Hsinchu 30013, Taiwan
***E-mail: radoong@mx.nthu.edu.tw**

The Au-Fe_3O_4 heterostructures by thermal decomposition of iron-oleate complex in the presence of Au nanoparticles (NPs) have been successfully fabricated using different sizes of Au NPs as the seeds for magnetically recyclable catalysis of *p*-nitrophenol reduction. The heterostructures exhibit bifunctional properties with high magnetization and excellent catalytic activity towards *p*-nitrophenol reduction. The epitaxial linkages in dumbbell- and flower-like heterostructures are different, leading to the change in magnetic and catalytic properties of the nanocatalysts. The pseudo-first-order rate constants for *p*-nitrophenol reduction are 0.61±0.08 and 0.32±0.04 min^{-1} for dumbbell- and flower-like Au-Fe_3O_4 heterostructures, respectively. In addition, the heterostructured nanocatalysts show good separation ability and reusability which can be repeatedly applied for nearly complete reduction of *p*-nitrophenols for at least 6 successive cycles. The kinetics of nitrophenol reduction was obtained by monitoring the absorbance at 400 nm by UV-Vis spectroscopy. The kinetic data could be well-described by the Langmuir-Hinshelwood model. These unique properties make Au-Fe_3O_4 heterostructures an ideal platform to study various heterogeneous catalytic processes which can be potentially applied in a wide variety of fields in purification, catalysis, sensing devices, and green chemistry.

Introduction

Nobel metal nanostructures have recently received much attention because of their unique optical, catalytic, and electrochemical properties which make them suitable materials for potential applications in various fields (*1*, *2*). Gold nanoparticles (Au NPs) have been found to play an important role in several catalytic processes including low temperature CO oxidation (*3*), reductive catalysis of chlorinated or nitrogenated hydrocarbons (*4–6*), and organic synthesis (*7*, *8*). Due to the high cost and limited supply, however, the improvement of the catalytic efficiency and the reduction of the used amounts are the top priorities for practical applications. The deposition of Au NPs onto porous supports such as TiO_2, SiO_2 and carbon is regarded as a conventional way to solve the problem by maximizing the loading of catalysts and to enhance the catalytic activity by well-tuning the surface functionality (*9–12*). However, the entrapment or immobilization of the nanocatalysts on solid supports normally results in a decrease in the catalytically active surface area and the reactivity of catalytic species (*13*).

Recently, the nanocatalysts with Janus/heterodimer geometry, which two or more materials are interconnected by junction regions with limited extension, have received increasing attention because they offer multi-functionality and enhanced physicochemical properties (*14–16*). Seh et al. (*17*) has introduced a facile method for the control synthesis of various Au-TiO_2 nanostructures with different geometry. The energetically stable Janus nanostructures exhibited the highest catalytic activity toward *p*-nitrophenol reduction because the exposed gold core offered high accessibility to reactants, while the TiO_2 coating on the other side imparted protection against aggregation. In addition, the catalysts could be reused over five cycles with no obvious reduction in activity. Sun et al. (*18*, *19*) has synthesized the Au-Fe_3O_4 dumbbell-like nanostructures (Au-Fe_3O_4 DBNPs) for CO oxidation as well as H_2O_2 reduction. Lin and Doong (*20*) has synthesized different morphology of Au-Fe_3O_4 heterostructure for catalytic reduction of *p*-nitrophenol. The combination of Au NPs with magnetic Fe_3O_4 NPs can not only provide catalytic activity but also be reclaimed via magnetic separation after use. Moreover, the magnetic property was enhanced via interfacial interaction (*21*). It is believed that electron transfer across the interface between these two NPs may lead to a dramatic change in physicochemical properties, thus offering an ideal platform to study the multifunctionality of nanomaterials (*22*, *23*). In addition, the Au-Fe_3O_4 heterostructures contain magnetically, optical and catalytically active NPs, which show highly potential applications to chemical catalysis, drug delivery and biomedical imaging (*24–26*).

In this study, the magnetically recyclable catalytic ability of Au-Fe_3O_4 heterostrctures including dumbbell-like nanoparticles (DBNPs) and flower-like nanoparticles (FLNPs) towards nirtophenol reduction is examined. The designed heterostructures show excellent dual functions which can not only undergo rapidly catalytic reduction of *p*-nitrophenols in the presence of $NaBH_4$ but also be easily recycled using external magnetic field. In addition, the kinetic and thermodynamic property of Au-Fe_3O_4 DBNPs in the reduction of *p*-nitrophenol are discussed. Effects of environmental parameters including initial concentration of *p*-nitrophenol and $NaBH_4$ on the reduction efficiency and rate of *p*-nitrophenol are investigated.

Materials and Methods

Chemicals

Iron chloride ($FeCl_3 \cdot 6H_2O$, 98%), oleylamine (>70%), oleic acid (90%), and 1-octadecene (90%) were purchased from Sigma-Aldrich. Sodium oleate (95%) was obtained from TCI. These compounds were used for the synthesis of Fe_3O_4 nanoparticles. Hydrogen tetrachloroaurate(III) trihydrate ($HAuCl_4 \cdot 3H_2O$) (Alfa Aesar) and tert-butylamine-borane complex (97%) (Sigma Aldrich) were used for the preparation of Au seeds. Tri-sodium citrate dehydrate (>99%), a phase transfer agent, was purchased from Ferak. In addition, sodium borohydride (95%) and *p*-nitrophenol, obtained from Riedel-de Haën and Fluka, respectively, were used for catalytic reaction. Solvents including ethanol absolute (99.8%) (Riedel-de Haën), cyclohexane (99.95%) (TEDIA), *n*-hexane (J. T. Baker) were used as received without further treatment.

Synthesis of Gold NPs with Different Sizes

The Au NPs with sizes of 4-5 nm were prepared by dissolving 40 mg of $HAuCl_4 \cdot 3H_2O$ in a mixture containing 4 mL of oleylamine and 4 mL of cyclohexane in air followed by magnetic stirring at 10 °C under a gentle stream of nitrogen gas. 0.2 mmole of tert-butylamine-borane complex was dissolved in 0.4 mL of oleylamine and 0.4 mL of cyclohexane, and then injected into the precursor solution. The solution color changed to deep red immediately after injecting the borane complex solution. The mixture was aged for 40 min at 10 °C followed by addition of 30 mL of ethanol to precipitate the Au NPs. The Au NPs was then harvested by centrifugation and re-dispersed in hexane. For preparation of 10-nm Au NPs, 40 mg $HAuCl_4 \cdot 3H_2O$ was dissolving in a mixture containing 4 mL of 1-octadecene and 4 mL of oleylamine in air. The resulting solution was put in an oil bath at 120 °C and reaction for 30 min under N_2 atmosphere. After reaction, the mixture was cooled down to room temperature and followed by addition of 30 mL of ethanol to precipitate the Au NPs. The product was centrifuged and re-dispersed in hexane.

Synthesis of Au-Fe_3O_4 Heterostructures

The synthetic method for the Au-Fe_3O_4 heterostructure is according to our previous work (*20*). A solution containing 0.5 mmole of oleic acid, 0.5 mmole of oleylamine, 1 mmole of $Fe(OL)_3$, 0.1 mmole of gold colloid dispersion, and 5 mL of octadecene was heated to 110 °C for 20 min. The solution was refluxed at 310 °C for 30 min. After cooling down to room temperature, the particles were separated by adding absolute ethanol, centrifugation and re-dispersion into hexane.

Preparation of Water-Soluble Au-Fe_3O_4 NPs

The Au-Fe_3O_4 NPs were washed with a mixture of hexane and ethanol (1:2) several times to remove excess capping agent on the surface of NPs. The heterostructured Au-Fe_3O_4 NPs was then dried and added into an aqueous solution containing 50 mM sodium citrate. After reaction of 10 min, the Au-Fe_3O_4 heterostructures were separated by magnet and wash with de-ionized water three times. The particles were then dissolved in de-ionized water.

Catalytic Reaction

The reduction of nitrophenol compound by water soluble Au-Fe_3O_4 NPs in the presence of $NaBH_4$ was carried out to examine the catalytic activity and recyclability of the Au-Fe_3O_4 nanocatalysts. 2 mL of de-ionized water, 40 µl of 10 mM nitrophenol, and 0.16 mL of 0.1 M $NaBH_4$ solutions were added into a quartz cuvette followed by addition of 2 mg of water soluble Au-Fe_3O_4 NPs to the mixture. The color of the solution changed gradually from yellow to transparent as reaction proceeded. UV-Vis spectrometry was used to record the change in absorbance at a time interval of 2 min.

It is known that the rate of surface-mediated degradation of organic pollutants at liquid-solid interface can be described by the Langmuir-Hinshelwood kinetic model:

$$r_0 = \frac{dC}{dt} = k_{app}\frac{K_F S_t C}{1+K_F C} \tag{1}$$

where r_o is the reaction rate of *p*-nitrophenol reduction, C is the aqueous concentration of *p*-nitrophenol, k_{app} is the limiting-step rate constant of reaction at maximum coverage under the given conditions, S_t is the total reaction sites of Au-Fe_3O_4 NPs, and K_F is the adsorption coefficient of *p*-nitrophenol. When the *p*-nitrophenol concentration (C) is low, eq (1) can be simplified to the pseudo-first-order kinetics:

$$\ln(\frac{C}{C_o}) = -k_{obs}t \tag{2}$$

where k_{obs} is the pseudo-first-order rate constant for *p*-nitrophenol (min^{-1}).

Characterization

Transmission electron microscopy (TEM) images were obtained on a JEOL 2011 microscope operated at 120 kV. High-resolution transmission electron microscopy (HR-TEM) was carried out on a JEOL JEM-2010 microscope at 200 kV. The samples were prepared by suspension in hexane. Wide-angle XRD patterns were recorded on a Bruker D8 X-ray diffractometer with Ni-filtered Cu Kα radiation (λ = 1.5406 Å) and operated at a generator voltage and an emission current of 40 kV and 40 mA, respectively. A Hitachi U-3010 UV–Vis spectrophotometer using a 1-cm path length quartz cuvette was used to identify the change in concentration of *p*-nitrophenol over a wavelength range from 200 to 600 nm. Magnetic measurements were carried out using a superconducting quantum interference device magnetometer (SQUID MPMS5, Quantum Design Inc.) with a maximum applied continuous field of 10,000 Gauss at room temperature.

Results and Discussion

Characterization of Au-Fe_3O_4 Heterostructures

The heterostructured Au-Fe_3O_4 nanoparticles were prepared by thermal decomposition of iron-oleate complex in the presence of different sizes of Au NPs. The morphology of these heterostructures is highly dependent on the size of Au seeds. Figure 1 shows the TEM images of Au-Fe_3O_4 heterostructures synthesized by using different sizes of Au seeds ranging between 5 and 10 nm (Figure 1a, c). The nanostructured nanoparticles show a dumbbell-like structure when small-sized Au NPs are used as seeds (Figure 1b). Using large Au NPs of 7-13 nm as seeds, flower-like structures with sizes of 20-28 nm are formed. A previous study depicted that the crystallinity of Au seeds controlled the nucleation process, with one iron oxide leaf nucleated per monocrystalline domain of gold (*27*). In this study, large Au NPs provide large surface areas and multiple monocrystalline domains for Fe_3O_4 to nucleation, resulting in the production of flower-like heterostructures.

The crystallinity of Au-Fe_3O_4 heterostructures is characterized by XRD. Figure 2a shows the XRD patterns of different morphologies of Au-Fe_3O_4 NPs. Five resolved peaks at 30.10°, 35.54°, 43.09°, 56.98°, and 62.58° 2θ, which can be assigned as the fcc Fe_3O_4, are observed. In addition, peaks at 38.18°, 44.39°, 64.58°, and 77.55° 2θ are common patterns for fcc-structured Au. The XRD patterns of Au-Fe_3O_4 NPs match well with those corresponding JCPDS standards of Au and Fe_3O_4 (JCPDS 04-0784; JCPDS 65-3107), clearly indicating the nature of heterodimer structures of Au-Fe_3O_4 NPs. In addition, the epitaxial linkage in heterostructures has a significant effect on the change in optical properties of Au NPs (*28*). The pure Au NPs show a surface plasmon resonance peak at 512 and 520 nm for 5 and 10 nm Au NPs, respectively (Figure 2b). After conjugation with Fe_3O_4 NPs, the peak is broadening and red-shifts to 543 nm in dumbbell-like structure and 606 nm in flower-like structures (Figure 2b). The relatively weak absorbances of both Au-Fe_3O_4 DBNPs and FLNPs are primarily attributed to the dilution effect of Fe_3O_4 on Au NPs in the heterostructures (*9*). Moreover, the magnetic measurement shows that Au-Fe_3O_4 NPs are superparamagnetic at room temperature (300 K) (Figure 2c). The hysteresis loops of heterostructures indicate that the saturation magnetization is 52 emu/g for Au-Fe_3O_4 DBNPs and 63 emu/g for Au-Fe_3O_4 FLNPsat 300K after normalization to the unit weight of Fe_3O_4. It is noteworthy that saturation magnetization obtained in this study is lower than that of bulk magnetite (90 emu/g) (*29*). However, these values are higher than those reported data prepared by the similar procedure after normalization to the unit weight of Fe_3O_4 (*30*, *31*).

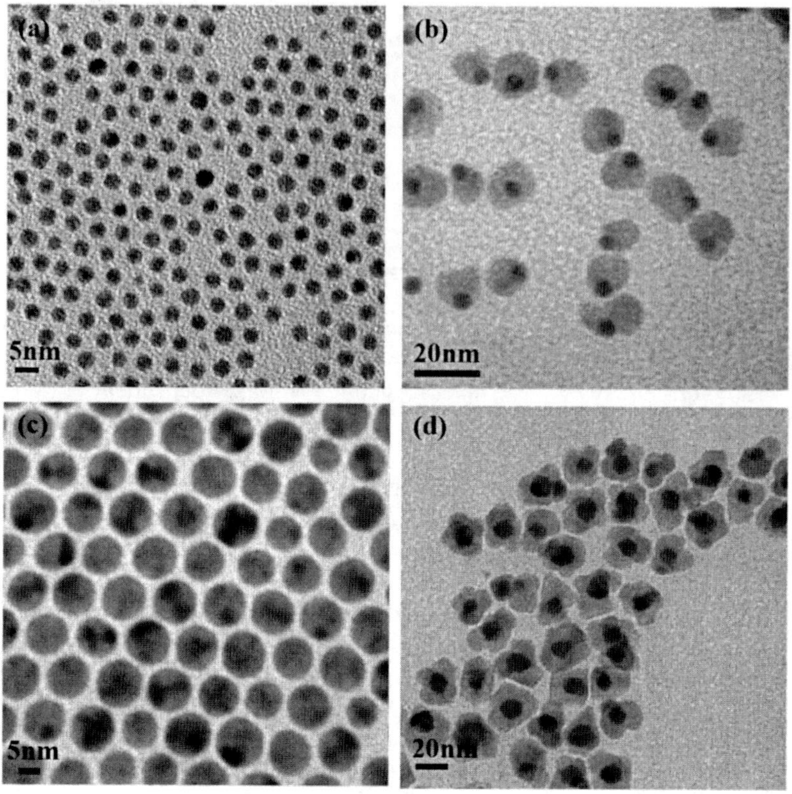

Figure 1. TEM images of (a) 5-nm Au NPs, (b) dumbbell-like Au-Fe_3O_4 heterostructures, (c) 10-nm Au NPs, and (d) flower-like Au-Fe_3O_4 heterostructures.

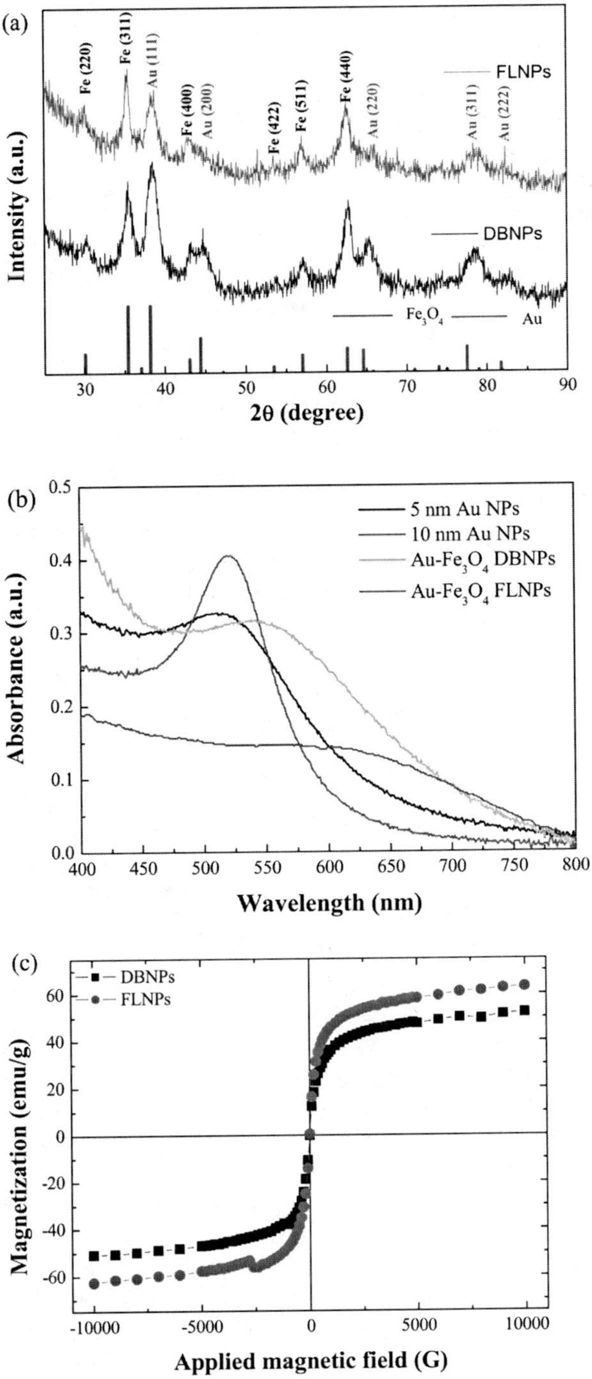

Figure 2. (a) XRD patterns, (b) UV-Vis spectra , and (c) magnetic hysteresis loops of dumbbell- and flower-like Au-Fe_3O_4 heterostructures.

Application of Au-Fe_3O_4 Heterostructures for Catalytic Reduction of *p*-Nitrophenols

The catalytic reduction of *p*-nitrophenol to their corresponding daughter derivatives, *p*-aminophenol, in the presence of $NaBH_4$ was chosen as a model reaction to investigate the bifunctionality of Au-Fe_3O_4 heterostructures. Such a reaction catalyzed by Au catalysts has been reported because this reaction can be rapidly and easily characterized (*32–34*). Figure 3(a) shows the typical UV-Vis spectra and concentration change of nitrophenol compounds in the presence of Au-Fe_3O_4 heterostructures and $NaBH_4$. The original absorption peak of *p*-nitrophenol is centered at 317 nm, and changes to 400 nm after addition of freshly prepared $NaBH_4$ solution, indicating the formation of *p*-nitrophenolate ions (*35*). This peak starts to decrease when the reduction proceeds in the presence of Au-Fe_3O_4 nanocatalysts. Addition of $NaBH_4$ in the absence of Au-Fe_3O_4 NPs has little effect on the change in absorbance at 400 nm, confirming that the reduction is mainly catalyzed by the Au-Fe_3O_4 NPs. In addition, the absorption peak at 400 nm decreases with the concomitant increase in peak intensity at 300 nm within 10 min after addition of Au-Fe_3O_4 catalysts (Figure 3a).

The pseudo-first-order kinetics can be applied to evaluate the rate constants for nitrophenol reduction because the concentration of $NaBH_4$ is higher than those of nitrophenols and can be considered as a constant during the reaction period. The concentration of *p*-nitrophenol at time t is denoted as C_t and the initial concentration of nitrophenols at t = 0 is regarded as C_0. The C_t/C_0 is measured from the relative intensity of absorbance (A_t/A_0). The linear relationship of $\ln(C_t/C_0)$ versus time (t) indicates that the reduction of nitrophenols by Au-Fe_3O_4 heterostructures follows the pseudo-first-order kinetics (eq. (2)). The rate constants for *p*-nitrophenol reduction are 0.61 ± 0.08 min^{-1} for dumbbell-like Au-Fe_3O_4 nanocatalysts and 0.32 ± 0.04 min^{-1} for flower-like Au-Fe_3O_4 NPs (Figure 3b). The catalytic efficiency as well as the rate constants for nitrophenol reduction by both dumbbell- and flower-like Au-Fe_3O_4 are higher than those previously reported values obtained from the catalysis of *p*-nitrophenol by Au-based materials (*36*, *37*). Deng et al. have synthesized Fe_3O_4@SiO_2-Au@mSiO_2 nanocomposites for catalytic reduction of *p*-nitrophenol (*36*). The rate constants were in the range of 0.20-0.35 min^{-1} with a catalyst dose of *ca.* 3mg. Chang et al. have synthesized Fe_3O_4-Au nanocomposites for catalytic reduction and found that the rate constants for catalytic reduction of *p*-nitrophenol were between 0.01 and 0.313 min^{-1}, depending on the amount of Au in nanocomposites (*37*). This result clearly indicates that Au-Fe_3O_4 heterostructures are superior nanocatalysts which can enhance the catalytic efficiency and minimize the used amounts of catalysts for reaction, especially only when trace amounts of Au catalysts are used (0.38-0.96 mg Au) for reduction. It is noteworthy that the catalytic efficiency of flower-like Au-Fe_3O_4 is lower than that of dumbbell-like structures, presumably due to that the Au surfaces in flower-like structures are mainly occupied by the Fe_3O_4 leaves, and thus suppress the reaction rate of nitrophenols. Therefore, the dumbbell-like Au-Fe_3O_4 heterostructures are selected as the model nanocatalysts for further experiments.

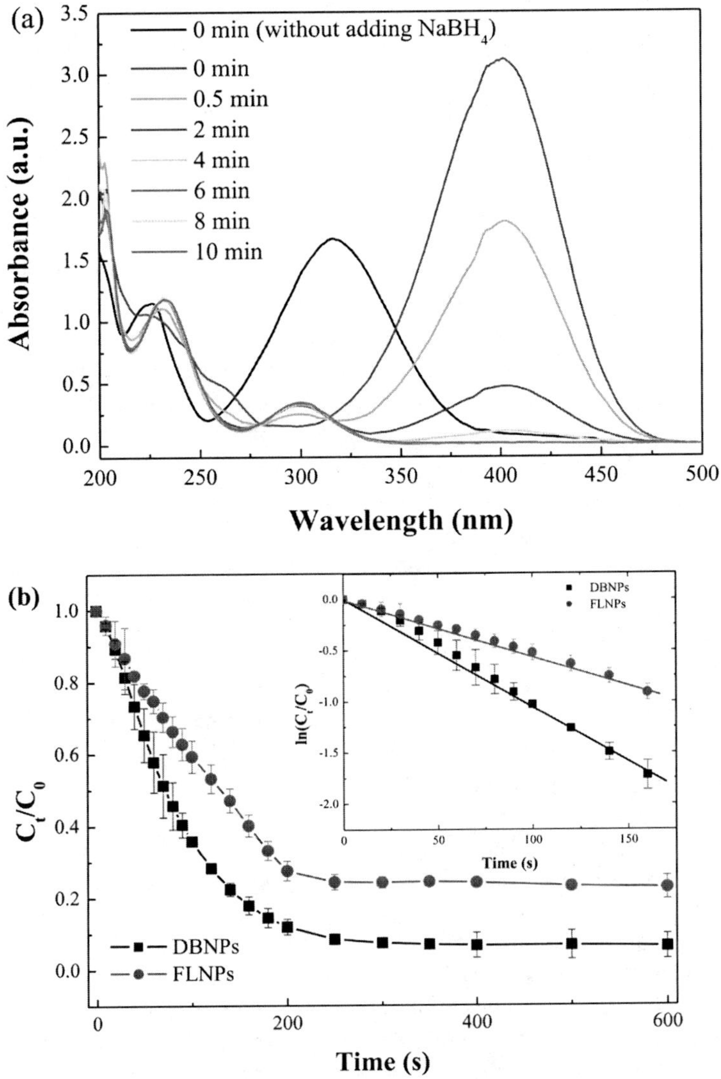

Figure 3. (a) Time-dependent UV-Vis spectral changes in p-nitrophenol catalyzed by Au-Fe_3O_4 heterostructures and (b) concentration change in nitrophenol compounds (C_t/C_0) in the presence of Au-Fe_3O_4 nanocrystals. Insets in Figures (b) is linear relationship of ln(C_t/C_0) as a function of time for p-nitrophenol. (Reproduced with permission from ref (20). Copyright 2011, J. Phy. Chem. C).

The as-prepared Au-Fe_3O_4 heterostructures show both catalytic and magnetic properties which can be easily recycled by an external magnet after the catalytic reduction. Figure 4a shows the magnetically recyclable reduction of nitrophenols in the presence of dumbbell-like Au-Fe_3O_4 nanocatalysts. The catalysts can be successfully recycled and reused for at least 6 successive cycles of reaction with stable conversion efficiency of around 100%. In addition, the same Au seeds used for synthesis of Au-Fe_3O_4 heterostructures were also employed for reduction of *p*-nitrophenol. The single-component Au NP catalysts were obtained by etching Fe_3O_4 away from the Au-Fe_3O_4 NPs in 0.5M H_2SO_4 solutions (*19*). The catalytic efficiency of *p*-nitrophenol by Au seeds is low when compared with that by dumbbell-like Au-Fe_3O_4 heterostructures, presumably due to the aggregation of Au seeds during etching step (Figure 4b). It is noteworthy that the k_{obs} for *p*-nitrophenol reduction decreased with the increase in recycling times, presumably due to the poison of *p*-aminophenol, the intermediate of *p*-nitropenol reduction, to Au-based nanomaterials.These results clearly demonstrate that the Au-Fe_3O_4 NPs are superior catalysts than Au itself and other supported Au catalysts, presumably attributed to the electronic junction effect of Au and Fe_3O_4 NPs (*18–21*). This electronic junction effect can also be observed in reduction catalysis of H_2O_2 by Au-Fe_3O_4 dumbbell-like structure (*19*). In addition, the Au NPs in dumbbell-like structure was stable against aggregation during harvest procedure, resulting in the enhanced catalysis of nitrophenol compounds. It is obvious that the presence of Fe_3O_4 NPs makes the dumbbell-like heterostructures a promising bifunctional probe for magnetically recyclable catalytic reduction. When the reduction is complete, the Au-Fe_3O_4 nanocatalysts can be separated easily and rapidly from the solution within 10 s by a magnet and then be re-dispersed into deionized water for the next cycle of catalysis (Figure 5).

In this study, $NaBH_4$ is used as reducing agent for the reduction of *p*-nitrophenol to *p*-aminophenol. The concentration effect of $NaBH_4$ on the reduction of *p*-nitrophenol is first examined to elucidate the reduction power needed for the catalytic reduction of nitrophenol by Au-Fe_3O_4 DBNPs. Figure 6 shows the concentration effect of $NaBH_4$ on p-nitrophenol reduction by Au-Fe_3O_4 DBNPs. The removal efficiency of *p*-nitrophenol by Au-Fe_3O_4 DBNPs increases upon increasing $NaBH_4$ concentrations. A nearly complete removal efficiency of p-nitrophenol is observed within 200 s when 3.3 mM $NaBH_4$ is added. The removal efficiency and rate of *p*-nitrophenol can be enhanced at high $NaBH_4$ concentration and only 100 s is needed to reach a nearly complete reduction of *p*-nitrophenol when the concentration of $NaBH_4$ is higher than 8.3 mM. The reduction of p-nitrophenol by Au-Fe_3O_4 DBNPs in the presence of various concentrations of $NaBH_4$ followed the pseudo-first-order rate kinetics.

The effect of initial *p*-nitrophenol concentration on the catalytic reduction efficiency and rate of *p*-nitrophenol by Au-Fe_3O_4 DBNPs is further examined. Figure 7 shows the removal efficiency of *p*-nitrophenol at various initial concentrations. The removal efficiencies of 4-NP by Au-Fe_3O_4 DBNPs are all higher than 99 % after 240 s of reaction when the initial concentration of *p*-nitrophenol are in the range 0.036 and 0.17 mM. The reduction of various *p*-nitrophenol also follows the pseudo-first-order kinetics and the k_{obs} for *p*-nitrophenol reduction decreases from 1.46 min^{-1} at 0.036 mM to 0.3 min^{-1} at 0.17 mM.

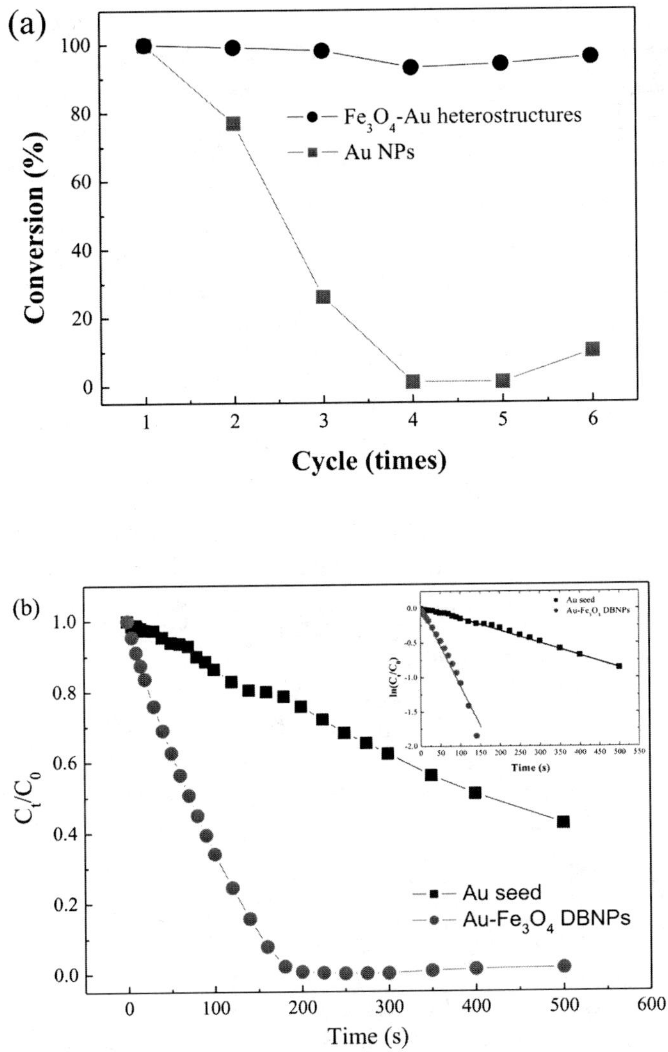

Figure 4. (a) Conversion efficiency of p-nitrophenol in 6 successive cycles of reduction by Au-Fe_3O_4 *and citrate-stabilized Au nanocatalysts. (b) The concentration change of p-nitrophenol (*C_t/C_0*) in the presence of Au seeds and Au-*Fe_3O_4 *DBNPs. Insets in Figures (b) is linear relationship of ln(*C_t/C_0*) as a function of time for Au seeds and Au-*Fe_3O_4 *DBNPs. (Reproduced with permission from ref (20). Copyright 2011, J. Phy. Chem. C).*

Figure 5. Pictures of catalytic reaction of p-nitrophenol by dumbbell-like Au-Fe_3O_4 heterostructures. (a) Yellow color of p-nitrophenol before the catalytic reaction, (b) transparent solution after the catalytic reduction, and (c) the recovery of Au-F_3O_4 heterostructures by a magnet. (Reproduced with permission from ref (20). Copyright 2011, J. Phy. Chem. C).

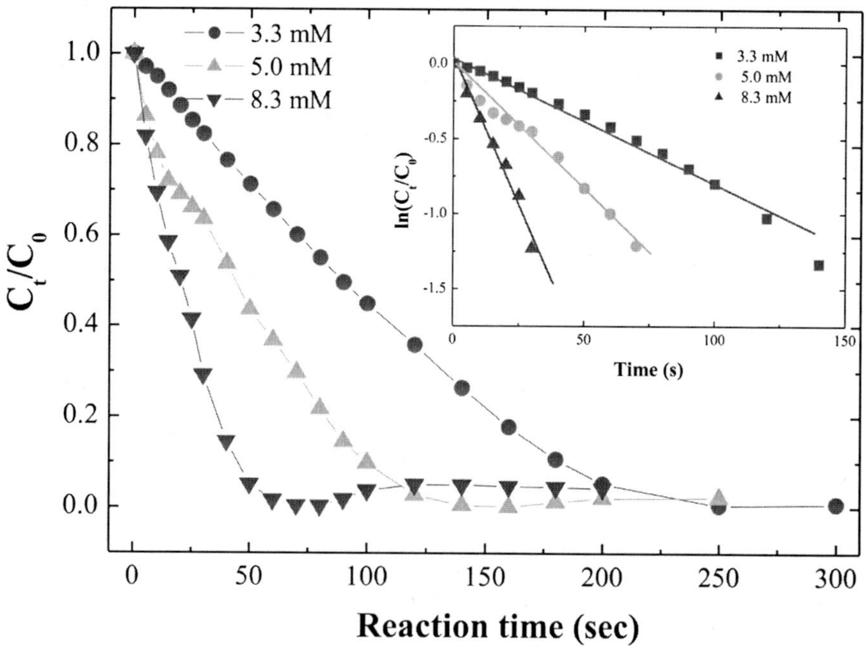

Figure 6. Concentration effect of $NaBH_4$ on the reduction of p-nitrophenol by Au-Fe_3O_4 DBNPs. Insets is linear relationship of $\ln(C_t/C_0)$ as a function of time under different concentration of $NaBH_4$.

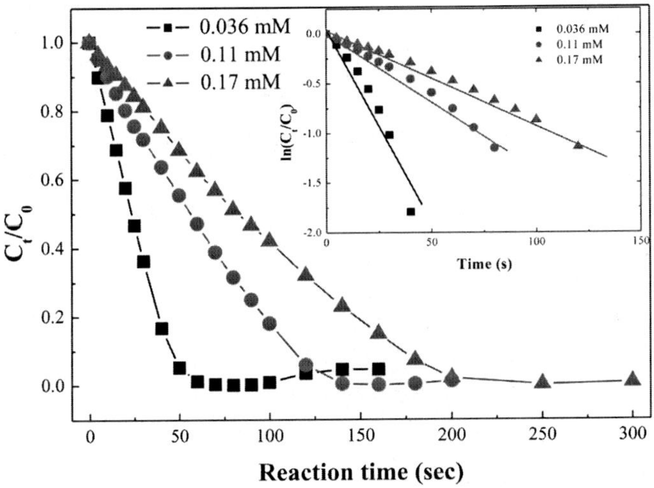

Figure 7. The change in concentration of p-nitrophenol with time at various initial p-nitrophenol concentrations ranging between 0.036 and 0.17 mM. Insets is linear relationship of $\ln(C_t/C_0)$ as a function of time under different concentration of p-nitrophenol.

Conclusions

In this study, we have demonstrated that the heterostructured Au-Fe_3O_4 nanocatalysts have excellent bifunctional characteristics for reusability and catalytic reduction. The size and morphology of Au-Fe_3O_4 nanocatalysts are highly dependent on the size of Au seeds. The catalytic performance of both dumbbell- and flower-like Au-Fe_3O_4 NPs is excellent for *p*-nitrophenol reduction in the presence of $NaBH_4$. In addition, the catalytic efficiency of dumbbell-like Au-Fe_3O_4 NPs is higher than that of flower-like heterostructures because of the high surface coverage of Au surface by Fe_3O_4 nanocrystals in flower-like heterostructures. The dumbbell-like nanoparticles also show good separability and reusability in 6 successive cycles of reduction. In addition, the catalytic efficiency and rate of *p*-nitrophenol is highly dependent on the initial concentration of *p*-nitrophenol and $NaBH_4$. The rate constant increases upon increasing $NaBH_4$ concentrations. In addition, the initial rate constant increases upon increasing initial *p*-nitrophenol concentrations. Results obtained in this study indicate that the interfacial interaction between Au and Fe_3O_4 in heterostructures can enhance the catalytic efficiency, and make them ideal platform to study the various heterogeneous catalytic processes.

Acknowledgments

The authors thank the National Science Council, Taiwan, for financial support under Contract No. NSC 99-2113-M-007-007-MY3.

References

1. Sau, T. K.; Rogach, A. L.; Jackel, F.; Klar, T. A.; Feldmann, J. *Adv. Mater.* **2010**, *22*, 1805–1825.
2. Jain, P. K.; Huang, X. H.; El-Sayed, I. H.; El-Sayed, M. A. *Acc. Chem. Res.* **2008**, *41*, 1578–1586.
3. Haruta, M.; Yamada, N.; Kobayashi, T.; Iijima, S. *J. Catal.* **1989**, *115*, 301–309.
4. Orlov, A.; Jefferson, D. A.; Macleod, N.; Lambert, R. M. *Catal. Lett.* **2004**, *92*, 41–47.
5. Praharaj, S.; Nath, S.; Ghosh, S. K.; Kundu, S.; Pal, T. *Langmuir* **2004**, *20*, 9889–9892.
6. Zeng, J.; Zhang, Q.; Chen, J.; Xia, Y. *Nano Lett.* **2010**, *10*, 30–35.
7. Gong, J. L.; Mullins, C. B. *Acc. Chem. Res.* **2009**, *42*, 1063–1073.
8. Hutchings, G. J. *Top. Catal.* **2008**, *48*, 55–59.
9. Xu, C.; Xie, J.; Ho, D.; Wang, C.; Kohler, N.; Walsh, E. G.; Morgan, J. R.; Chin, Y. E.; Sun, S. *Angew. Chem., Int. Ed.* **2008**, *47*, 173–176.
10. Chen, M. S.; Goodman, D. W. *Science* **2004**, *306*, 252–255.
11. Comotti, M.; Li, W. C.; Spliethoff, B.; Schuth, F. *J. Am. Chem. Soc.* **2006**, *128*, 917–924.
12. Tang, S. C.; Vongehr, S.; Meng, X. K. *J. Mater. Chem.* **2010**, *20*, 5436–5445.
13. Lim, C. W.; Lee, I. S. *Nano Today* **2010**, *5*, 412–434.
14. Li, Y. Q.; Zhang, G.; Nurmikko, A. V.; Sun, S. H. *Nano Lett.* **2005**, *5*, 1689–1692.
15. Buonsanti, R.; Grillo, V.; Carlino, E.; Giannini, C.; Gozzo, F.; Garcia-Hernandez, M.; Garcia, M. A.; Cingolani, R.; Cozzoli, P. D. *J. Am. Chem. Soc.* **2010**, *132*, 2437–2464.
16. Mamidala, V.; Xing, G. C.; Ji, W. *J. Phys. Chem. C* **2010**, *114*, 22466–22471.
17. Seh, Z. W.; Liu, S. H.; Zhang, S. Y.; Bharathi, M. S.; Ramanarayan, H.; Low, M.; Shah, K. W.; Zhang, Y. W.; Han, M. Y. *Angew Chem. Int. Ed.* **2011**, *50*, 10140–10143.
18. Wang, C.; Yin, H. F.; Dai, S.; Sun, S. H. *Chem. Mater.* **2010**, *22*, 3277.
19. Lee, Y. M.; Garcia, M. A.; Huls, N. A. F.; Sun, S. H. *Angew. Chem., Int. Ed.* **2010**, *49*, 1271.
20. Lin, F. H.; Doong, R. A. *J. Phys. Chem. C* **2011**, *115*, 6591–6598.
21. Lopes, G.; Vargas, J. M.; Sharma, S. K.; Be′ron, F.; Pirota, K. R.; Knobel, M.; Rettori, C.; Zysler, R. D. *J. Phys. Chem. C* **2010**, *114*, 10148–10152.
22. Costi, R.; Saunders, A. E.; Banin, U. *Angew. Chem., Int. Ed.* **2010**, *49*, 4878–4897.
23. Frey, N. A.; Phan, M. H.; Srikanth, H.; Srinath, S.; Wang, C.; Sun, S. *J. Appl. Phys.* **2009**, *105*, 07B502.
24. Xu, C. J.; Wang, B. D.; Sun, S. H. *J. Am. Chem. Soc.* **2009**, *131*, 4216–4217.

25. Jiang, J.; Gu, H. W.; Shao, H. L.; Devlin, E.; Papaefthymiou, G. C.; Ying, J. Y. *Adv. Mater.* **2008**, *20*, 4403–4407.
26. Choi, J. S.; Jun, Y. W.; Yeon, S. I.; Kim, H. C.; Shin, J. S.; Cheon, J. *J. Am. Chem. Soc.* **2006**, *128*, 15982–15983.
27. Wei, Y.; Klajn, R.; Pinchuk, A. O.; Grzybowski, B. A. *Small* **2008**, *4*, 1635.
28. Wang, C.; Xu, C.; Zeng, H.; Sun, S. *Adv. Mater.* **2009**, *21*, 3045–3052.
29. Cornell, M. R.; Schwertmann, U.*The Iron Oxides*; VCH: New York, 1996; p 117.
30. Park, J.; Lee, E.; Hwang, N. M.; Kang, M.; Kim, S. C.; Hwang, Y.; Park, J. G.; Noh, H. J.; Kim, J. Y.; Park, J. H.; Hyeon, T. H. *Angew. Chem., Int. Ed.* **2005**, *44*, 2872–2877.
31. Zhen, G.; Muir, B. W.; Moffat, B. A.; Harbour, P.; Murray, K. S.; Moubaraki, B.; Suzuki, K.; Madsen, I.; Agron-Olshina, N.; Waddington, L.; Mulvaney, P.; Hartley, P. G. *J. Phys. Chem. C* **2011**, *115*, 327–334.
32. Wu, Y. P.; Zhang, T.; Zheng, Z. H.; Ding, X. B.; Peng, Y. X. *Mater. Res. Bull.* **2010**, *45*, 513–517.
33. Dotzauer, D. M.; Dai, J. H.; Sun, L.; Bruening, M. L. *Nano Lett.* **2006**, *6*, 2268–2272.
34. Huang, J. F.; Vongehr, S.; Tang, S. C.; Lu, H. M.; Shen, J. C.; Meng, X. K. *Langmuir* **2009**, *25*, 11890–11896.
35. Pradhan, N.; Pal, A.; Pal, T. *Langmuir* **2001**, *17*, 1800–1802.
36. Deng, Y. H.; Cai, Y.; Sun, Z. K.; Liu, J.; Liu, C.; Wei, J.; Li, W.; Liu, C.; Wang, Y.; Zhao, D. Y. *J. Am. Chem. Soc.* **2010**, *132*, 8466–8473.
37. Chang, Y. C.; Chen, D. H. *J. Hazard. Mater.* **2009**, *165*, 664–669.

Chapter 17

Stabilization of Fe^0 Nanoparticles with Silica for Enhanced Transport and Remediation of Hexavalent Chromium in Groundwater

Yongchao Li,[1] Zongming Xiu,[2] Tielong Li,[*,1] and Zhaohui Jin[1]

[1]College of Environmental Science and Engineering, Nankai University, Tianjin 300071, P.R. China
[2]Department of Civil and Environmental Engineering, Rice University, Houston, Texas 77005, U.S.A.
***E-mail: litielong@nankai.edu.cn. Phone: 86-22-2350-4302.**

Two types of silica supported/coated Fe nanoparticles were prepared, characterized and studied for the reactivity of Cr (VI) removal. The silica fume supported-Fe nanoparticles (SF-Fe) were prepared using commercial silica fume as a support, and the mono-dispersed Fe nanoparticles ($Fe@SiO_2$) coated with SiO_2 shell was fabricated using a facile one-step method developed by aqueous reduction method combined with modified Stöber method. In this reaction, borohydride was acted not only as a reductant for iron salt but also a catalyst for hydrolysis and polycondensation reaction of tetraethylorthosilicate (TEOS). The feasibility of using SF-Fe and $Fe@SiO_2$ for reductive immobilization of Cr (VI) was investigated through batch reduction tests. Compared with unsupported Fe, SF-Fe and $Fe@SiO_2$ were significantly more active in Cr (VI) removal. The highest removal capacity of $Fe@SiO_2$ can reach 467 mg Cr/g Fe at an initial Cr (VI) concentration of 70 mg/L under pH 6.0 ± 0.1. Silica was also found to inhibit the formation of Fe (III)/Cr (III) precipitation on Fe nanoparticles' surface, which enhanced the deactivation resistance of iron. Column tests showed that the SF-Fe could be readily transported in model soil, indicating the enhanced mobility by silica fume. Overall, the silica fume support /SiO_2 coating may improve the stability, reactivity and mobility of

Fe^0 nanoparticles, which has important implication for heavy metal contaminated sites remediation.

Keywords: Silica fume; Core-shell; Fe nanoparticles; Stability; Cr (VI) removal; Transport

Introduction

Chromium is widely detected in surface water and groundwater at sites associated with industrial and military activities (*1*). In natural waters, chromium exists in both valence states, Cr (VI) and Cr (III). Cr (VI) anions are highly soluble in aquatic systems and are severe contaminants to environment due to their carcinogenic, mutagenic, and teratogenic features in biological systems while Cr (III) species are relatively stable and have low solubility and mobility in soils and aquifers (*2*). Traditionally, Cr (VI) is removed from water through reduction of Cr (VI) to Cr (III) using a reducing agent (*3*). As a reductive material, nano-scale zero-valent iron particles (nZVI) are extensively used for removing all kinds of contaminants including Cr (VI) (*4–6*). Due to its larger specific surface area and more active sites, the reactivity of Fe nanoparticle has been significantly improved compared to micro-sized Fe (*7*). Compared to the traditional passive processes such as the "funnel and gate" or permeable reactive barrier (PRB) processes, the in situ injection technology offers a number of key advantages (*8*). For example, Fe nanoparticles (Fe NPs) can proactively attack contaminant plumes in the source zone and potentially reduce the remediation cost and time substantially. However, this promising technology has been held back by the key technical barrier that the nanoparticles tend to agglomerate and grow to micron scale or larger, thereby rapidly losing their soil mobility and chemical reactivity (*9*, *10*).

A stabilizer can enhance dispersion (or reduce agglomeration) of nanoparticles through (a) electrostatic repulsion and (b) steric hindrance. Therefore the agglomeration of magnetic metal nanoparticles was reduced (*11*). In addition, to weaken the physical interactions, encapsulating nanoparticles with select stabilizers can also passivate the highly reactive surface from reacting with the surrounding media such as dissolved oxygen (DO) and water. Therefore, several stabilizers such as resin (*12*), carbon (*13*), poly acrylic acid (*14*), starch (*15*) and sodium carboxymethyl cellulose (*16*) have been commonly used as dispersants for preparing Fe NPs in aqueous media.

However, a good support should be cheap, widely available and able to disperse the metals to form feasible particle size. At the same time, the support must be suitable to stimulate the reaction on its surface. Silica fume is generated during silicon metal production as very fine dust of silica from a blast furnace and historically considered a waste product. It can be successfully used to improve the resistance of concrete against chloride penetration (*17*). However, no studies have been reported on silica fume in environmental remediation. The silica fume is chosen as a support because it is cheap and easy to obtain, and because it also can provide a high surface area upon which nZVI particles can be dispersed. Furthermore, silica surface had been shown to strongly bind Fe (III) and Cr (III) via surface complexation (*18*, *19*) which may be important for Cr (VI) removal. Therefore, whether this commercial available silica fume would enhance the transport of nZVI in the subsurface environment is worth of study.

Recently, the deposition of friendly SiO_2 shells on nanoparticles are a class of materials widely used in many fields of colloid and materials science. Previous investigations demonstrated that SiO_2 shell can prevent agglomeration and oxidation of the nanoparticles including Fe_3O_4, and Fe (*20*, *21*). Also, nanoporous SiO_2 shell facilitates the mass transfer between Fe NPs and pollutant. However, such metal/inorganic type core-shell nanostructures are often generated by using two sequential synthetic methods (*20*, *21*). First, the nanoparticles are prepared and transferred to an ethanol/ammonia mixture. Then under continuous stirring, tetraethylorthosilicate (TEOS) is slowly added to this dispersion, in several hours later, silica is formed on the surface of nanoparticles through hydrolysis and condensation of TEOS. However, pure metal particles are considered not to be suitable for direct SiO_2 coating due to their low chemical affinity to SiO_2 (*22*). Amino-terminated silane coupling agents (*23*), gelatin (*20*), and sodium dodecylsulfate (*24*) are often used as the primer to make the Fe surface vitreophilic for subsequent SiO_2 coating. Therefore, it is significant to develop a simple method to coat Fe NPs with well-defined SiO_2 shells.

This work aimed to investigate the feasibility of using silica fume as a novel support for preparing physically stable and chemically reactive Fe NPs and one-pot synthesis of Fe@SiO_2 for the removal of Cr (VI). The specific objectives were to (1) prepare and characterize the SF-Fe and Fe@SiO_2, (2) investigate the effect of silica fume support and SiO_2 shell on Cr (VI) reduction by Fe NPs, (3) investigate the preliminarily test the mobility of the SF-Fe in soil.

Materials and Methods

Tetraethylorthosilicate (TEOS) and potassium borohydride (KBH_4) were purchased from Guangfu Fine Chemical Institute, Tianjin, China. Potassium dichromate ($K_2Cr_2O_7$) was obtained from Fuchen Chemical Reagent Factory, Tianjin, China. Anhydrous ethanol (CH_3CH_2OH) and anhydrous ferric chloride ($FeCl_3$) were purchased from First Chemical Reagent Factory, Tianjin, China. 10-phenanthroline ($C_{12}H_8N_2 \cdot H_2O$) was obtained from North Tianyi Chemical Reagent Factory, Tianjin, China. Silica fume (60-180nm, 85% SiO_2) chosen as the support was supplied by Guangyuan Ferroalloy co., ltd, Inner Mongolia, China. All the chemicals used in this research were analytical grade or higher and were used as received.

Preparation of SF-Fe

HCl (0.0045 M) was added into 13 g/L silica fume suspension while stirring. After 2 hours, 50 mL serum was withdrawn and centrifuged. In a typical SF-Fe synthesis, 0.1453 g of $FeCl_3$ was dissolved in 50 mL of 30% technical grade ethanol, 70% deionized water (v/v). Support material (centrifuged silica fume) was added to this mixed solution while stirring. KBH_4 (1.8 g) solution was slowly added to the mixture and ferric iron was reduced according to the following reaction:

$$4Fe^{3+} + 3BH_4 + 9H_2O \rightarrow 4Fe^0_{(S)} + 3H_2BO_3 + 12H^+_{(aq)} + 6H_{2(g)} \quad (1)$$

After addition of all of the KBH_4, the mixture was stirred for 30 min. The resulted SF-Fe was separated with magnet, and then washed twice with ethanol, effectively substituting ethanol for the water in the mixture. The whole process was carried out in a nitrogen atmosphere. The unsupported nano iron was prepared in similar procedures as described before expect adding support.

Preparation of Fe@SiO$_2$

The Fe@SiO_2 nanocomposite was prepared by a KBH_4 reduction method combined with a modified Stöber method. Briefly, 0.0436g $FeCl_3$ was first dissolved 30 mL 70% ethanol solution. To this solution, the predetermined amount of TEOS was added for the synthesis of SiO_2 shell. The resulting solution was settled for 0.5 h under stirring. Next, 5 mL KBH_4 aqueous solution with precisely adjusted injection speed were added to the mixture with vigorous stirring. After 1 hour at room temperature, the resulting Fe@SiO_2 were separated by a magnet and then washed repeatedly with deionized water and ethanol to remove nonmagnetic by-products. In the general preparation of core-shell nanoparticles using the Stöber method, alcohol type, volume ratio of ethanol/water, amount of ammonia aqueous and TEOS were crucial parameters. In this research, the KBH_4 was acted as the ammonia because the pH of reactant increased with the addition of KBH_4. Therefore, in order to obtain mono-dispersed core-shell nanospheres and to avoid homogeneous nucleation of silica in bulk solution, the effect of TEOS amount and KBH_4 injection speed on the synthesis of Fe@SiO_2 were studied. The preparation conditions and properties of the samples are summarized in Table 1. Pure SiO_2 without Fe NPs was also prepared in a similar procedure as described above except addition of the $FeCl_3$.

Characterization of SF-Fe and Fe@SiO$_2$

Particle size, morphology, and crystal structure were characterized using transmission electron microscopy (TEM, JEOL JEM-100CX) and X-ray diffraction (XRD, PhilipsD/Max-2500). The binding energies were measured by a Kratos Axis Ultra DLD X-ray photoelectron spectroscope (XPS).

Table 1. Characterization of Fe NPs Coated with SiO_2 Shell

Sample	*TEOS amount (mL)*	*KBH_4 injection speed (mL/min)*	*KBH_4 amount (g)*	*Average Fe size (nm)*	*Average shell thickness (nm)*	*Morphology of Fe@SiO_2*
1	0.05	5	0.145	27	3	Single Fe core with thin shell
2	0.1	5	0.145	25	9	Single Fe core
3	0.2	5	0.145	23	18	Single Fe core with thick shell
4	0.1	2.5	0.145	26	10	Assembly consisted of several Fe cores
5	0.1	10	0.145	22	6	Single SiO_2 without Fe core occurred

Batch Experiments

Batch experiments for the reduction of Cr (VI) in water were carried out in polytetrafluoro ethylene (PTFE)-sealed serum bottles covered with aluminum foil at atmospheric pressure. To each bottle, 100 mL of an aqueous Cr (VI) solution in water and a certain amount of Fe nanoparticles were added. All reactors were agitated on a shaker at a rate of 220 rpm during the experiment. At timed intervals, samples were taken by a 1 mL-syringe, filtered through a filter (0.22 μm), and tested for Cr (VI) concentration using the diphenylcarbohydrazide method. To assure data quality, all experiments were performed in duplicate.

Transport Behavior of SF-Fe in Soil

As a preliminary test of the mobility of the SF-Fe and unsupported Fe in soil, simple column breakthrough and elution tests were carried out with quartz sand (0.8-1.0 mm diameter) packed in a 20 mL glass column. The porosity of the packing was 0.35. After a 10 mL of 2.5 g/L Fe suspension was fed, the top of each column was flushed with 6 pore volumes of DI water at a flow rate of 14.0 mL/min. The effluent was collected at the bottom of the filter bed, and the particles not retained in the filter column were digested with HCl, and analyzed for total iron using 1, 10-phenanthroline spectrophotometry.

Results and Discussion

Characterization of SF-Fe

Figure 1a showed that approximately spherical Fe nanoparticles (black one) with size ranging from 20 nm to 110 nm were dispersed well on the surface of silica fume (gray one). Figure 1b represented a histogram of SF-Fe particle size obtained by analysis of TEM micrographs of 196 composite particles. It was shown that the SF-Fe particle size range was on the order of 0.15-0.45 μm, which is the optimal particle size for transport through soil (*25*). In contrast, in the absence of a support, many research indicated that Fe nanoparticles sizes did not appear as discrete nano-scale particles, but formed chain-like aggregates. The size of some denser flocs could be greater than 10 μm (*26*).

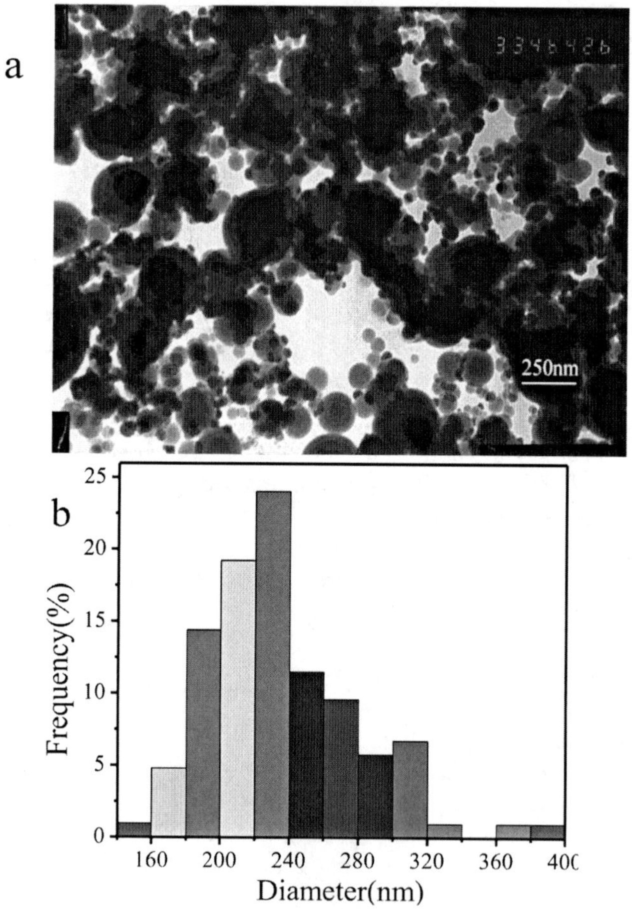

Figure 1. (a) TEM image of SF-Fe composite particle. (b) Particle size distribution of SF-Fe based on TEM analysis.

Moreover, the SF-Fe exhibited superior stability against aggregation in water. As seen in Figure 2d and f, unsupported Fe precipitated and settled at the bottom of tube in 10 min. This was due to the strong *van der waals* attraction associated with the high Hamaker constant (*27*) of the magnetite shell as well as the magnetic attractions (*28*) between the particles. However, SF-Fe suspension was well dispersed and remained as a slightly gray-colored solution for up to 48h (Figure 2g), indicating that the strong *van der waals* attraction that exits between the Fe nanoparticles were overcome by the electrostatic repulsions from the great negatively charged silica fume (-48.1mv, pH 6.5).

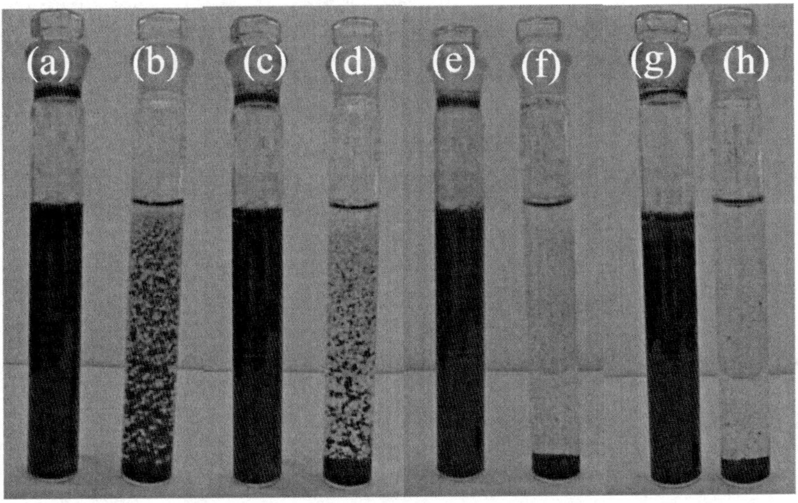

Figure 2. Tubes containing unsupported Fe and SF-Fe at various times: (a) and (b) SF-Fe and unsupported Fe after 1 min, (c) and (d) SF-Fe and unsupported Fe after 5 min, (e) and (f) SF-Fe and unsupported Fe after 10 min, (g) and (h) unsupported Fe and SF-Fe after 2 days.

In addition, the presence of the Fe peak at 45° in the X-ray diffraction data of Figure 3 showed that both unsupported Fe and SF-Fe were in poor crystallinity.

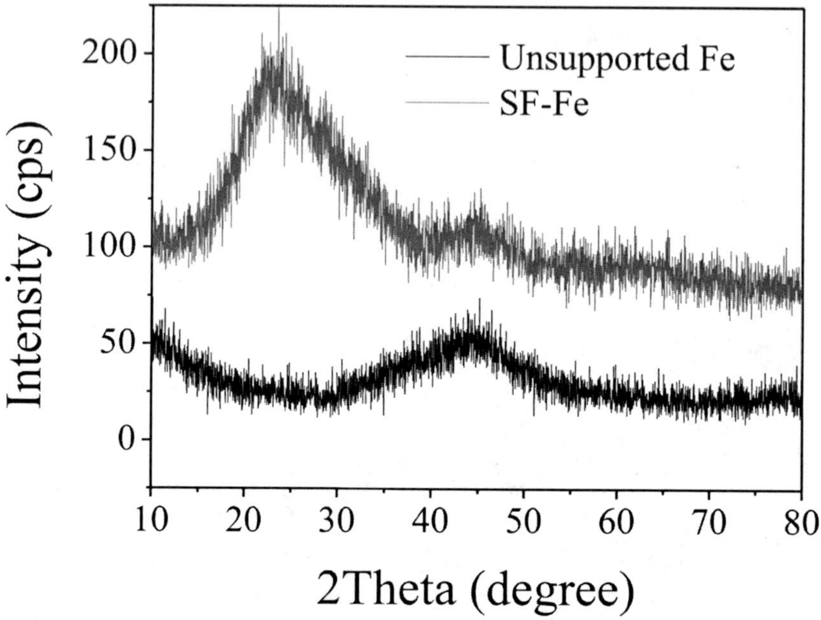

Figure 3. XRD spectra of unsupported Fe and SF-Fe.

Characterization of Fe@SiO_2

XRD Pattern of Fe@SiO_2

Figure 4 shows the XRD patterns of all the prepared nanocomposites. The broad diffraction peaks around 26° was ascribed to amorphous SiO_2 (*24*). The positions of broad diffraction peaks at 44° showed that all the Fe NPs were poorly ordered and amorphous and no other iron oxides was identified. Although the XRD results demonstrated that the SiO_2 already combined with Fe NPs, the composite structure of products was not clear.

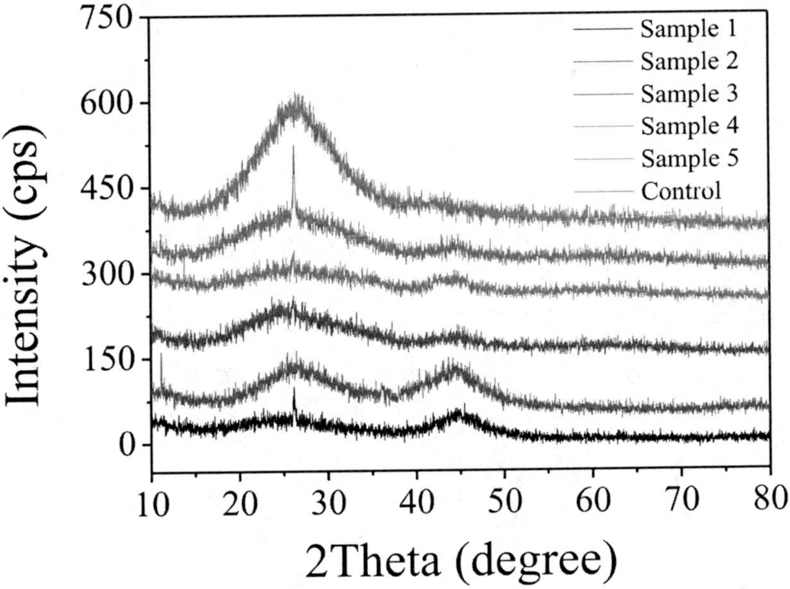

Figure 4. the XRD patterns of all the prepared Fe@SiO_2 nanocomposites.

TEM Image of Fe@SiO_2

TEM analysis of samples prepared under various preparation conditions is presented in Figures 5 a-e. It was evident that all of the synthesized nanoparticles had a clearly distinguished core/shell structure: the dark sphere was attributed to Fe and the grey shell can be attributed to SiO_2. The pores of SiO_2 shells appeared to be small but they were clearly visible. Such results demonstrated the high efficiency of this preparation method despite its simplicity, as compared with those reported by Wang and Harrison (*20*), Yang et al (*22*, *23*), and Yuan et al. (*24*). Figure 5f represents a distribution map of originally Fe particle size obtained by analysis of multiple TEM micrographs of all the composite particles. It was clearly seen that the particle size range was on the order of 10-60 nm, and most of particles size concentrated in 20-30 nm. The calculated average diameters were all about 25 nm as shown in Table 1. Compared with chain-like aggregates of uncoated Fe NPs, both the small particle size and the mono-dispersity of Fe core in Fe@SiO_2 indicated that SiO_2 shell suppressed the growth of the Fe NPs and maintained a higher surface area of the particles. Although the KBH_4 injection speed and TEOS concentration had little effect on the original Fe particles size, they were important for the morphology of the nanocomposites.

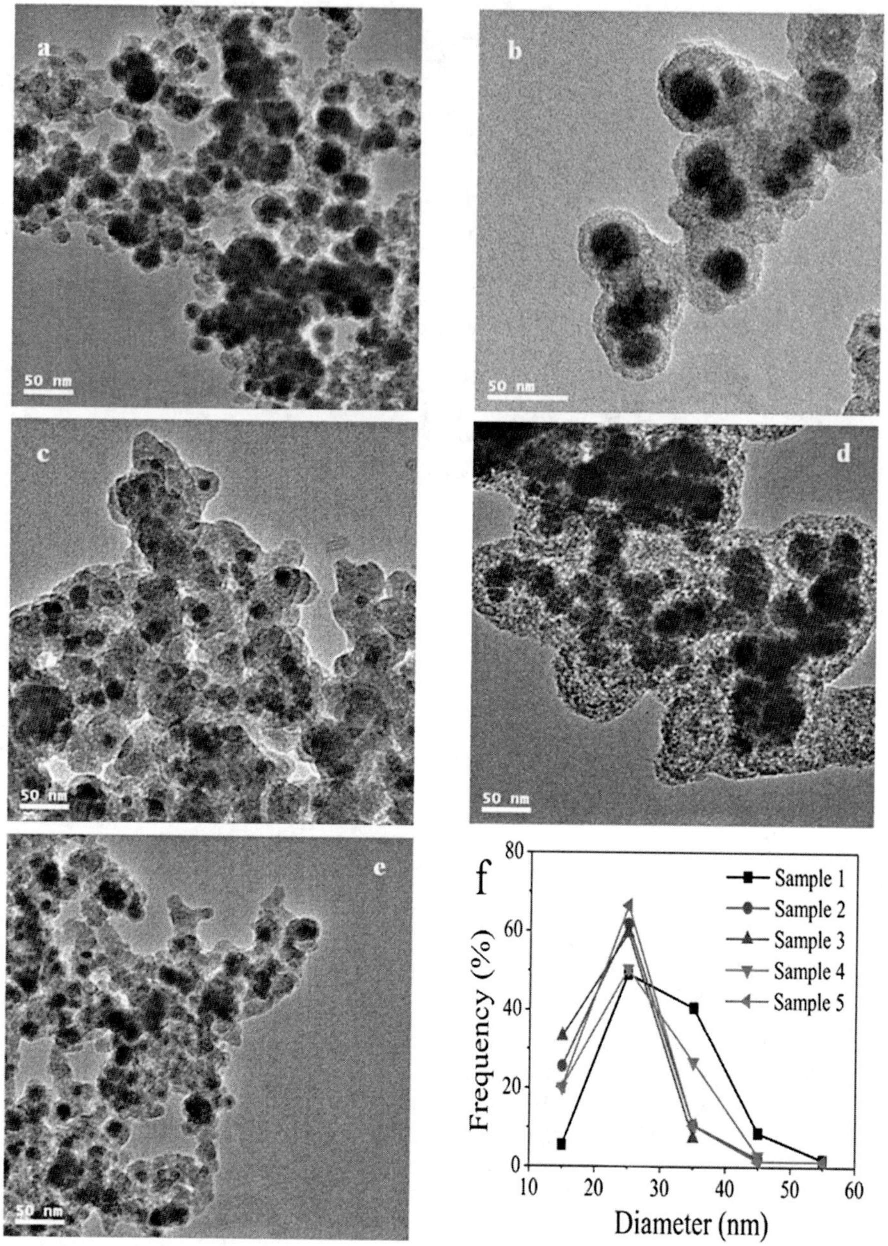

Figure 5. TEM images of Fe NPs encapsulated in SiO_2 beads: (a) sample 1, (b) sample 2, (c) sample 3, (c) sample 4, (d) sample 5 as described in Table 1. (e) Size distribution of Fe NPs in Fe@SiO_2 synthesized at various conditions.

To study the influence of the TEOS amount on the formation of Fe@SiO_2, different amounts of TEOS including 0.05, 0.1 and 0.2 mL were added and the other parameters were kept constant, as shown in Table 1. Figure 5a shows that Fe NPs (sample 1 in Table 1) coated with a 3 nm SiO_2 shell could be produced by adding a low dose (0.05 mL) of TEOS. TEM images shown in Figure 5b and c revealed that the thickness of SiO_2 shell on the surface of Fe NPs increased gradually with the increasing of TEOS amount. The average thickness of SiO_2 shell increased to about 18 nm when 0.2 mL TEOS was added. Furthermore, with the increase of the thickness of SiO_2 shell, Fe NPs became more mono-dispersed as a result of the reduction in the relatively size distribution. The same phenomenon had also been reported by Yang et al. (*22*).

To study the influence of the KBH_4 injection speed on the formation of Fe@SiO_2, different KBH_4 injection speed including 2.5, 5 and 10 mL/min were used and the amount of other reactants were kept constant, as shown in Table 1. As can be seen in Figure 5d, the SiO_2 shell had a higher porosity and consisted of several aggregated Fe NPs (sample 4) because the SiO_2 shell can not be formed timely at a KBH_4 injection speed of 2.5 mL/min. When KBH_4 injection speed increased to 5.0 mL/min, the agglomeration of Fe NPs was prevented and resulted in single Fe NP encapsulated into SiO_2 shell. However, in the case of higher KBH_4 injection speed (10 mL/min for sample 5), bare SiO_2 beads without Fe NPs were simultaneously present, as shown in Figure 5e. This was attributed to rapid condensation/deposition of TEOS during coating process leading to SiO_2 nuclei formed quickly. Hence, to produce single Fe NP coated with a uniform SiO_2 shell, the KBH_4 injection speed should be controlled to avoid the formation of SiO_2 nuclei and Fe NPs aggregation.

XPS Spectra of Fe@SiO_2

Sample 2 was analyzed by high-resolution XPS. The application of XPS in this work helps to answer which elements are present at the particles surface and what chemical or valence states of these elements are present. Figure 6 presents the whole region scan of the Fe@SiO_2 surface. It can be seen that, the principal elements at the surface were oxygen, silicon, carbon, and boron. Most importantly, there were no characteristic peaks for iron when the binding energy was between 700~740 eV. Silicon and oxygen peaks resulted from the TEOS. Carbon peak was likely due to the carbon dioxide contamination during the sample preparation and transfer. Boron was a residue from the borohydride oxidation. The XPS just analyzed the element composite of the product surface with less than 3 nm, whereas the average SiO_2 shell thickness of sample 2 was about 9 nm. Therefore, no characteristic peaks of iron were found, confirming a complete surface SiO_2 coating of Fe particles.

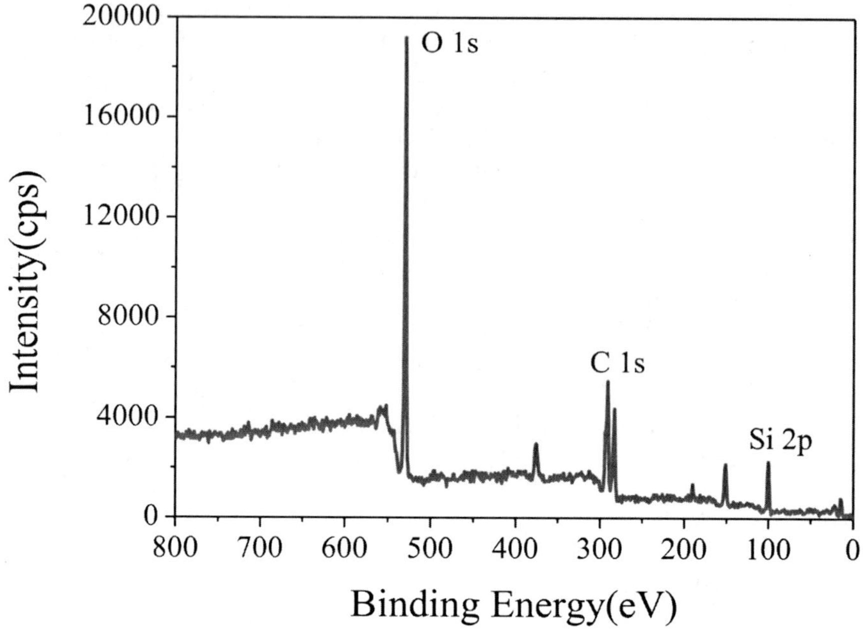

Figure 6. XPS spectrum of fresh Fe@SiO₂.

Reduction and Immobilization of Cr (VI) by Different Fe NPs

Figure 7 compared the reactivity of Cr (VI) reduction using SF-Fe and unsupported Fe. The normalized residual concentration (C/C_0) was used to describe the removal rate. At the Fe nanoparticles doses of 0.4 g/L, 88.00% of Cr (VI) were removed by SF-Fe, however, only 65.45% of Cr (VI) were removed by unsupported Fe in 120 min. Apparently, the presence of silica fume prevented the resulted iron particles from agglomeration, and thus maintained the high surface area and reactivity of the SF-Fe. Whereas, micrometer-size aggregate lead to low surface and reactivity of the unsupported Fe. In addition, silica fume could inhibit the formation of Fe (III)–Cr (III) precipitation on the Fe nanoparticles surface may also have contributed to the enhanced Cr (VI) removal (*18*, *19*). Silica fume without Fe was used as a control and almost no Cr (VI) was removed from the solution, so it was likely that Cr (VI) didn't adsorb to the silica fume.

The reactivity of the Fe@SiO_2 was also investigated by testing the removal of Cr (VI). Figure 8 shows the batch kinetic data during reduction of Cr (VI) by uncoated Fe NPs and Fe@SiO_2 at a Fe dose of 0.15 $g \cdot L^{-1}$ and initial pH of 6.0 ± 0.1. After 180 min reaction, the pH increased to about 10.5 as a result of the release of OH^- (Eq. (2)).

$$Cr_2O_7^{2-} + 2Fe^0 + 7H_2O \rightarrow 2Cr^{3+} + 14OH^- + 2Fe^{3+} \quad (2)$$

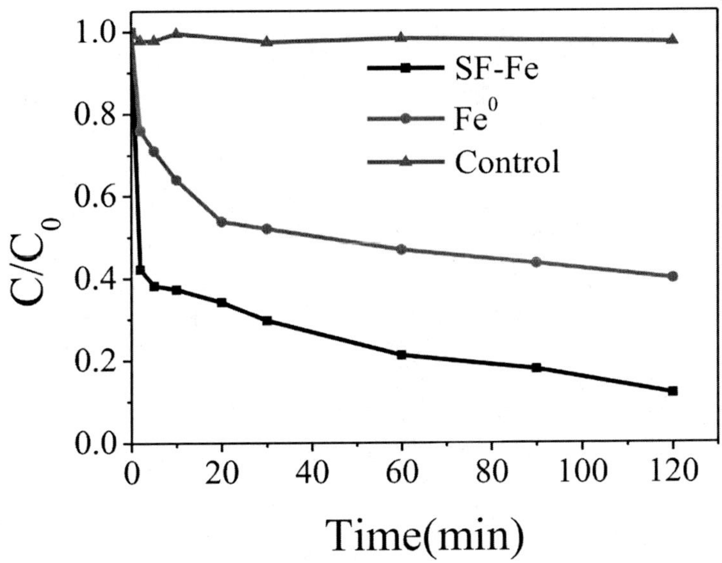

Figure 7. Cr (VI) removal from solution for unsupported Fe and SF-Fe. Initial Cr (VI) concentration: 40 mg/L; Fe dose: 0.4 g/L.

As shown in Figure 8, around 100% and 16.36% of 70 $mg{\cdot}L^{-1}$ Cr (VI) was reduced by Fe@SiO_2 and uncoated Fe NPs in 180 min, respectively. The pure SiO_2 (without Fe NPs) was investigated as a control. The control reactor showed no loss of Cr (VI) during the whole experiment, which proved that Fe NPs were responsible for all the losses of Cr (VI). The presence of SiO_2 shell prevented the growth and agglomeration of the resultant iron particles. Moreover, Cr (VI) ions can readily approach Fe surface since the SiO_2 shell was porous, and thus maintained the high surface area and reactivity of the Fe NPs. Therefore, a significant increase of the Cr (VI) removal capacity by Fe@SiO_2 was obtained. The removal capacity of Cr (VI) ions over Fe@SiO_2 synthesized in this work (466.67 mg Cr/g Fe) was approximately 267 times that of micron-Fe and 1.83 times that of reported CMC-stabilized Fe NPs. On the whole, the main advantage of utilizing SiO_2 shell was not to remove Cr (VI) directly but to prevent the aggregation of Fe NPs and to prevent the product aggregation on Fe NPs surface.

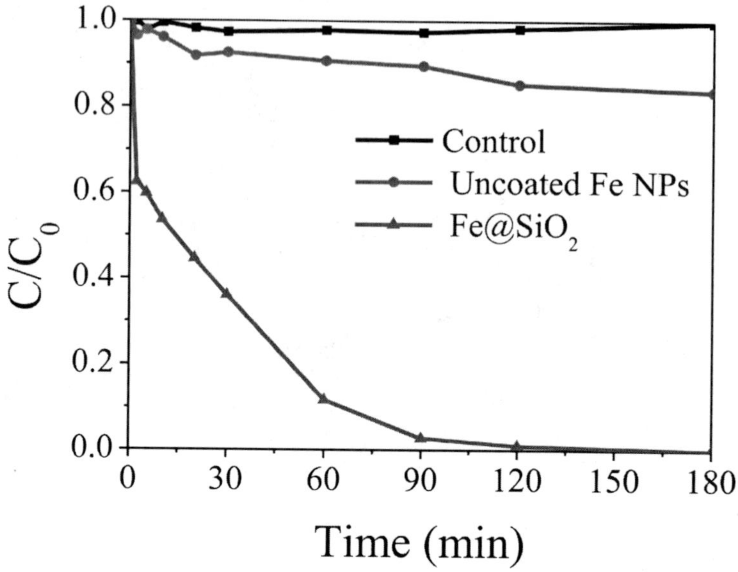

Figure 8. Removal of Cr (VI) using Fe@SiO$_2$ and uncoated Fe NPs. Initial Cr (VI) concentration: 70 mg/L; Fe dose: 0.15 g/L; pH: 6.0±0.1; temperature: 25°C.

Fate of SF-Fe, Fe@SiO$_2$, and Reduced Cr (VI)

SF-Fe and unsupported Fe treated with Cr (VI) were recovered and analyzed by using a Kratos Axis Ultra DLD multi-technique X-ray photoelectron spectroscopy (XPS). According to Eq. (2), Fe was primarily oxidized to Fe (III) and Cr (VI) was reduced to Cr (III). At the end of the experiment, SF-Fe, Fe@SiO$_2$ and unsupported Fe treated with Cr (VI) were recovered and analyzed by XPS. Detailed XPS surveys on the region of Fe2p and Cr2p are presented in Figures 9a, b, c. All the particles exhibited Fe 2p$_{3/2}$ and 2p$_{1/2}$ binding energies of about 710.5 and 723.5 eV, which were indicative for Fe (III) species (*29*). The photoelectron peaks for Cr 2p$_{3/2}$ and 2p$_{1/2}$ centered at 575.5 and 585 eV respectively, indicating that Cr (III) was the predominant chromium species on the surface (*30*). According to the inductively coupled plasma atomic emission spectroscopy (ICP-AES) analysis, after the reduction by Fe@SiO$_2$, dissolved iron and total chromium in water was only 0.0172, 0.01167 mg respectively. This result indicated that most of the iron and chromium formed a precipitate on the Fe@SiO$_2$ surface because silica surface can strongly bind Fe (III) and Cr (III) via surface complexation. Therefore, Fe (III)/Cr (III) precipitation on the SiO$_2$ shell surface occurred.

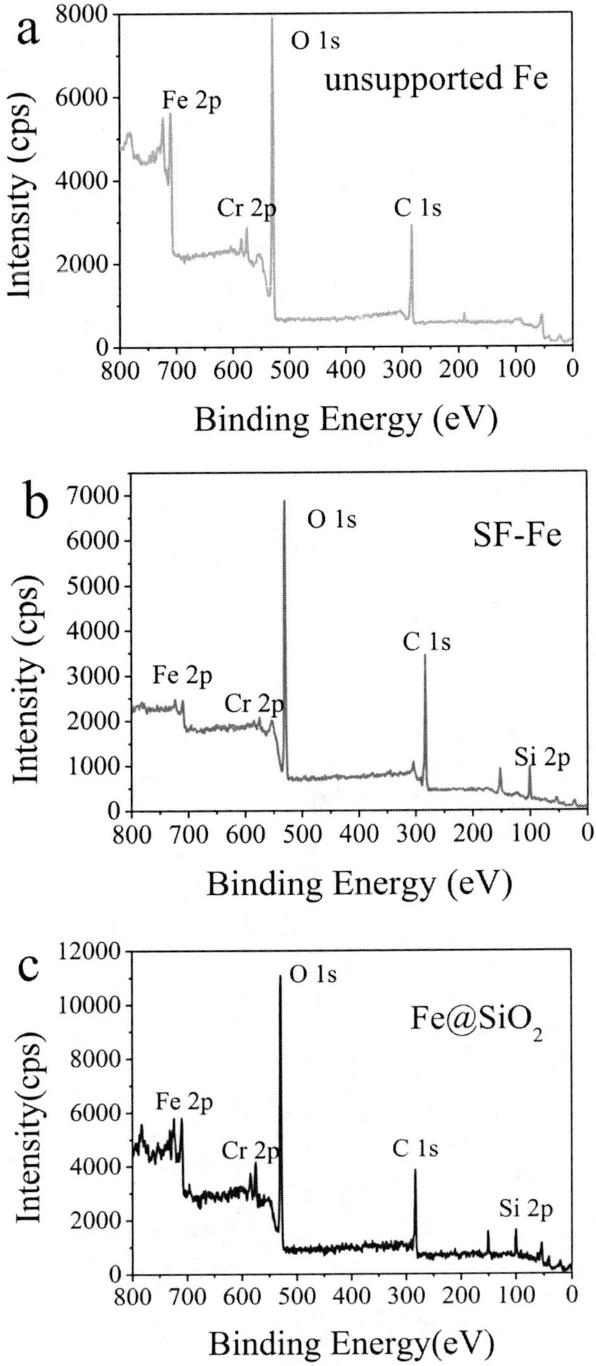

Figure 9. XPS survey spectrum of (a) unsupported Fe, (b) SF-Fe, (c) Fe@SiO₂ after reaction.

Mobility Characteristics

Figure 10a illustrated the elution results in vertical column, with panels (i) indicating results for the SF-Fe and panels (ii) indicating results for the unsupported Fe. The elution results indicated that most of the unsupported Fe was trapped within the first few centimeters of the column, and visible penetration did not exceed the middle of the column. In marked contrast to this poor transportability, SF-Fe reached the column bottom and eluted efficiently with collection of the particles in the beakers as shown in panel (i). Figure 10b showed photographs of the horizontal columns with panel (i) depicting the column containing unsupported Fe after 12PV of water flushing and panel (ii) depicting the column containing SF-Fe after 6PV of water flushing. Agglomerates and large clusters were observed at the column inlet for unsupported Fe, while SF-Fe were more uniformly located in small clusters throughout the column and appeared to have adsorbed to the sand surfaces.

Breakthrough curves indicated that almost 51.50% and 38.29% of the SF-Fe were eluted from the vertical and horizontal sand column respectively under the specified conditions. In contrast, no significant elution of unsupported Fe was observed in both columns. SF-Fe (0.15-0.45 μm) was in the optical size for transport could be reason for this phenomenon, but it was also possible that the stabilizer silica fume improved elution efficiency by decreasing Fe adhesion to the sand grains. Whereas for unsupported Fe, it appeared in the form of dendritic flocs rather than discrete particles that may result in pore plugging and thus retard filtration. Both vertical and horizontal column experiments indicated potential of the SF-Fe, however it didn't necessarily indicate applicability to field conditions. A variety of groundwater and soil conditions had to be considered before field application became a realistic possibility.

a

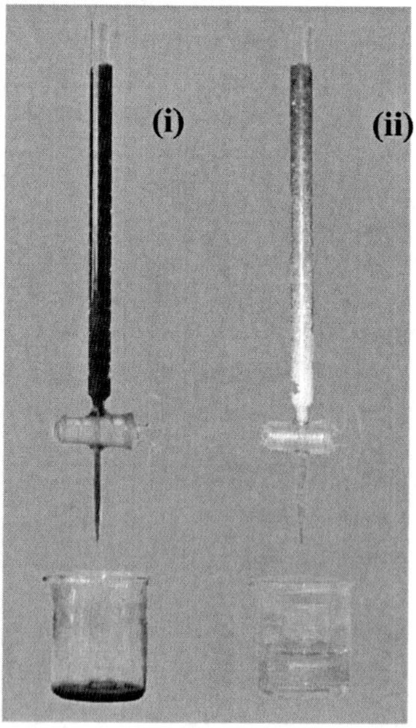

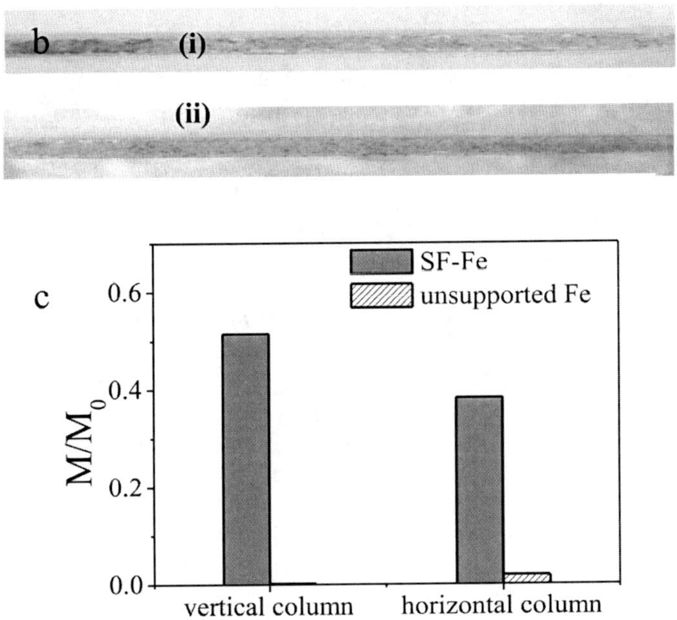

Figure 10. (a) Elution characteristics of SF-Fe and unsupported Fe in vertical columns. Clear transport of SF-Fe was seen in panel (i). Panel (ii) illustrated accumulation of unsupported Fe at the top of column. (b) Elution characteristics of SF-Fe and unsupported Fe in horizontal columns. Panel (i) showed unsupported Fe retained in the column after flushing with water for 12PV. Panel (ii) showed the SF-Fe has moved through the horizontal column partially after 6 PV of flushing. (c) Elution profiles of SF-Fe and unsupported Fe in vertical and horizontal columns. M/M_0 represents the fraction of particles that are eluted.

Conclusions

Commercial silica fume was used as a support for synthesizing Fe nanoparticles. The synthesized Fe nanoparticles were evenly disposed on the surface of silica fume. Reserving Fe NPs in the SiO_2 shell may reduce the toxicity of nanoparticles on the indigenous dechlorinating bacteria community (*31*, *32*) and reduce the risk of exposure to human and animals. Moreover, the SiO_2 shell can be an ideal reaction media for producing non-agglomerated and mono-dispersed Fe nanoparticles. We developed a simple one-step method to obtain core-shell Fe@SiO_2 nanocomposites without using surface-capping agents. Compared to unsupported Fe, SF-Fe was more reactive in reduction of Cr (VI). Column transport experiments confirmed that the SF-Fe moved more effectively through a sand filter applied as model soil than unsupported Fe.

Acknowledgments

This work was supported by the National Natural Science Foundation of China (nos. 40971254 and 41173102) and the National Youth Science Foundation of China (no. 20907023).

References

1. Palmer, C. D.; Wittbrodt, P. R. Processes affecting the remediation of chromium-contaminated sites. *Environ. Health Perspect.* **1991**, *92*, 25–40.
2. Fendorf, S.; Wielinga, B. W.; Hansel, C. M. Chromium transformations in natural environments: The role of biological and abiological processes in chromium (VI) reduction. *Int. Geol. Rev.* **2000**, *42* (8), 691–701.
3. Costa, M. Potential hazards of hexavalent chromate in our drinking water. *Regul. Toxicol. Pharmacol.* **2003**, *188* (1), 1–5.
4. Kanel, S. R.; Manning, B.; Charlet, L.; Choi, H Removal of arsenic (III) from groundwater by nano scale zero-valent iron. *Environ. Sci. Technol.* **2005**, *39*, 1291–1298.
5. Wang, C. B.; Zhang, W. X. Synthesizing nanoscale iron particles for rapid and complete dechlorination of TCE and PCBs. *Environ. Sci. Technol.* **1997**, *31*, 2154–2156.
6. Cao, J; Zhang, W Stabilization of chromium ore processing residue (COPR) with nanoscale iron particles. *J. Hazard. Mater.* **2006**, *132*, 213–219.
7. Ponder, S. M.; Darab, J. G.; Mallouk, T. E. Remediation of Cr (VI) and Pb (II) aqueous solutions using supported nanoscale zerovalent iron. *Environ. Sci. Technol.* **2000**, *34*, 2564–2569.
8. Mackay, D. M.; Cherry, J. A. Groundwater contamination: Pump and treat remediation. *Environ. Sci. Technol.* **1989**, *23* (6), 630–636.
9. Li, X. Q.; Elliott, D. W.; Zhang, W. X. Zero-valent iron nanoparticles for abatement of environmental pollutants: materials and engineering aspects. *Crit. Rev. Solid State Mater. Sci.* **2006**, *31* (4), 111–122.
10. Saleh, N.; Phenrat, T.; Sirk, K.; Dufour, B.; Ok, J.; Sarbu, T.; Matyjaszewski, K.; Tilton, R. D.; Lowry, G. V. Adsorbed triblock copolymers deliver reactive iron nanoparticles to the oil/water interface. *Nano Lett.* **2005**, *5* (12), 2489–2494.
11. Cushing, B. L.; Kolesnichenko, V. L.; O'Connor, C. J. Recent advances in the liquid-phase syntheses of inorganic nanoparticles. *Chem. Rev.* **2004**, *104*, 3893.
12. Park, H.; Park, Y. M.; Yoo, K. M.; Lee, S. H. Reduction of nitrate by resin-supported nanoscale zero-valent iron. *Water Sci. Technol.* **2009**, *59*, 2153–2157.
13. Schrick, B.; Hydutsky, B. W.; Blough, J. L.; Mallouk, T. E. Delivery vehicles for zerovalent metal nanoparticles in soil and groundwater. *Chem. Mater.* **2004**, *16* (11), 2187–2193.
14. Yang, G. C. C.; Tu, H. C.; Hung, C. H. Stability of naniron slurries and their transport in the subsurface environment. *Sep. Purif. Technol.* **2007**, *58* (1), 165–172.

15. He, F.; Zhao, D. Preparation and characterization of a new class of starch-stabilized bimetallic nanoparticles for degradation of chlorinated hydrocarbons in water. *Environ. Sci. Technol.* **2005**, *39* (9), 3314–3320.
16. He, F.; Zhao, D.; Liu, J.; Roberts, C. B. Stabilization of Fe-Pd nanoparticles with sodium carboxymethyl cellulose for enhanced transport and dechlorination of trichloroethylene in soil and groundwater. *Ind. Eng. Chem. Res.* **2007**, *46* (1), 29–34.
17. Gesoglu, M.; Guneyisi, E Strength development and chloride penetration in rubberized concretes with and without silica fume. *Mater. Struct.* **2007**, *40*, 953–964.
18. Fendorf, S. E.; Lamble, G. M.; Stapleton, M. G.; Kelley, M. J.; Sparks, D. L. Mechanisms of chromium (III) sorption on silica. 1. Cr (III) surface-structure derived by extended X-ray-absorption fine-structure spectroscopy. *Environ. Sci. Technol.* **1994**, *28*, 284–289.
19. Oh, Y. J.; Song, H.; Shin, W. S.; Choi, S. J.; Kim, Y. H. Effect of amorphous silica and silica sand on removal of chromium (VI) by zero-valent iron. *Chemosphere* **2007**, *66*, 858–865.
20. Wang, G. H.; Harrison, A. Preparation of iron particles coated with silica. *J. Colloid Interface Sci.* **1999**, *217*, 203–207.
21. Li, Y. S; Church, J. S.; Woodhead, A. L.; Moussa, F. Preparation and characterization of silica coated iron oxide magnetic nano-particles. *Spectrochim. Acta, Part A* **2010**, *76*, 484–489.
22. Yang, P.; Ando, M.; Murase, N. Various Au nanoparticle organizations fabricated through SiO_2 monomer induced self-assembly. *Langmuir* **2011 a**, *27*, 895–901.
23. Yang, T. I.; Brown, R. N. C.; Kempel, L. C.; Kofinas, P. Controlled synthesis of core–shell iron-silica nanoparticles and their magneto-dielectric properties in polymer composites. *Nanotechnology* **2011 b**, *22*, 105601–105609.
24. Yuan, M. L.; Tao, J. H.; Yan, G. J.; Tan, M. Y.; Qiu, G. Z. Preparation and characterization of Fe/SiO_2 core/shell nanocomposites. *Trans. Nonferrous Met. Soc. China* **2010**, *20*, 632–636.
25. Zhan, J. J.; Zheng, T. H.; Piringer, G.; Day, C.; McPherson, G. L.; Lu, Y. F.; Papadopoulos, K.; John, V. T. Transport characteristics of nanoscale functional zerovalent iron/silica composites for in situ remediation of trichloroethylene. *Environ. Sci. Technol.* **2008**, *42* (23), 8871–8876.
26. Sun, Y. P.; Li, X. Q.; Zhang, W. X.; Wang, H. P. A method for the preparation of stable dispersion of zero-valent iron nanoparticles. *Colloid Surf., A* **2007**, *308* (1−3), 60–66.
27. Saleh, N.; Sirk, K.; Liu, Y.; Phenrat, T.; Dufour, B.; Matyjaszewski, K.; Tilton, R. D.; Lowry, G. V. Surface modifications enhance nanoiron transport and NAPL targeting in saturated porous media. *Environ. Eng. Sci.* **2007**, *24* (1), 45–57.
28. Zhang, L.; Manthiram, A. Chains composed of nanosize metal particles and identifying the factors driving their formation. *Appl. Phys. Lett.* **1997**, *70* (18), 2469–2471.
29. Nurmi, J. T.; Tratnyek, P. G.; Sarathy, V.; Baer, D. R.; Amonette, J. E.; Pecher, K.; et al. Characterization and properties of metallic iron

nanoparticles: Spectroscopy, electrochemistry, and kinetics. *Environ. Sci. Technol.* **2005**, *39*, 1221–1230.

30. Fiol, N.; Escudero, C.; Villaescusa, L. Chromium sorption and Cr (VI) reduction to Cr (III) by grape stalks and yohimbe bark. *Bioresource Technol.* **2008**, *99*, 5030–5036.
31. Xiu, Z. M.; Jin, Z. H.; Li, T. L.; Mahendra, S.; Lowry, G. V.; Alvarez, P. J. J. Effects of nano-scale zero-valent iron particles on a mixed culture dechlorinating trichloroethylene. *Bioresource Technol.* **2010**, *101* (4), 1141–1146.
32. Xiu, Z. M.; Gregory, K. B.; Lowry, G. V.; Alvarez, P. J. J. Effect of bare and coated nanoscale zerovalent iron on tceA and vcrA gene expression in *Dehalococcoides* spp. *Environ. Sci. Technol.* **2010**, *44* (19), 7647–7651.

Chapter 18

Synthesis, Characterization, and Cyanide Photodegradation Over Cupric Oxide-Doped Zinc Oxide Nanoparticles

Abdulaziz Bagabas,[*,1] Mohamed F. A. Aboud,[2,3] Reda M. Mohamed,[4] Zeid AL-Othman,[2] Ahmad S. Alshammari,[1] and Emad S. Addurihem[1]

[1]Petrochemcials Research Institute, King Abdulaziz City for Science and Technology (KACST), P.O. Box 6086, Riyadh 11442, Saudi Arabia
[2]Department of Chemistry, College of Science, King Saud University (KSU), P.O. Box 2455, Riyadh 11451, Saudi Arabia
[3]Department of Mining, Metallurgy, and Petroleum Engineering, Al-Azhar University, Cairo 11371, Egypt
[4]Department of Chemistry, Faculty of Science, King Abdulaziz University (KAU), P.O. Box 80203, Jeddah 21589, Saudi Arabia
***E-mail: abagabas@hotmail.com**

A simple, rapid, inexpensive, room-temperature wet chemical route for synthesizing CuO-doped ZnO nanoparticles was established. Undoped and CuO-doped ZnO photocatalysts, with 1, 2, 3, and 4 wt.% of CuO, were prepared by using cyclohexylammine for coprecipitation of the metals from their aqueous solutions, followed by calcination at 500°C. The photocatalysts were characterized by ICP elemental analysis, UV-Vis absorption spectroscopy, XRD, SEM, EDX, and TEM. The synthesized materials were tested for oxidative photodegradation of cyanide in aqueous medium at room temperature. The undoped material exhibited an activity of 56% cyanide removal efficiency. Doping of ZnO with 1 and 2 wt.% CuO dramatically enhanced the photocatalytic activity to 89% and 93 %, respectiviely, while to 97% for both 3 and 4 wt.%. This improvement in photocatalytic activity can be attributed to the reduction of ZnO band gap and the increase in charge separation on the surface of photocatalyst particles.

Introduction

Zinc oxide (ZnO) is a II-VI semiconductor with broad range of applications due to its unique properties. It can be also used as a model for applying quantum mechanics principles in the nanometric regime (*1*). Its direct, wide band gap energy (E_g) of 3.37 eV (*2*) enables its use in optoelectronic devices. The high sensitivity of the ZnO surface conductivity with adsorbed species nominates its utilization for gas detection (*3*). The abnormal radiation hardness makes it applicable in space applications (*4*).

The ZnO large exciton binding energy of 60 meV (*2*), compared with that of GaN of ~ 25 meV (*5*), enhances its luminescence efficiency of the emitted light even at room temperature and higher. The visible photo-luminescence (PL) emission around ~ 2.5 eV, originated from intrinsic defects (*6*), makes ZnO suitable for applications in field emission and vacuum fluorescent displays. Its reported room temperature ferromagnetism upon transition metal doping motivates its applications for spintronic devices (*7*). In addition, its relatively low cost, superior chemical and mechanical stability (*8*), the availability of large-area substrates with desirable c-axis preferential growth nature and technological compatibility with the conventional silicon process (*9*) make it very strong competitor for many applications.

Alteration of ZnO specifications, electronic and optical properties in particular, can be made by its doping with transition metals such as manganese, iron, cobalt, nickel, copper and lanthanides such as europium, erbium, and terbium (*1*).

The Cu-doped ZnO semiconductor research was mainly directed to catalytic applications such as methanol synthesis (*10*), production of hydrogen by partial oxidation of methanol (POM) (*11*), carbon monoxide oxidation (*12*), degradation of textile dye pollutants within aqueous solutions (*13*) and dilute magnetic semiconductors for spintronic devices (*14*, *15*).

Copper forms in different bonding states within ZnO lattice such as metallic (Cu^0), monovalent ($Cu^I{}_2O$) and divalent ($Cu^{II}O$), depending on the annealing conditions (temperature and oxygen pressure), where the fully oxidized divalent state Cu^{2+} is favored when the above conditions are promoted, otherwise other states would be present.

Various methods for doping ZnO with CuO have been reported in the literature such as hydrothermal (*1*, *16*), sol-gel (*17*, *18*), and co-precipitation (*19*). Hydrothermal method applies both pressure (150 kPa) (*1*) and temperature (190°C) (*16*). Furthermore, it resulted in secondary phase of CuO above 3% (*1*) and maximum solubility of 1% for CuO (*16*). It also led to surface-enrichment with copper, and hence, inhomogeneous distribution of CuO within ZnO (*16*). The sol-gel method, on the other hand, uses a template of organic material, which requires high temperature 430°C for its removal from the desired oxide product (*18*). However, this organic sometimes could not be removed and resulted in presence of mixed states of copper (*17*). The co-precipitation uses a strong base such as sodium hydroxide at moderate temperature 80°C and results in segregated phases of CuO and ZnO (*19*).

We are reporting herein a co-precipitation method, avoiding the drawbacks of the above reported procedures, using a moderate base of cyclohexylamine (CHA) without using any organic template or surfactant at ambient temperature and pressure. Our method, in addition, resulted in copper-substituted ZnO wurtzite lattice structure nanoparticles, regardless of doping concentration (1, 2, 3 and 4 weight percents of copper oxide).

Experimental

Materials

Copper nitrate trihydrate (98-103%, Fluka), zinc nitrate hexahydrate (pure, POCH), and cyclohexylamine (GC>99%, Merck) were commercially available and were used without further purification. Deionized water (18.2 MΩ.cm) was obtained from a Milli-Q water purification system (Millipore).

Preparation Method

Undoped ZnO

As described previously (*20*), zinc nitrate hexahydrate was mixed with cyclohexylamine in water in 1:2 mol ratio at room temperature to prepare undoped ZnO precipitate, which was calcined at 500°C for three hours.

CuO-Doped ZnO

Calculated amounts of copper nitrate trihydrate, zinc nitrate hexahydrate, and cyclohexylamine were mixed according to the mol ratios as shown in Table I. For each mixture, metal nitrate precursors were first mixed and dissolved in 500 ml of deionized water at room temperature, under continuous magnetic stirring. The addition of cylcohexylamine resulted in a very light bule precipitate. Depth of blue color increased with increasing the copper content (Figure 1). The reaction mixtures were left stirring for one week. The precipitates were filtered off through F-size fritted filters, and then were copiously washed with deionized water. The precipitates were dried under vacuum for one day. After drying, the precipitates were mixed with 300 ml water and were magnetically stirred for one day for the removal of any impurity. The precipitates were filtered off, air-dried, and then calcined at 500°C for three hours. Brown solids were obtained after calcination. The depth of brown color increased with increasing the copper content (Figure 1).

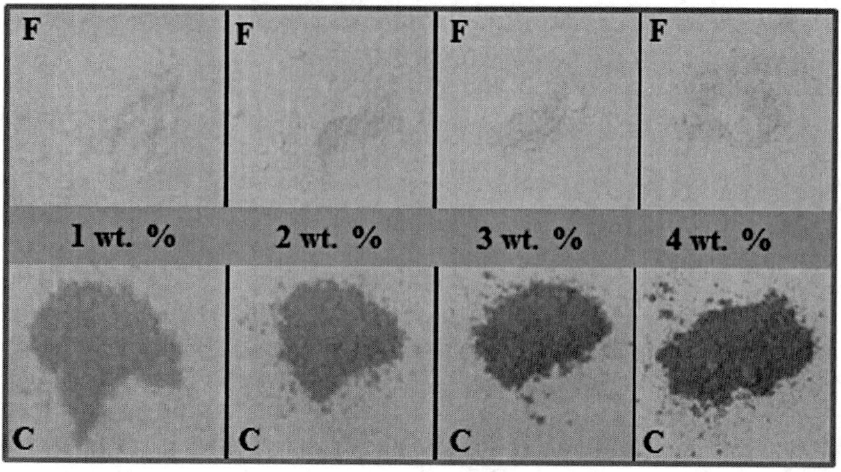

Figure 1. Catalyst color depth change of the Cu^{2+}-doped ZnO samples before (F) and after calcination (C).

Table I. Mole Ratio of Copper Nitrate to Zinc Nitrate to CHA for Preparing the Catalyst Precursors

Precursor (wt %)	*Cu^{2+}:Zn^{2+}:CHA mol ratio*
1	1:117:236
2	1:47:97
3	1:31:65
4	1:23:48

Materials Characterization

Inductively-coupled plasma (ICP, Varian Vista-MPX) was used to determine the copper and zinc component in the clacined CuO-doped ZnO catalysts, obtained at 500°C.

X-ray diffraction (XRD) patterns were recorded for phase analysis using Philips X pert pro diffractometer, operated at 40 mA and 40 kV by using CuK_{α} radiation and a nickel filter, in the 2 theta range from 2 to 80° in steps of 0.02°, with a sampling time of 1s per step. XRD patterns were recorded for Cu^{2+}-doped ZnO materials before and after calcination.

The morphology (size and shape) was investigated using a field emission scanning electron microscope (FE-SEM model: FEI-200NNL) and a high resulotion transmission electron microscope (HRTEM model: JEM-2100F JEOL). Carbon-coated copper grids were used for mounting the samples for HRTEM analysis. Elemental microanalysis of the surface was performed by energy dispersive X-ray spectroscopy (EDX), which is coupled to FE-SEM.

UV-Vis absorption spectra for Cu^{2+}-doped ZnO materials before and after calcination were recorded on a Perkin Elmer Lambda 950 UV/Vis/NIR spectrophotometer, equiped with 150 mm snap-in integrating sphere for capturing diffuse and specular reflectance.

Photocatalytic Evaluation

All the experiments were carried out using a horizontal cylinder annular batch reactor. A black light-blue florescent bulb (F18W- BLB) was positioned at the axis of the reactor to supply UV illumination. Reaction suspension was irradiated by UV light of 365 nm at a power of 18 W. The experiments were performed at room temperature by suspending 0.02 wt% of CuO-doped ZnO catalyst sample into 300 ml, 100 ppm potassium cyanide at pH 8.5, adjusted by ammonia solution. This specific pH value was chosen on the basis of previous investigation, revealed the preferred adsorption of OH^- ion over CN^- ion at higher pH values (*21*), while hydrogen cyanide, HCN, elevates at pH≤7 according to the following equation:

$$CN^-_{(aq)} + H_2O_{(l)} = HCN_{(g)} + OH^-_{(aq)}$$

The reaction was carried out isothermally at 25 °C and a sample of the reaction mixture was taken after 120 minutes. The CN^- content in the solution after reaction time was analyzed by volumetric titration with $AgNO_3$. The removal efficiency of CN^- has been measured by applying the following equation:
% Removal efficiency = $(C_o - C)/C_o \times 100$, where C_o the original cyanide content and C the retained cyanide in solution.

Results and Discussion

Catalyst Characterization

The physical observation of the Cu^{2+}-doped ZnO samples before and after calcination (Figure 1) was evident for increasing the copper content through the increase in color depth.

Table II confirms that the theoretical and experimental, obtained by ICP, CuO content in the calcined samples were in good agreement.

Figure 2 shows the XRD spectra of the CuO/ZnO samples with varying the copper concentration from 1 up to 4 weight percent before and after calcination. No indication of any copper secondary phases were observed upon incorporation of copper for all the samples. The absence of secondary phases could be attributed either to the complete solubility of copper within ZnO, which is higher than the reported solubility limit value of 1% (*16*) and the 3% (*1*). The low solubility was attributed to the high covalent character in Cu-O bonding, resulted from the high localization of 3d state on Cu (*15*). It is also likely that the high dispersion of copper phases within ZnO phase prevented its detection by XRD technique (*22*). However, this possibility was excluded on the basis of XRD results, explained below, and on the basis of the TEM results. The other possibility of not detecting any copper phase due to its amorphous nature was also ruled out on the basis of the TEM results, which showed the crystal planes.

Table II. Theoretical and Experimental Results of CuO Content in the CuO-Doped ZnO Catalysts

Sample	*Theoretical (wt.%)*	*Experimental (wt.%)*
1% CuO/ZnO	1	0.7
2% CuO/ZnO	2	1.9
3% CuO/ZnO	3	2.9
4% CuO/ZnO	4	3.9

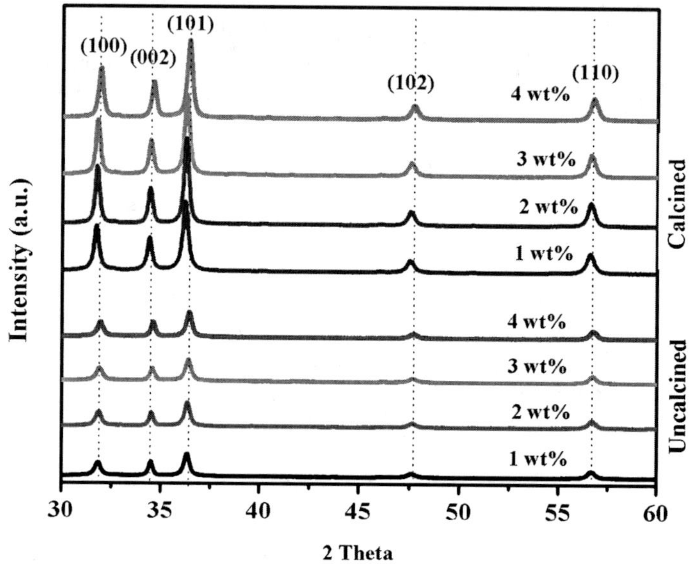

Figure 2. XRD patterns of the Cu^{2+}-doped ZnO samples with varying the copper concentration from 1 to 4 weight percent before and after calcination.

The shift of the ZnO peaks (for instance 100, 002, and 101) to higher 2θ is a result of replacing Zn^{2+} (0.06 nm) by the smaller Cu^{2+} ions (0.057 nm) (*23*) in the wurtzite lattice . The shift in the uncalcined samples was more pronounced than the calcined samples due to the decrease in defects, resulting from copper substitution upon calcinations (Figure 2). No copper phases were detected by XRD for the uncalcined or the calcined samples which confirms complete substitution of Cu^{2+} in the ZnO wurtzite lattice even at room temperature.

The importance of having complete solubility in our CuO-doped ZnO catalyst systems is to produce a coupled system instead of having two independent ZnO and CuO composits such as those systems used in hetrojunctions (*24*) that may not be in harmony with each other and minimize the charge transfer from one to another.

Figure 3 (a-d) shows the SEM micrographs for the calcined CuO-doped ZnO samples. The same morphology was observed in all samples irrespective of CuO wt%. The particles are agglomerated in rice-like shape. The elemental microanalysis of the surface by EDX (Figure 4) confirmed the purity of the calcined samples and the presence of copper, zinc, and oxygen on their surfaces. However, the surfaces are rich in zinc, which is consistent with the oxygen-deficiency for the *n*-type ZnO.

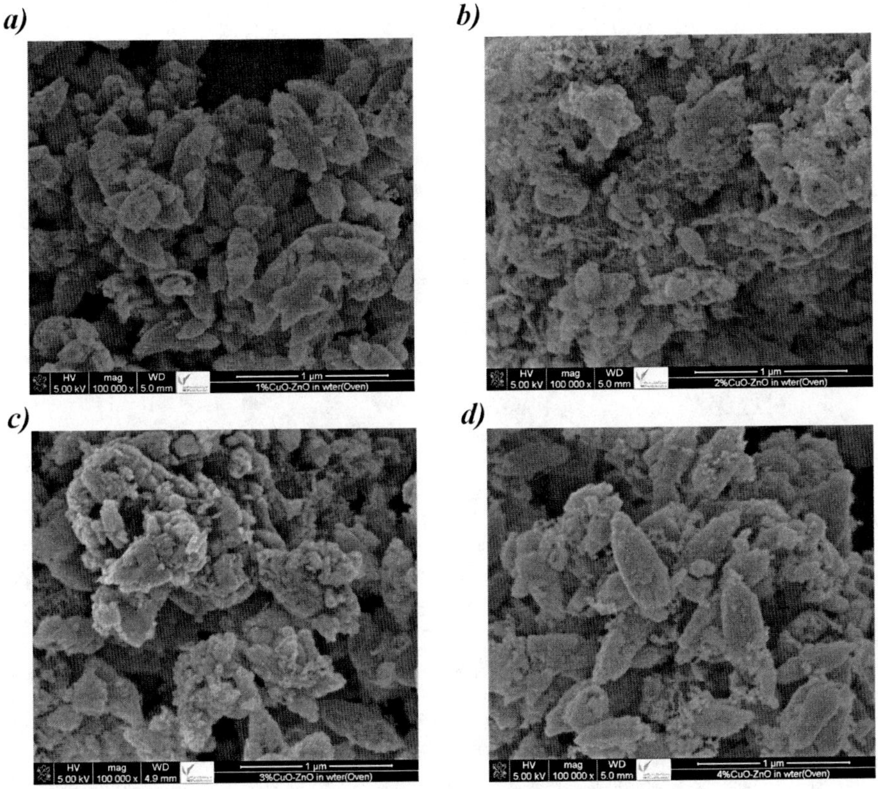

Figure 3. SEM micrographs for the calcined CuO-doped ZnO samples a) 1 wt.%, b) 2 wt.%, c) 3 wt.%, and d) 4 wt.%.

Figure 5 shows high resolution TEM micrographs for the calcined CuO-doped ZnO samples. All the samples show similar morphology and particle size irrespective of CuO wt% doping. The particles have different shapes such as rectangular- and round-like. The average size of the particles ranges from 5-20 nm. The lattice fringes, in addition, match those of ZnO only, which supports the results obtained from XRD, indicating the replacement of Zn^{2+} ions by Cu^{2+} ions in the wurtzite lattice of ZnO.

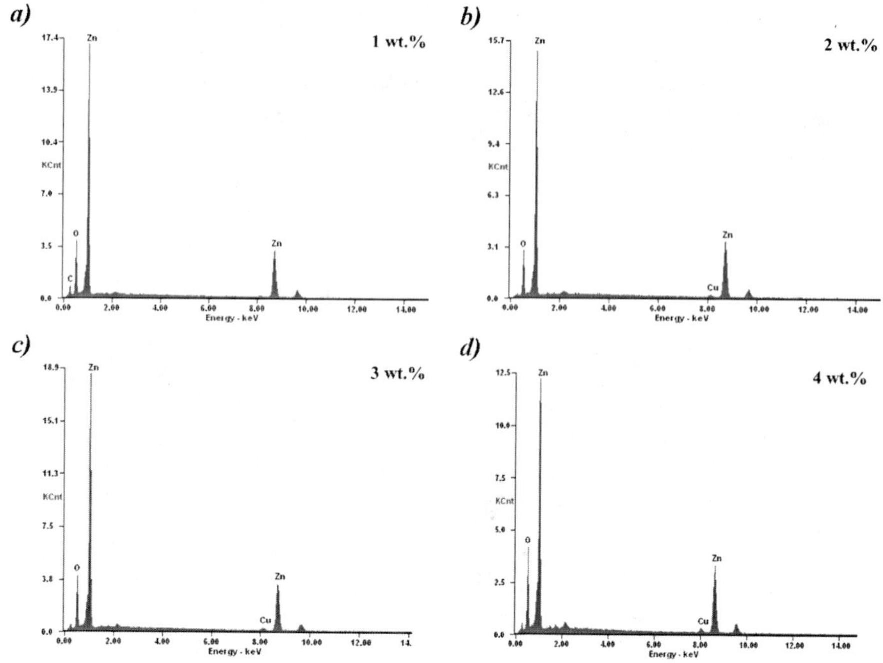

Figure 4. EDX for calcined CuO-doped ZnO samples a) 1 wt.%, b) 2 wt.%, c) 3 wt.%, and d) 4 wt.%.

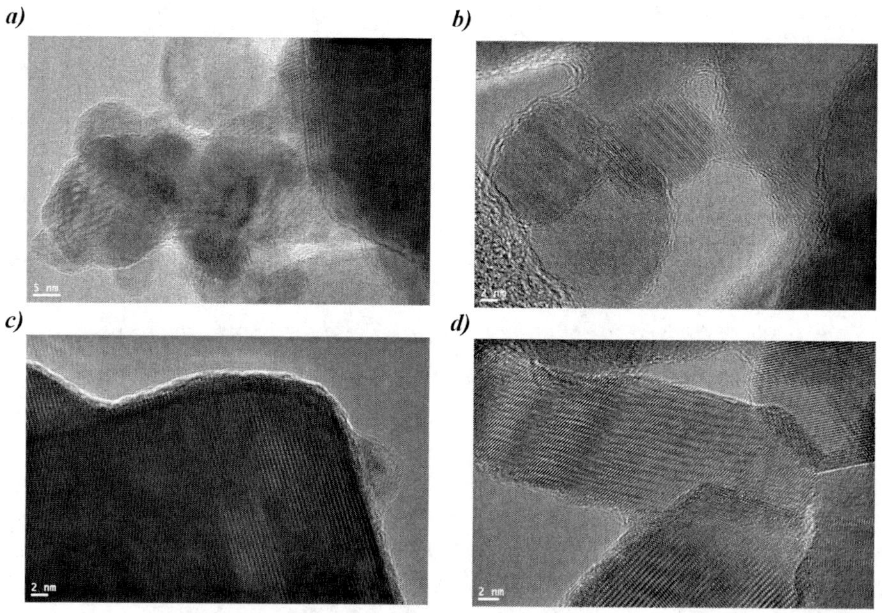

Figure 5. high resolution TEM micrographs for the calcined CuO-doped ZnO samples a) 1 wt.%, b) 2 wt.%, c) 3 wt.%, and d) 4 wt.%.

Figure 6 shows the UV-Vis absorption spectra for the copper oxide doped ZnO samples before and after calcination. The incorporation of Cu^{2+} is responsible for ZnO E_g reduction, i.e. red shift. As shown in Table III, a slight shift in the E_g of ZnO with increasing the content of CuO from 1 to 4 wt.% was obsereved.

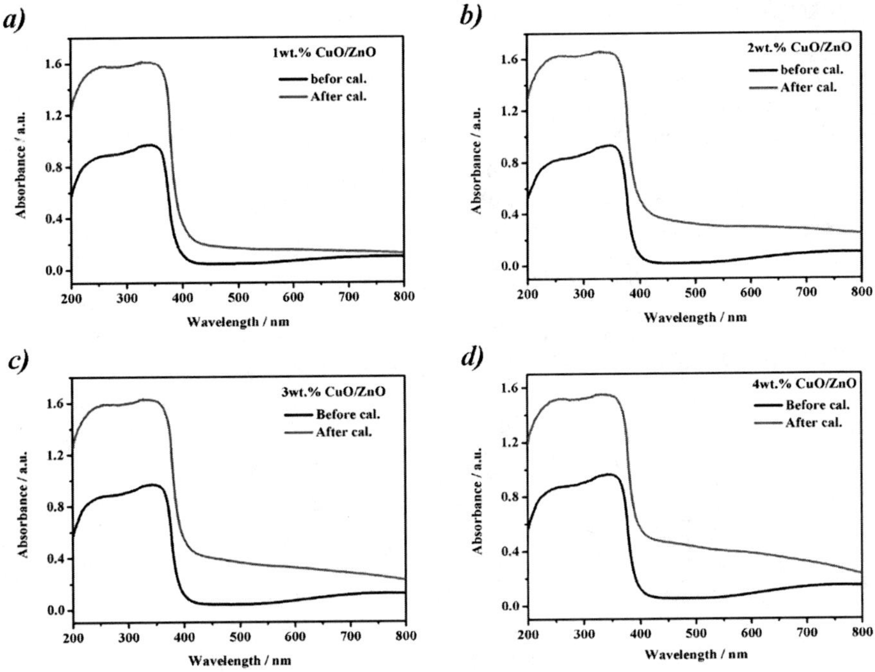

Figure 6. UV-Vis absorption spectra for the copper oxide doped ZnO samples before and after calcination a) 1 wt.%, b) 2 wt.%, c) 3 wt.%, and d) 4 wt.%.

Table III. Band Gap Energy of the Cu^{2+}-Doped ZnO Samples before and after Calcination

Sample	*E_g / eV (Uncalcined)*	*E_g / eV (Calcined)*
1% CuO/ZnO	3.24	3.21
2% CuO/ZnO	3.22	3.20
3% CuO/ZnO	3.22	3.19
4% CuO/ZnO	3.21	3.19

The E_gs of the unclacined and calcined samples were almost comparable. This result might imply that Cu^{2+} substituted Zn^{2+} in ZnO wurtzite lattice at room temperature, which was also supported by X-ray results. However, the small red shift in the E_gs of the calcinced samples could be due to the enhancement of copper substitution upon calcinations.

The red shift in the calcined samples was less compared to the reported E_g (2.9 eV) for CuO/ZnO nanocomposite, which was prepared physically by wet impregnation method and attributed to the stoichiometry deficiency of ZnO due to impregnated CuO (*13*). The E_g reduction (band offsets) in our solid solution system may be attributed to the following effects (*25*):

1) The strong *d-p* coupling between copper and oxygen moves O 2*p* orbital up, that narrows the direct fundamental E_g of ZnO.
2) Creation of impurity energy band, especially at higher concentrations, above the ZnO valance band maximum (VBM) which creates a mixture of direct and indirect transitions.

Effect of CuO Doping on Photocatalytic Activity

Table IV shows that increasing the CuO content from 0 to 3 wt% led to increase in the photocatalytic activity of cyanide degradation. However, increasing the CuO content to 4 wt% did not enhance the photocatalytic activity. This finding might be due to the identical E_g of both 3% and 4 wt% CuO-doped ZnO catalysts (Table III). The enhancement in photocatalytic degradaion of cyanide ion with increasing the CuO wt.% content could be attributed to the inhibition of electron-hole pair recombination and efficient separation of the charges (*26*). Such easy transfer of electrons from CuO to ZnO is due to the close match of work function between CuO and ZnO (5.3eV) (*13*).

Table IV. Effect of CuO wt.% on Photocatalytic Activity

Photocatalyst	*% of cyanide degradation*
ZnO	56
1% CuO/ZnO	89
2% CuO/ZnO	93
3% CuO/ZnO	97
4% CuO/ZnO	97

Conclusion

Our CuO-doped ZnO solid-solution system showed enhancement in the photocatalytic performance due to E_g reduction. The narrow E_g of CuO (1.7, 1.33 eV) (*26*, *27*) results in efficient separation of charges, which were photo-generated in the copper oxide/zinc oxide under UV-vis light illumination and suppressed their recombination. The mechanism of the photo-generated charges separation is due to their transfer between the two semiconductor materials (*p*-type copper oxide/*n*-type zinc oxide) as follows; the photogenerated electrons transfer from the conduction band (CB) of CuO to that of ZnO, while the photogenerated holes immigrate in the opposite direction from the valance band (VB) of ZnO to that of CuO. Consequently, more electrons are accumulated in (CB) of ZnO and consumed for reduction of pollutants.

References

1. Sharma, P. K.; Dutta, R. K.; Pandey, A. C. *J. Magn. Magn. Mater.* **2009**, *321*, 4001–4005.
2. Zhao, L.; Lu, P. F.; Yu, Z. Y.; Guo, X. T.; Shen, Y.; Ye, H.; Yuan, G. F.; Zhang, L. *J. Appl. Phys.* **2010**, *108*, 113924–113930.
3. Heo, Y. W.; Tien, L. C.; Norton, D. P.; Kang, B. S.; Ren, F.; Gila, B. P.; Pearton, S. J. *Appl. Phys. Lett.* **2004**, *85*, 2002–2004.
4. Ju, S.; Lee, K.; Janes, D. B.; Dwivedi, R. C.; Awuah, H. B.; Wilkins, R.; Yoon, M. H.; Facchetti, A.; Mark, T. J. *Appl. Phys. Lett.* **2006**, *89*, 073510–073512.
5. Johnson, J. C.; Choi, H. J.; Knutsen, K. P.; Schaller, R. D.; Yang, P.; Saykally, R. J. *Nat. Mater.* **2002**, *1*, 106–110.
6. Liu, J.; Zhao, Y.; Jiang, Y. J.; Lee, C. M.; Liu, Y. L.; Siu, G. G. *Appl. Phys. Lett.* **2010**, *97*, 231907–231909.
7. Sofer, Z.; Sedmidubsky, D.; Huber, S.; Hejtmanek, J.; Maryško, M.; Jurek, K.; Mikulics, M. *J. Cryst. Growth* **2011**, *314*, 123–128.
8. Look, D. C. *Mater. Sci. Eng. B 80* **2001**, *383*, 1–3.
9. Lee, J.-B.; Lee, H.-J.; Seo, S.-H.; Park, J.-S. *Thin Solid Films* **2001**, *399*, 398–399.
10. Bao, J.; Liu, Z.; Zhang, Y.; Tsubaki, N. *Catal. Commun.* **2008**, *9*, 913–918.
11. Schuyten, S.; Guerrero, S.; Miller, J. T.; Shibata; Wolf, E. E. *Appl. Catal., A* **2009**, *352*, 133–144.
12. Taylor, S. H.; Hutchings, G. J.; Mirzaei, A. A. *Catal. Today* **2003**, *84*, 113–119.
13. Sathishkumar, P.; Sweena, R.; Wu, J. J.; Anandan, S. *Chem. Eng. J.* **2011**, *171*, 136–140.
14. Kim, C. O.; Kim, S.; Oh, H. T.; Choi, S. H.; Shon, Y.; Lee, S.; Hwang, H. N.; Hwang, C. C. *Phys. B* **2010**, *405*, 4678–4681.
15. Wang, X.; Xu, J. B.; Cheung, W. Y.; An, J.; Ke, N. *Appl. Phys. Lett.* **2007**, *90*, 212502–212504.
16. Wang, X.; Xu, J.; Zhang, B.; Yu, H.; Wang, J.; Zhang, X.; Yu, J.; Li, Q. *Adv. Mater.* **2006**, *18*, 2476–2480.
17. Bao, J.; Liu, Z.; Zhang, Y.; Tsubaki, N. *Catal. Commun.* **2008**, *9*, 913–918.
18. Fernandes, D. M.; Silva, R.; Hechenleitner, A. A. W.; Radovanovic, E.; Melob, M. A. C.; Pineda, E. A. G. *Mater. Chem. Phys.* **2009**, *115*, 110–115.
19. Li, B.; Wang, Y. *Superlattices Microstruct.* **2010**, *47*, 615–623.
20. Bagabas, A.; Mohamed, R.; Aboud, M. F. A.; Mostafa, M. M.; Alshammari, A.; AL-Othman, Z. U.S. Patent filed, 2012.
21. Chiang, K.; Amal, R.; Tran, T. *J. Mol. Catal. A: Chem.* **2003**, *193*, 285–297.
22. Ma, Z.; Zaera, F. Characterzation of Heterogenoues Catalysis. In *Surface and Nanomolecular and Nanomolecular Catalysis*; Richards, R. M., Ed.; Taylor & Francis: Boca Raton, FL, 2006; pp 1–26.

23. Jolly, W. L. *Modern Inorganic Chemistry*, 2nd ed.; McGraw-Hill Book Co.: Singapore, 1991; pp 592–597.
24. Hu, Y.; Zhou, X.; Han, Q.; Cao, Q.; Huang, Y. *Mater. Sci. Eng.* **2003**, *99*, 41–43.
25. Ahn, K.-S.; Deutsch, T.; Yan, Y.; Jiang, C.-S.; Perkins, C. L.; Turner, J.; Al-Jassim, M. *J. Appl. Phys.* **2007**, *102*, 023517–023522.
26. Wei, S.; Chen, Y.; Ma, Y.; Shao, Z. *J. Mol. Catal. A: Chem.* **2010**, *331*, 112–116.
27. Wong, L. M.; Chiam, S. Y.; Huang, J. Q.; Wang, S. J.; Pan, S. J.; Chim, W. K. *J. Appl. Phys.* **2010**, *108*, 033702–033707.

Chapter 19

Scientific and Regulatory Issues in Exposure Assessment of Conventional Metals vs Nanosized

Najm Shamim* **and Siroos Mostaghimi**

U.S. Environmental Protection Agency, Office of Consumer Safety and Pollution Prevention, Office of Pesticides Program, Antimicrobials Division, Risk Assessment & Science Support Branch, Arlington, Virginia, 22202
***E-mail: Shamim.Najm@epa.gov**

Metals and salts are ubiquitously present in the environmental media, and some of the metals are an integral part of biological systems. Application list of metals and salts is too large to enumerate. Traditionally pesticides, whether agricultural, antimicrobials or biopesticides, are organic based (mostly synthetic) or biochemicals (naturally occurring molecules are modified). In the last 40 years or so metal and salts, and even organometallics, have been applied as pesticides in agricultural sector (as fungicides, algaecides), in the antimicrobial sector as antifoulants and as wood preservatives to prevent rotting of wood. To conduct risk assessment, risk characterization, and ultimately to manage risks for organic pesticides, there are more defined steps, developed over a period of time, that are available as opposed to for metals. For example generation of data for mammalian toxicity or ecotoxicity etc. is carried out by well-tested methods for organic pesticides. With metals and salts when used as pesticides, there are inherent difficulties that exist due to the fact that each metal has unique physical /chemical characteristics, ecotoxicity pathways are more difficult to quantify. Use of nanometals and nanometal oxides enhances the difficulties in data generation as well as the quantitative determination of risks.

This chapter provides an over view of the scientific and regulatory difficulties that various agencies and organizations, particularly those that are both regulatory/ scientific organizations and conduct risk assessments, risk characterizations, and ultimately risk management of chemicals--be it as general purpose chemicals or designer chemicals like pesticides. The chapter also briefly discusses some of the manufacturing processes that are being utilized to increase the use of green chemistry principals to synthesize/manufacture nanometals/ nanometal oxides to increase sustainability, reduce pollution, and decrease some risks. This would likely help not only the pesticide industry in the manufacture of the products but also help agencies like EPA to explore and possibly adopt a new paradigm of reduced risks for some of the nanomaterials. The chapter provides examples in the fate and ecotoxicity disciplines only and related copper as an example.

Introduction

Metals have found applications in a wide variety of ways in our civilization from making small tools to constructing complicated machines. These naturally occurring materials could also be used as 'designer chemicals' like agricultural and antimicrobial pesticides.

This chapter is an overview of the application of metals/ metal oxides, and also some metal salts as pesticides, and their possible effects on the environment and human health. Exposure to metals/ metal oxides when used in the agricultural field can be high for terrestrial and aquatic organisms, environmental media like water, soils and sediments, as well as birds etc. Since there are many established use sites and use application scenarios for the antimicrobials, exposure in the antimicrobial setting is use site dependent. The antimicrobial uses may be indoor or outdoor uses. For example use of an antimicrobial pesticide as a disinfectant or a sanitizer on the kitchen counter tops may result in migration of the pesticide into food and may trigger a dietary risk and thus a dietary assessment of a pesticide would be conducted. On the other hand, metals/ metal oxides could be used as antifouling agent when applied on ship hauls and boat bottoms. This application could likely result in high exposure to sediments, and aquatic organisms. The use of metals/metal oxides as wood preservatives could also likely result in exposure to the environmental media.

The Office of Pesticides Programs within the Office of Chemical Safety and Pollution Prevention (OSCPP) of Environmental Protection Agency (EPA) is mandated through FIFRA (Federal Insecticide, Fungicide and Rodenticide Act) not only to review all data on each product submitted for registration, but also to conduct a complete risk assessment and risk characterization for the environment and human health for all products which go through registration with OPP, before granting a license interstate commerce.

There are a number of Federal Agencies in the United States and in other countries around the world which have a mandate as regulatory Agencies. In the following section we have referred some of the agencies along with their website links which provide details about each statute.

FIFRA (Federal Insecticide, Fungicide and Rodenticide Act) (*1*). This statute is for pesticides regulations under EPA and Office of Chemical Safety and Pollution Prevention (OCSPP)

FFDCA (Federal Food, Drug and Cosmetic Act (FFDCA) (*2*). This statute is in the purview of the Food and Drug Administration (FDA).

REACH (Registration, Evaluation, Authorization and Restriction of Chemical Substances) (*3*) is the statute under European Community Regulations of Chemicals and their safe use.

TSCA: (Toxic Substance Control Act) (*4*): An EPA statute and it is within the Office of Chemical Safety and Pollution Prevention (OSCPP), It regulates all chemicals other than pesticides, and food additives. Note that under OSCPP, EPA has two major statutes: TSCA and FIFRA to regulate the chemicals that are likely to hit the market place. Some substances that are excluded from this statute are: drugs, food, and pesticides, which are covered under FFDCA and some under FIFRA.

It should be noted that the TSCA covers the largest share of chemicals, industrial scale (high production volume) to small scale production of chemicals. FIFRA, on the other hands, cover the largest share of designer chemicals like agricultural pesticides, antimicrobials and biopesticides. Within the FIFRA scenario, metals, and metal oxides have a small share of applications, but present a great challenge for risk assessment and risk characterization. With the advent of nanotechnology and nanoproducts, some with a pesticidal claim present additional issues for risk assessment and risk characterization processes.

All the regulations and their subsets, within the context of each statute, endeavor to ascertain the safety of chemicals. These regulations aim at the environmental protection and human health safety. In addition, the fate of these chemicals after use (as solid waste, in aqueous medium, in soils, or sediments) is important to understand and efforts are made to identify, minimize the hazards of these chemicals in the environmental media and to protect human health. Classical examples of the use of chemical substances and their hazards through improper disposal and misuse were brought forth in Rachael Carson's Silent Spring (*5*).

Organics vs Metal/Metal Oxides/Salts as Pesticides

Most of the registered pesticides are organic and are made for specific purposes. These are synthesized for use as fungicide or algaecide or insecticide or rodenticide etc. Risk assessment and risk characterization processes for organics pesticide have a set of parameters that are reasonably understood or streamlined. However, with metals, each metal, or metal oxide or any metal salt has unique physical and chemical characteristics, and each differ in behavior in the environmental media, and living systems and is also different from organic pesticides which makes the risk assessment/ risk characterization processes for metals different and difficult, if not impossible Also, as indicated above, metals and metal oxides as nanoproduct present additional scientific and regulatory challenges.

EPA's document 'Framework for Metal Risk Assessment' at length discusses the unique characteristics of physical / chemicals of metals and has shown the uniqueness of metals as summarized below (*6*):

* "Metals are universally, naturally or through anthropogenic processes present in the environment as mixtures and some of the metals are essential for humans, animals, plants and microorganisms.
* Metals are not created or destroyed by any chemical or biochemical process. Metals undergo transformation from one form to other forms; in particular transition metals show versatility for these kinds of transformations.
* Each metal shows a specific pathway for absorption, distribution, transformation and excretion within an organism, which is dependent on the organism's ability to regulate or store the metal.
* Fate and transport of metals is driven by a number of factors: speciation, and possible kinetic steps are involved in this process, partitioning from one environmental medium to another, deposition and possible resuspension in the same environmental medium or other medium, bioaccumulation of metals is more complicated than organic pesticides and goes through processes like bioaccessibility, bioavailability, or it shows a tendency of being essential to an organism and through an uptake of a metal and distribution in an organism. Generally metals do not show biomagnification like some organic pesticides.
* For risk assessment processes some factors that will have to be taken into account are: metal speciation: each has unique pathways. Transition metals show a greater tendency of speciation.
* Partitioning from one media to the other; each metal has a unique way of achieving this.
* Possible deposition or resuspension in the same environmental media. Also unique to metals are:

    - Bioaccessibility
    - Bioavailability
    - Bioaccumulation etc. All are unique for each metal.
    - Uptake, distribution and excretion from an organism"

What Are Nano Materials?

Since the advent of National Nanotechnology Initiative (NNI) which defines a nanomaterials as: "any engineered material between 1 to 100 nm in any one dimension is a nano material," various scientific and regulatory organizations, national and international, have been making serious efforts to come up with a definition of their own. This definition includes all types of nanomaterials, carbon nanotubes and metals, metal oxides, or sulfides etc. that are engineered to nano-scale materials. Following examples show the complexities introduced by nanotechnology, even for defining a nanomaterial: It provides a flavor of numerous approaches.

a. Health Canada has put forward an interim Policy Statement on Working Definition for nanomaterial: Any manufactured product, material, substance, ingredient, device or structure is a nanomaterial **if**: it is within a nanoscale in at least one spatial dimension or: it is smaller or larger than the nanoscale in all spatial dimension and exhibits one or more nanoscale phenomenon. The definition of nanoscale is the same as in NNI. The Nanoscale phenomena indicate that the material as identified above, its size has properties that are distinguishable from the chemical or physical properties of individual atoms, individual molecules and bulk materials (*7*).
b. Food and Drug Administration (FDA): It recently issued a draft guidance on nanomaterials for cosmetics: "As described in this draft guidance, **if** a FDA-regulated product contains nanomaterials or otherwise involves the application of nanotechnology, FDA will ask: (1) whether an engineered material or an end product has at least one dimension in the nanoscale range (approximately 1 nm to 100 nm); or (2) whether an engineered material or end product exhibits properties or phenomena, including physical or chemical properties or biological effects, that are attributable to its dimension(s), even if these dimensions fall outside the nanoscale range, up to one micrometer. Once the guidance is finalized, the agency intends to apply these considerations broadly to all FDA-regulated products, including cosmetic products (*8*)." This tentative definition has the underlying assumption that is possible that some of the engineered material may have unique characteristics different from the regular materials, even up to a micrometer scale, and thus such engineered materials will be closely scrutinized by FDA.
c. EU (European Union) has put forward this nanomaterial definition: "Nanomaterial" means a natural, incidental or manufactured material containing particles, in an unbound state or as an aggregate or as an agglomerate and where, for 50 % or more of the particles in the number size distribution, one or more external dimensions is in the size range of 1) 1 nm - 100 nm. In specific cases and where warranted by concerns for the environment, health, or safety or competitiveness, 2) the size distribution threshold of 50 % may be replaced by a threshold between 1 and 50 %. By definition from point 2, fullerenes, graphene flakes and single wall carbon nanotubes (SWCNT) with one or more external dimensions below 1 nm should be considered as nanomaterials.
   For the purposes of point (2), "particle", "agglomerate" and "aggregate" are defined as follows:

   - "Particle" is a minute piece of matter with defined physical boundaries;

- "Agglomerate" means a collection of weakly bound particles or aggregates where the resulting external surface area is similar to the sum of the surface areas of the individual components;
- "Aggregate" is particle comprising of strongly bound or fused particles.

Where technically feasible and requested in specific legislation, compliance with the definition in point (2) may be determined on the basis of the specific surface area by volume. A material should be considered as falling under the definition in point (2) where the specific surface area by volume of the material is greater than 60 m^2 / cm^3. However, a material which, based on its number size distribution, is a nanomaterial should be considered as complying with the definition in point (2) even if the material has a specific surface area lower than 60 m^2/cm^3 (*9*). Admittedly EU has included in their definition many parameters and factor that may possibly make some natural, engineered or even incidental material to have unique characteristics different from bulk materials.

d. FIFRA has similar issues as far as a definition of a nanomaterial is concerned. EPA at this time has no formal definition of a nonmaterial A recent Federal Register Notice amplifies the issues faced by EPA (*10*). All definitions, or lack thereof, are, to say the least, is a tentative approach, and will likely evolve into a concrete base over a period of time as the nanoscience evolves and finds its firm foundation.

There is a general consensus that any new chemical methods and approaches that can reduce risks to the environmental and human health are beneficial to all stakeholders. This is critically important for nanoproducts too. Thus a simultaneous development of two important new fields, namely: green chemistry and sustainability warrants a discussion in this paper before we go to the known scientific issues for nanomaterials.

Sustainability and Nanotechnology

The word sustainability is no longer used lightly word to merely put an emphasis on a scientific argument. Sustainability is a use of resources in such a way that these resources can prolong economic, social and environmental stability. Kates. R, et al. (*11*) have stated that sustainability science has emerged as a new scientific discipline. Sustainability science, like nanotechnology encompasses all disciplines of natural, biological, social as well as engineering and medical sciences. Sustainability itself utilizes the concepts of sustainable development and environmental science. Ideally a combination of sustainability, nanotechnology, and using some of the green chemistry approaches can go a long way to reduced waste, reduced exposure and hazards and thus reduced risks from the new class of nanoproducts like nanosensors, nanopesticides etc.

Rajender Varma's review on the greener syntheses using mechano-chemical mixing, microwave and ultraviolet radiation processes amply shows sustainability of resources can be enhanced by finding and using less expensive and less hazardous routes for the synthesis of organic chemicals (*18*). In addition his published work clearly shows that using a naturally occurring chemical substance like vitamin B_2 can be used as a capping or dispersing agent for the synthesis of gold and platinum nanoparticles. Vitamin B_2 also acts as a reducing agent, has high water solubility, is easily biodegradable, and has inherent low toxicity (*19*). In another work, Varma synthesized nanocomposites of metallic and bimetallic (Pt-IN, Ag-Pt, Pt-Fe, Cu-Pd, Pt-Pd, and Pd-Fe) by reacting 3% by weight PVA (polyvinyl alcohol) with microwave radiation maintaining the reaction temperature at 100 °C (*20*).

In another recent work, K. Onubogu et al. have shown that using gum arabic to synthesize silver, gold, and platinum nanoparticles results in mono dispersed and highly crystalline and stable nanoparticles (*21*). An excellent review by Paul Anastas has outlined the more recent trends in Green Chemistry and Sustainability (*22*). Another recent review by Palacio and Bhushan on ionic liquids shows by the applications of green methods and nanotechnology can result in effective molecular lubricants (*23*).

Nanotechnology is advancing at a rapid pace, more on the side of applications les so on the development of its scientific structure. Thus nanotechnology industry is moving forward to develop applications and quickly market these new nanoproducts. This creates a host of difficulties for international scientific and regulatory agencies.

This chapter endeavors to point out some difficulties for such agencies which have both scientists and regulatory managers to tackle the incoming nanoproducts for registration. The next section briefly outlines some of the major scientific issues that both scientists and regulatory communities have faced, and are still facing.

No scientific discipline advances and develops in isolation, and nanotechnology is a discipline that interconnects all disciplines of science. Nanotechnology has not matured yet as a science and still needs and uses the old scientific concepts while also developing new scientific frontiers. Nanotechnology is constantly exploring new ways to make, or engineer new nanoproducts for different uses. A quick review of recent literature on nanotechnology shows, that making use of ionic liquids for manufacturing new, unique, and in high purity nanoproducts result in less waste and a step in the right direction to sustain resources. Thus ionic liquids could become an important tool for the advancement of manufacturing new and novel nanoproducts and advancement of sustainability approach.

Ionic Liquids

Ionic liquid is defined in more than way; however, in general, ionic liquids are defined as: "liquids that are exclusively made in such a way that they exhibit ionic conductivity. Some traditionally known molten salts are also included in this definition. Ionic liquids have low melting points (< 100 C), and they are typically organic salts or eutectic mixtures of an organic salt with inorganic salt (*12*)."

The field of ionic liquids is relatively new, particularly for the synthesis of new types of substances. Numerous new and unique routes of manufacturing, synthesizing have been discovered through the applications of ionic liquids.

What makes ionic liquids different from other ionic compounds (organic or inorganic) Generally ionic liquids have special characteristics that make them different form other liquids: for example, ionic liquids, besides the characteristics noted above, have high thermal stability, liquid range is high (> 200 C), and a large number of these substances have very low vapor pressure, are moderately polar, and act as excellent solvents because of a wide liquid range. Because ionic liquids have a wide liquid range, most of them can be recycled and used again. Thus a major principle of green synthesis is upheld (*13*). In other words, use of ionic liquids the sustainability of a resource is enhanced. Many types of ionic liquids have been synthesized and investigated for their applications; Most common ones are: a) imidazolium (cations) with a variety of organic and inorganic anions, b) pyridinium (cations) also with many anions, organic and inorganic, c) quataranized phosphonium and ammonium (cations) with a wide range of inorganic and organic anions (*14*); d) benzothiazole-based quaternary ammonium salts, which in turn can be made into ionic liquids (*15*). In the past ionic liquids were touted to be benign substances, some recent publications indicate serious environmental, toxicity and ecotoxicity issues related to the uses of these chemicals (*16*).

Determination of octanol/water coefficient for ionic liquids remains a challenge because of the presence of both anion and cationic parts an active and contributing species in the overall characteristics of ionic liquids. This is an important parameter as it is a good, reliable indicator for bioaccumulation, and bioconcentration of chemical substances into various aquatic organisms. A few papers have been published for designing new approaches for the determination of Log Kow (*17*).

Ionic liquids have been used in various fields: such as designing new chemical reactions, synthesis of new drugs, in lubrication processes, synthesis of antiviral drugs, application as antimicrobials, catalysis, and in the synthesis of new nanomaterials like alloy nanoparticles, microspheres, carbon nanotube, metal oxide nanoparticles.

Scientific Issues

There are a number of major scientific issues that all regulatory and scientific agencies are facing regarding the nanomaterials. It is not possible to discuss in detail all of them, but some major ones are:

A **A new nanoproduct when it is engineered as a technical (pristine) Product.**

i). **Characterization**: Complete characterization is critically important for this new technical nanoproduct. Full characterization of nanoproducts is a tedious, and expensive process. For example, just one element (for example nanosilver) can:

- Result in the formation of many different nanoparticles based on morphology (spherical, cylindrical, circular),
- Result in nano particles that are of different sizes (1-10nm, 10-20, 20-40 nm etc.), each will have different size distribution,
- Have stability dependent on the type of dispersing agents used; the product could be monodispersed, or polydispersed; it could be a colloidal solution, having a mixture of sizes. But when it is put for application such as if it is coated in solid form on a matrix (example sox), or is converted into a nanocomposite in silica matrix, or is embedded in an organic polymer, etc. the question arises if it has the same characteristics as the pristine one? This will entail a separate characterization.

ii). **Characterization methods**: What analytical methods would give best possible characterization of a nano product? Is it DLS (if a colloidal solution), or UV-Visible spectra, or methods like SEM, TEM, AFT, x-ray diffraction (XRD), Raman spectrometry, and C-13 NMR if it is a carbon nanotube or is graphene, or use near IR for surface characterization etc. It is clear that a single analytical method is not going to provide full characterization.

iii). **Reproducibility of Characterization from one lab to another**. This is a major issue that still remains unresolved and is mentioned in all scientific meetings all over the world. It creates a major uncertainty in data evaluation as well and providing robust risk assessments.

iv). **Data on environmental fate** in environmental media as well as ecotoxicity is still sparse thus hampering the risk assessment/ risk characterization processes.

B. **When the nanomaterial is put in the product form as a pure substance like a colloidal suspension, in a formulation, as a composite in matrices like a polymer, or silica, or as a coating,** does it:

i) Migrates or leaches out from a composite
ii) What is the identity of the material that leaches out: is it still nano of the same size as the original? Or is it aggregated or has it de-aggregated (reduced in size). Each possibility introduces

an uncertainty for risk assessment and risk characterization processes.

iii) Has the morphology of the nanomaterial that has leached out changed?

iv) How do the environmental fate, ecotoxicity and mammalian toxicity of the leached/migrated material compared the parent or technical.

v) If the nanomaterial has changed during the migration, its full characterization of this material is needed.

C. **Leached out nanomaterial as a waste:**

i) If it reaches wastewater streams: is it more or less toxic to the microbes in water than the pristine or the conventional counterpart.

ii) Types the changes it goes through in this medium: does it become bigger size or smaller size, more hazardous than the pristine product?

iii) What is the Impact on the aquatic organisms?

iv) Is it more or less toxic to the sediment organisms?

vi) Does it the same identity in these matrices?

D. **Waste incineration: at the last stage of its life cycle, if it goes through incineration:**

i) What is the pathway of the atmospheric transport of this nanomaterial (if it is still nanosized)?

ii) What is the Identity of the material in the air?

iii) Exposure impact to humans (through inhalation) and to birds and insects, that is, is it more or less toxic than the pristine material?

The listed factors do not represent a complete set of scientific issues that regulatory organizations would face. This is true for nanomaterials used as pesticides or industrial chemicals or drugs, or as cosmetic products etc.

Regulatory Issues

The fact that each scientific and regulatory Agency has defined nanomaterial slightly differently (or has not defined it) indicates that product characterization is still a major issue. And risk assessment and risk characterization is a major hurdle for such nano products and perhaps in many cases ill-defined products. Regulatory issues are totally dependent on the scientific issues and or their resolutions.

A number of publications have appeared, with concrete ideas and suggesting reasonable steps to move forward with the risk assessment and risk characterization of nanomaterials which could possibly result in a new paradigm.

A. A German Report recently published indicate their thinking about regulating nanomaterials This approach basically divides the nanomaterials in to ‘ no cause for concern-criteria’ and ‘concern-criteria.’ If there is a cause of concern about a specific nanoproduct then based on risk assessment, it can be classified into risk is probable, possible or not likely (*30*).

B. More recently a paper has been published (*31*) on a risk assessment framework for assessing metallic nanomaterials on the environmental, specifically addressing the aquatic exposure and behavior. The paper has pointed out the facts that limited data is available for such materials, there is a huge variability in the available data, and with so many uncertainties, the methodology developed in the paper is primarily qualitative, and if the uncertainty factors are removed, this approach can result in a semi-quantitative assessment of aquatic exposure from these nanomaterials (three nanomaterials were considered: nanosilver, nanocerium dioxide, and nanotitanium dioxide).

   The paper also suggests that the added factors for the nanomaterials exposure (even for aquatic environment only) are not represented in the conventional metals risk assessment approach, and some of these additional factors are:

   - Possibility of aggregation (unique to nanomaterials)
   - Adsorption
   - Persistence
   - Accumulation
   - Mobility in different environmental media like water.

   To this a number of material and environmental factors have to be taken into account that have been suggested to influence the environmental behavior of the nanomaterials. One inherent limitation with this approach is that it is built around Irish aquatic waterways. Also the possibility that some of these or other nanometallic or nanometallic oxides may find agricultural uses, which will create an entirely new scenario and a very different conceptual model, will have to be developed. In our opinion this is the first attempt after the EPA’s publication of Framework on Metals Risk Assessment. It is noted that no consideration was given to the nano aspect in the EPA document on of Framework on Metals Risk Assessment.

C. Nano Risk Framework (*32*). This effort was the result of stakeholders cooperating to design a new approach for a new field of nanotechnology. The risk framework was summarized in six steps:
   1) A general description of the new nanomaterial should be developed; 2) The description can lead to generating three sets of profiles: properties of the nanomaterial, hazard profile and exposure pathways of the nanomaterial through its life cycle; 3) Evaluate or assess the risks and characterize risks of the nanomaterial due to its application, and identify

the gaps in the lifecycle profiles, and fill the gaps through new data generation or making 'reasonable worst case' assumptions or values; 4) Assess risk management: results from step 3 will provide directions to take for the management of possible risks which may result due to the use of the nanomaterials; 5) Results from the first four steps should be enough to decide whether to continue the development of the product and move forward for the production. Relevant information may be shared with external and internal stakeholders; 6) the nanomaterial profile should reviewed on a regular basis, and any new triggers that come into existence should be taken into account (adapted). The new factors may change the risk factors, and therefore risk characterization and risk management processes would have to be altered.

D. Gwinn and Tran's work (*33*). This work also points out the major issues regarding the risk management of nanomaterials which start with the uncertainties of characterization of nanomaterials from their pristine state (when they are made) when in use, and when disposed off, which lead to ambiguous risk assessment process for exposure scenarios in various environmental media. In addition, risk management requires human toxicity and eco- effects data. To top it off, regulations specifically addressing nanomaterials are lacking and traditional regulatory statutes (TSCA, REACH, FIFRA, CWA, etc.) are not equipped to address these major that lead to risk assessment, risk characterization and risk management.

On the other hand, American Bar Association's Paper on Nanotechnology (*34*) has stated that FIFRA and TSCA are adequate statues to deal with nanoproducts.

To bring home the point that dealing with metals and their salts for their use as pesticides invite additional factors that organic pesticides do not. Copper has been used in agricultural sector but we have chosen the use of copper as antimicrobials and also for only one use scenario. The following section points out the scientific and regulatory factors that come into play for this one limited use scenario. The example of copper nanoproduct is for the sake of discussion only. There are no test guidelines for such a scenario, and discussion below is based on pure assumptions which may fit exactly in a real life situation.

Copper as a Nanoproduct

Nano-metals and nano- metal oxides have found a number of applications in various fields and applications of these into specialized field like pesticides are not lagging. Nanosilver applications as an antimicrobial on socks (treated article) have become a classical example of its application. Recent publications of some the environmental issues related to nanosilver have appeared in literature (*24*, *25*)

Nanocopper, and to an extent micronized copper is touted for its application as an antimicrobial pesticide: Conventional copper (regular ionic copper, copper salts) has been used as a pesticide: fungicide, algaecide use in agricultural sector but as a wood preservative, and antifoulant uses in the antimicrobial sector are known. Nanosized copper may find its use as a pesticide. It has been shown that the particle composition of metallic nanomaterials and their toxic effects on the aquatic organisms (*26*). due to the unique properties of nanometals affect their efficacy against fungi, and termites when they are used as wood preservatives; or the question has been raised that if the leach ability of nanometals from treated wood is impacted due the nanosized materials (*28*)?; A few studies on the determination of toxicity , dispersion, and removal from municipal wastewaters or from natural river waters also have been published. (*27*, *29*).

General Discussion on Problem Formulation

It is prudent to design a problem formulation to critically examine and assess a chemical based on a use scenario or use scenarios. The purpose of any problem formulation before performing risk assessment on a chemical is to gain information and an understanding of the environmental fate, transport and ecological effects of that chemical according to its uses. The discussion in this section is a blue print of the general approaches/steps that may be taken to initiate and complete risk assessment and risk characterization processes for any product, including nanoproducts.

The created document on the Problem Formulation would contain the following informations:

A. Basic Information:

1. Major uses of the chemical
2. Active ingredient and pertinent formulas,
3. Fate characteristics such as solubility, hydrolysis, etc.
4. General information about the uses of the chemical and information about how the chemical is applied, and number of applications.
5. Information about the regulatory actions related to the chemical.
6. Any risk assessment conducted for this chemical before?
7. List of registered products which this chemical is in their ingredient.
8. The description of the methodologies as well as models which will be used for risk assessment of this chemical.

   It should be noted that items 1 through 4 are in integral part of data submissions by the registrants. Items 5 through 8 are obtained from the agency databases.

B. Fate and transport of the chemical on the environment
   The issues explored include:

1. Mode of action (moa), which discusses how this chemical effects the environment and its effect on human, plants and animals. The bioavailability of the chemical is evaluated.
2. An overview of the pesticide use and a past history of the labels used for this chemical and a general description of the minimum and maximum use of this chemical according to the registered labels.
3. Use categories- Which could be either agricultural or non-agriculture uses and if there are any direct water applications of this chemical. The urban uses of the chemical would also be explored.
4. Specific use of the chemical as antimicrobial which could be used for control of microbial growth such as wood preservatives, antifouling materials used on ships hauls for preventing algae and slime growth as well as roof coatings on shingles for preventing algal growth.
5. A conceptual model could be developed to explore the receptors of the stress from this chemical in the environment as well as changes that the stress will bring on individual organisms, food chain and general habitat. The conceptual model also explores the stress put on terrestrial plants and animals because of the different uses of this chemical on the environment.

After gathering all information as discussed above a risk hypothesis may be developed, and then the potential adverse effects of the use of the chemical on the environment is estimated. To measure the exposure real data as well as different environmental models are used to estimate the environmental concentrations (EECs) of the chemical. These concentrations would be compared with the eco toxicity data available for this chemical.

Estimation of Concentration of Chemical for a Special Use Scenario

In this section the process used for estimating concentration of a chemical, is discussed. For example copper may be used on roof shingles. In order to be able to estimate the concentration of the chemical which leaches out of shingles and it may end up in a body of water the following information would be needed:

1. Physical and chemical characteristics of the chemical.
2. How much of the chemical is applied to the shingles and by what methodology (is the chemical imbedded in the shingles, or applied after the shingles are installed?)
3. Number of houses or structures which have shingles that are treated with the chemical.
4. Average dimensions of the roof for each house.
5. Annual weather data for the area

6. Leaching rate of the chemical from the shingles.

Methodology

Using the above information the concentration of the chemical leached from shingles could be estimated by using the generic steps outlined below. Once again these assumptions and real life scenario could be different from the process presented here.

1). To consider a conservative scenario, the highest one event rainfall in a year is used to calculate the amount of water which would go through the gutters in each house.

2). For a screening level estimation of copper that leaches out, it could be assumed that there are 10 houses per hectare (Ha). It may be assumed that roofs of the homes for this screening level estimation are treated with copper. A ratio of 10 to 1 is used for calculating the concentrations. The number of houses in a 10 hectare land would be used to calculate the amount of the water from rainfall which comes out of each roof through gutters. It may be assumed that water from 10-hectare land would end up into a one-hectare pond with 20 million liters of water. It is assumed that the chemical is mixed with water evenly. It should be noted that the fate of the chemical in water is not considered in this process. The estimated concentrations of the chemical in the pond water would be calculated. This concentration could be used to compare with eco-toxicological data of risk quotients (RQs) for the same chemical to perform exposure assessment for that chemical.

If it could be assumed that the product is nanocopper applied to shingles because it shows greater activity and better efficiency, then all the factors listed in the preceding sections for the scientific data for nanomaterials would be to taken into account before even a screening level estimation of nanocopper is made. For example:

- Does nanocopper leach out as nanocopper of the same size as when it was incorporated into the shingles?
- Does it stay as nano, or aggregates when it leaches out; what is size of aggregate?
- When it reaches the wastewater streams what is the size of this material?
- Does it affect the microbes of the surface water?
- What exposure model would be used for the determination of leached copper?
- What exposure models would be used for the determination of risk quotients: BLM, MEMPAC etc.
- None of them were created to estimate nano copper. Thus the results will be flawed.

Conclusions

Nanoscience and nanotechnology are developing in big strides, more in the application field. Full and unambiguous characterization of a nanomaterial when it is made in the pristine state, and when it is applied in a product, and when it is disposed off still have gaps in the knowledge base. The fate and transport methodologies for identification of nanomaterials in the environmental media are still in semi-developed state. The situation becomes more complicated with nanometals due to the added unique characteristics of nanometals.

Sustainability and green chemistry principles could provide the scientists insights into designing new nanoproducts which would help reduce pollution, and reduce risks.

Acknowledgments

Disclaimer: Views expressed in this chapter are those of the authors and do not represent the current thinking or policy of the U.S. Environmental Protection Agency.

We would like to thank the Office of Pesticides Programs for encouraging out reach activities. Special thanks to Joan Harrigan-Farrelly and Dr. Jennifer McLain, respectively, Director and Associate Director of the Antimicrobials Division. Many thanks to OPP's Nano Group Members: Drs. Jed Costanza (OPP Nano Group Lead), Jonathan Chen, and Jenny Tao for providing useful discussion and valuable suggestions.

References

1. Federal Insecticide, Fungicide, and Rodenticide Act, Code of Federal Regulation Title 40, 2010, Parts 150 through 189 Codification of General and Permanent Rules Related to the Protection of Environment Pertaining to Pesticides. http://www.epa.gov/compliance/monitoring/programs/fifra/.
2. Federal Food, Drug, and Cosmetic Act, Code of Federal Regulation Title 21, Codification of General Rules Related to the Food, Drugs and Cosmetics. www.fda.gov/RegulatoryInformation/Legislation/FederalFoodDrugand.
3. REACH: Registration, Evaluation, Authorization and Restriction of Chemical Substances. 2007; Statute under European Community Regulations of Chemicals and Their Safe Use. ec.europa.eu/environment/chemicals/reach/reach_intro.htm.
4. Toxic Substance Control Act, Code of Federal Regulations Title 15, 1976, Parts 2601 through 2692. www.epa.gov/lawsregs/laws/tsca.html.
5. Carson, R. *Silent Spring*; Printing Houghton Mifflin Co.: New York, 1994.
6. Framework for Metals Risk Assessment: EPA Document No. 120/R-07/001, March 2007. www.epa.gov/raf/metalsframework/index.htm.
7. Interim Policy Statement on Health Canada's Working Definition for Nanomaterials, October 2011. www.hc-sc.gc.ca/sr-sr/pubs/nano/pol-eng.php.
8. FDA Draft Guidance On Use of Nanotechnology in Cosmetics. www.insidecosmeceuticals.com/blogs/regulatory-rundown/2012/04/fda.

9. EU Definition of Nanotechnology. ec.europa.eu/environment/chemicals/nanotech/index.htm.
10. Federal Register, June 2011.
11. Kates, R.; Clark, W.; Corell, R.; Hall, J.; Jaeger, C.; Lowe, I.; McCarthy, J.; Schellnhuber, H.-J.; Bolin, B.; Dickson, N.; Faucheux, S.; Gallopin, G.; Grubler, A.; Huntley, B.; Jager, J; Jodha, N.; Kasperson, R.; Mabogunje, A.; Matson, P.; Mooney, H. Sustainability Science. *Science* **2001**, *292* (5517), 641–642.
12. Freemantle, M. *An Introduction to Ionic Liquids*; RSC Publishing: London 2010, 1
13. Earle, M. J.; Seddon, K. R. *Pure Appl. Chem* **2000**, *72* (7), 1391–1398.
14. Wells, A. S.; Coombe, Y. T. *Org. Proc. Res. Dev.* **2006**, *10*, 794–798.
15. Nadeem, S.; Munawar, A. M.; Ahmad, S.; Smiglak, M.; Drab, D. M.; Malik, K. I.; Amjad, R.; Ashraf, C. M.; Rogers, R. D. *ARKIVOC* **2010**, *VII*, 19–37.
16. Pham, T. P. T.; Cho, C. W.; Yun, Y.-S. *Water Res.* **2010**, *44*, 352–372.
17. Lee, S. H.; Lee, S. B. *J. Chem. Technol. Biotechnol.* **2008**, *84*, 202–207.
18. Varma, R. *Green Chem. Lett. Rev.* **2007**, *1* (1), 1–14.
19. Nadagouda, M. N.; Varma, R. S. *Green Chem.* **2006**, *8*, 516.
20. Nadagouda, M. N.; Varma, R. S. *Macromol. Rapid Commun.* **2007**, 465.
21. Onubou, K.; Medina-Ramirez, I.; Basher, S; Luo, Z.; Liu, J. *Int. J. Green Nanotechnol.* **2011**, *3* (2), 1–10.
22. Anastas, P. T.; Beach, E. S. *Green Chem. Lett. Rev.* **2007**, *1* (1), 1–19.
23. Palacio, M.; Bhushan, B. *Tribol. Lett.* **2010**, *40*, 247–268.
24. Farkas, J.; Peter, H.; Christian, P.; Urrea, J. A. G.; Hassellov, M.; Tuoriniemi, J.; Gustafsson, S.; Olsson, E; Hylland, K.; Thomas, K. V. *Environ. Int.* **2011**, *37*, 1057–1062.
25. Li, Y.; Chen, D. H.; Yan, I.; Chen, Y.; Mittelstaedt; Zhang, Y.; Biris, A. S.; Heflich, R. H.; Chen, T. *Mutat. Res.* **2012**, *745*, 4–10.
26. Griffitt, R. J.; Luo, J; Gao, J.; Bonzongo, J.-C.; Barber, D. S. *Environ. Toxicol. Chem.* **2008**, *27* (9), 1972–1978.
27. Gao, J.; Youn, S.; Hovsepyan, A.; LLanzea, V. L.; Wang, Y.; Bitton, G.; Bonzongo, J.-C. *Environ. Sci. Technol.* **2009**, *43*, 3322–3328.
28. Clausen, C. A. *Int. Biodeterior. Biodegrad.* **2009**, *63*, 490–495.
29. Ganesh, R.; et al. *Environ. Sci. Technol.* **2010**, *44*, 7808–7813.
30. Okopol GmbH Institut fur Okologie und Politic recently published a report on the assessment tool for Nanomaterials, which is the result of the German NanoCommission' work and the stakeholder Dialogue " Risk Management in the Nano World". *Assessment Tools for Nanomaterials*, April 2012.
31. O'Brien, N. J.; Cummins, E. J. *Risk Anal.* **2011**, *31* (5), 706–726.
32. NANO Risk Framework, 2007, through the collaborative efforts of Environmental Defense and DuPont Nano Partnership Group.
33. Gwinn, M. R.; Tran, L. *Wiley Interdiscip. Rev.: Nanomed. Nanobiotechnol.* **2010**, *2*, 130–137.
34. American Bar Association, 2006. http://www.americanbar.org/content/dam/aba/migrated/environ/nanotech/pdf/TSCA.authcheckdam.pdf.

Editors' Biographies

Najm Shamim

Dr. Najm Shamim works as a chemist at the U.S. Environmental Protection Agency's Office of Consumer Safety and Pollution Prevention (OCSPP). He specializes in the environmental fate and transport studies of antimicrobials and conducts risk assessments of this class of pesticides. For his work at the EPA, he has received *two Bronze medals* and *two Excellence in Science Awards*. Prior to joining the EPA, he was a chemist at the National Institute of Standards and Technology (NIST) where he helped build and expand the NIST Mass spectral and vapor phase FT-IR databases. His experience also includes lecturership in the Chemistry Department of Howard University, Washington, DC.

Virender Sharma

Dr. Virender Sharma is a Professor of Chemistry and Director of the Center of Ferrate Excellence at the Florida Institute of Technology. His research activities include synthesis and properties of ferrates and formation and fate of nanoparticles in the aquatic environment. He has published *176* peer-reviewed journal articles, *32* book chapters and *44* proceedings and preprint articles. He has edited the book titled "Ferrates: Synthesis, Properties, and Applications in Water and Wastewater Treatment" (Oxford University Press) and is also the sole author of the book titled "*Reactive Intermediates in the Oxidation of Amino Acids, Peptides, and Proteins*". His scholarly research activities have been recognized with honors given by state and national institutions and organizations. Florida Tech honored him with the "*Award for Excellence in Research*" in 2008. The Orlando section of the American Chemical Society gave him an *outstanding chemist award.* Other awards include *certificate of merit and best symposium awards* by the ACS. Lately, he has been honored with the *Excellence in Review Award* by the editor of the Environmental Science & Technology. During the past ten years, he has been involved in the International Forum on Sustainability and is a regular invited speaker at this forum where he has been emphasizing the sustainability of water supplies in the growing world.

Indexes

Author Index

Subject Index

D

E

F

G

H

I

M

N

P

R

S

T

W